Nelson Maths

Teacher's Book

Karen Morrison
Lisa Greenstein

OXFORD
UNIVERSITY PRESS

Great Clarendon Street, Oxford, OX2 6DP, United Kingdom

Oxford University Press is a department of the University of Oxford.

It furthers the University's objective of excellence in research, scholarship, and education by publishing worldwide. Oxford is a registered trade mark of Oxford University Press in the UK and in certain other countries.

© Cloud Publishing Services CC and Lisa Greenstein 2022
The moral rights of the authors have been asserted.

First published 2022

British Library Cataloguing in Publication Data

Data available

ISBN: 978-1-382-01020-7

1 3 5 7 9 10 8 6 4 2

Paper used in the production of this book is a natural, recyclable product made from wood grown in sustainable forests. The manufacturing process conforms to the environmental regulations of the country of origin.

Printed by CPI Group (UK) Ltd, Croydon CR0 4YY

Acknowledgements

The publisher and authors would like to thank the following for permission to use photographs and other copyright material:

Cover: Matthieu Nivesse. **Photos: p20:** Oxford University Press.

Artwork by Joseph Wilkins, John Haslam, Q2A Media, Integra Software Services, Pantek Media, and Oxford University Press.

Every effort has been made to contact copyright holders of material reproduced in this book. Any omissions will be rectified in subsequent printings if notice is given to the publisher.

Cover activities

The following activities are based on the Level 6 Pupil book and Workbook cover image. You can use these stimulus questions according to children's learning to date.

Number
Arrays: See whether the children can use their knowledge of arrays to tell you how many windows there are on certain buildings, without counting each window. Encourage the children to record their calculations.

Geometry
Shapes around us: Ask the children how many different 2D and 3D shapes they can see. Encourage children to describe the properties of the shapes. Can they identify types of angles in the picture? How many parallel and perpendicular lines can the children find? Are there any lines of symmetry? Ask children to tell you as many shape facts as they can about one building at a time.

Patterns around us: Can children see any patterns? Encourage them to look for repeating, growing and tessellating patterns. Can they explain the rule and tell you what the next term would be? What about the third or tenth term?

Statistics
Charts and graphs:

- Can the children spot anything that looks like a chart or a graph? Support them to find the representation of the pie chart and line graph in the fireworks, and the pie charts and bar charts on the signs on top of the buildings.
- Ask the children what they think each graph shows. Agree that there isn't enough information on the graphs to know what they're about, and discuss what information is missing.

Contents

How Nelson Maths Works

LEVEL	PUPIL BOOK	WORKBOOKS	TEACHING SUPPORT	DIGITAL CONTENT ON OXFORD OWL
STARTER LEVEL				For all levels: • Digital versions of the Pupil Books, Workbooks and Teacher's Books • Assessment support • Parent notes • Curriculum mapping and planning guides • Vocabulary support
1				
2				
3				
4				
5				
6				

Oxford OWL Access the digital content for this course online at **www.oxfordowl.co.uk**

How to use Nelson Maths

Nelson Maths is a comprehensive maths programme for children aged 4–11. The course covers the following five strands: Number, Measure, Geometry, Statistics and Algebra to ensure full coverage of your chosen curriculum. The course is also full of opportunities to develop children's problem-solving skills through carefully designed questions and activities.

Nelson Maths is made up of seven levels (one for each year group) and Level 6 contains 15 units, which should ideally be taught in order. The units have been arranged in a careful progression, designed to support children to build their skills and knowledge and make connections across the different strands. Your children may progress through some units more quickly than others, but it is recommended that you spend approximately two weeks on each unit.

Children each have a **Pupil Book** and write-in **Workbook**. The main lesson activities are included in the **Pupil Book**, and the **Workbook** includes extra practice and consolidation activities. The **Teacher's Book** contains detailed teaching support for each unit, which will help you to introduce new concepts, revise prior learning, deepen children's understanding of a topic and encourage mathematical conversations and discussion.

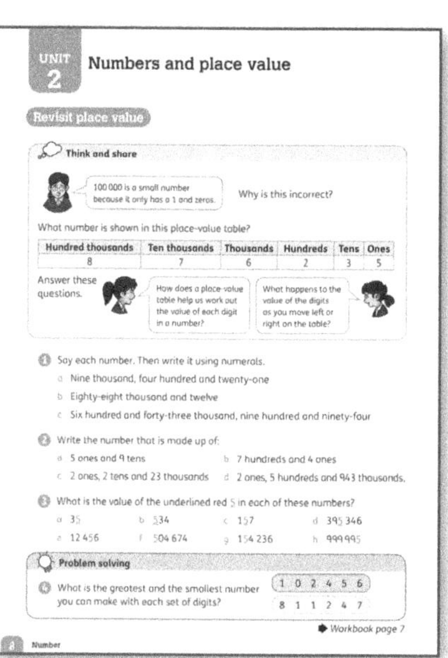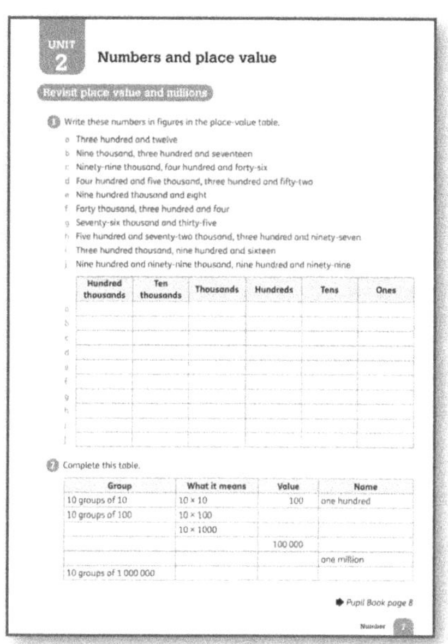

Each unit in the **Pupil Book** is supported by a corresponding unit in the **Workbook**.

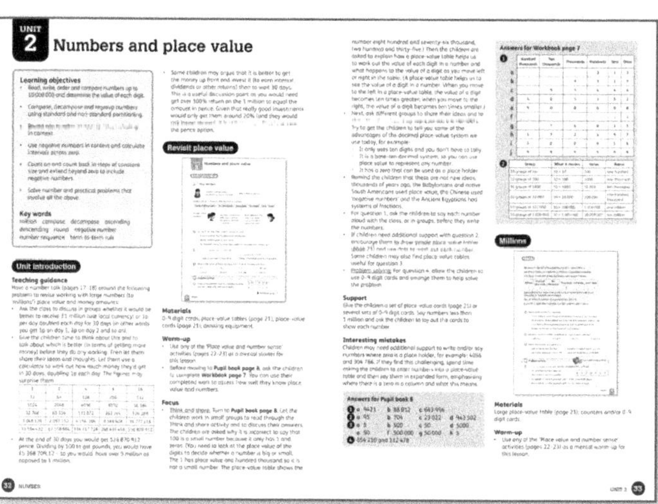

The **Teacher's Book** provides detailed teaching guidance for each unit and answers to the activities.

Read the relevant unit of the Teacher's Book before you begin teaching it. It's important to familiarise yourself with the learning objectives, key vocabulary and resources you need before each lesson. The following pages will give you a good understanding of how to use the Pupil Book, Workbook and Teacher's Book together.

In addition to the printed materials, you can also find lots of supplementary digital resources online on Oxford Owl. Please see page 4 for more information.

Using the Pupil Books

The **Pupil Book** units are designed to be worked through in order, following the discussion prompts and activities suggested in the **Teacher's Book**, and using the linked **Workbook** pages to help children to consolidate their understanding through independent practice. The following features of the **Pupil Books** are designed to help you get the most out of your lessons.

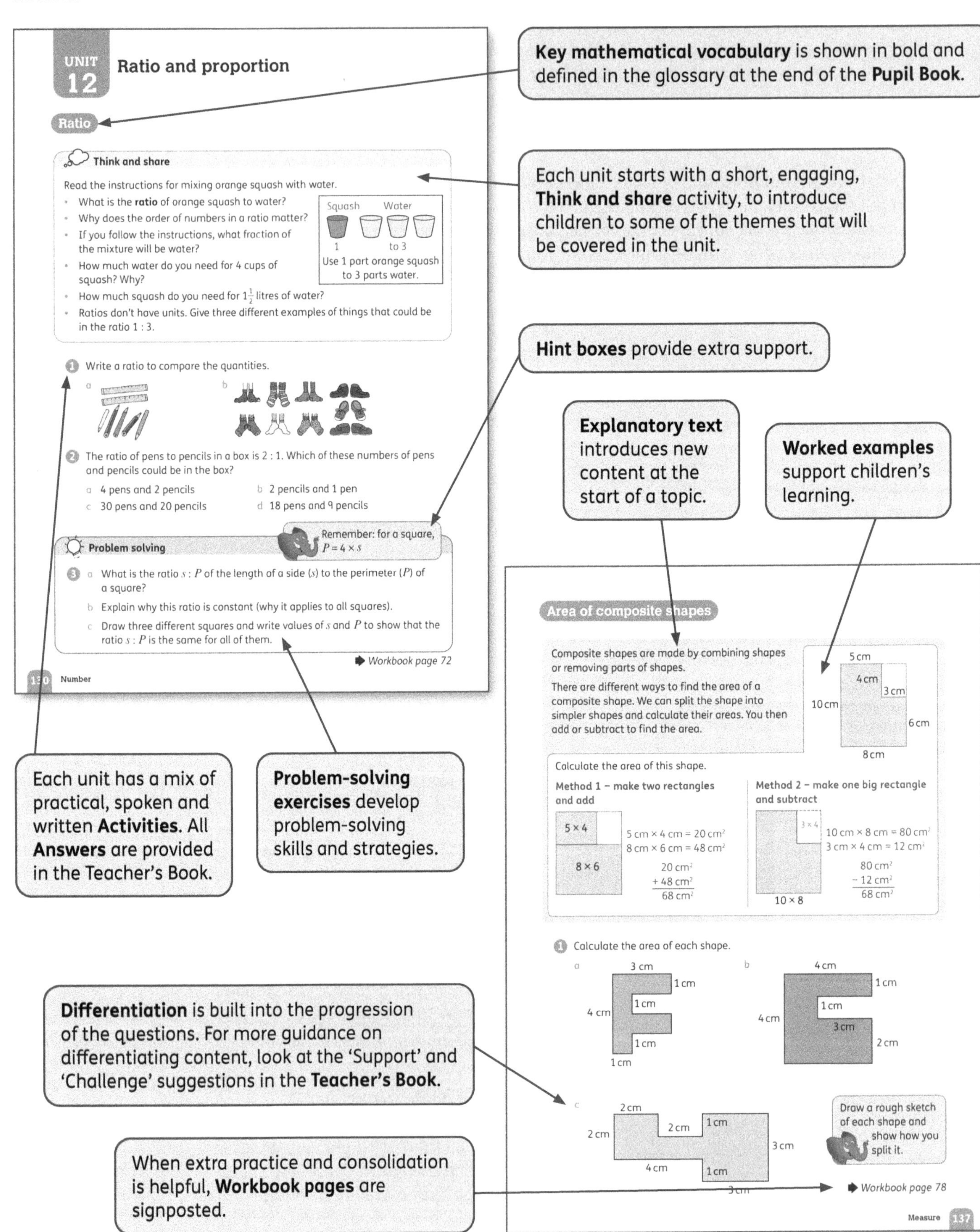

Key mathematical vocabulary is shown in bold and defined in the glossary at the end of the **Pupil Book**.

Each unit starts with a short, engaging, **Think and share** activity, to introduce children to some of the themes that will be covered in the unit.

Hint boxes provide extra support.

Explanatory text introduces new content at the start of a topic.

Worked examples support children's learning.

Each unit has a mix of practical, spoken and written **Activities**. All **Answers** are provided in the Teacher's Book.

Problem-solving exercises develop problem-solving skills and strategies.

Differentiation is built into the progression of the questions. For more guidance on differentiating content, look at the 'Support' and 'Challenge' suggestions in the **Teacher's Book**.

When extra practice and consolidation is helpful, **Workbook pages** are signposted.

The first unit in each Level is called **Think maths**. It encourages a growth mindset and builds resilience by teaching children that mistakes are a positive part of their learning journey. It prepares children to think mathematically and make connections across maths.

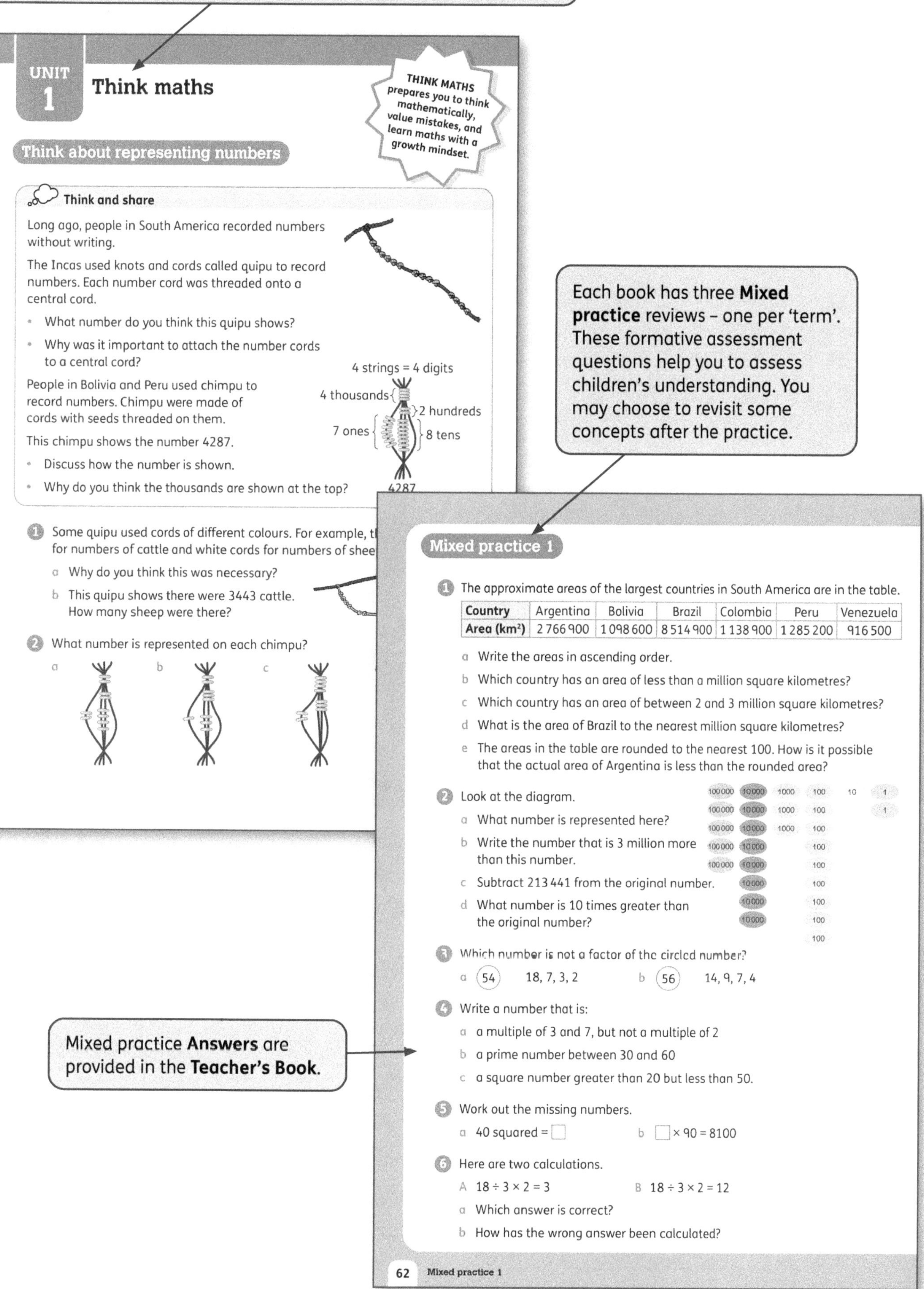

Country	Argentina	Bolivia	Brazil	Colombia	Peru	Venezuela
Area (km²)	2 766 900	1 098 600	8 514 900	1 138 900	1 285 200	916 500

Each book has three **Mixed practice** reviews – one per 'term'. These formative assessment questions help you to assess children's understanding. You may choose to revisit some concepts after the practice.

Mixed practice **Answers** are provided in the **Teacher's Book.**

Using the Workbooks

The **Workbooks** provide essential practice to help children consolidate their learning. They provide new questions and activities linked to the concepts introduced in the **Pupil Books**. They are designed for independent use, making them ideal for homework or additional classwork. We recommend that children complete each page in the **Workbook** after they have completed the corresponding lesson in the **Pupil Book**. When extra practice and consolidation is helpful, **Workbook** pages are signposted at the end of the Pupil Book lesson.

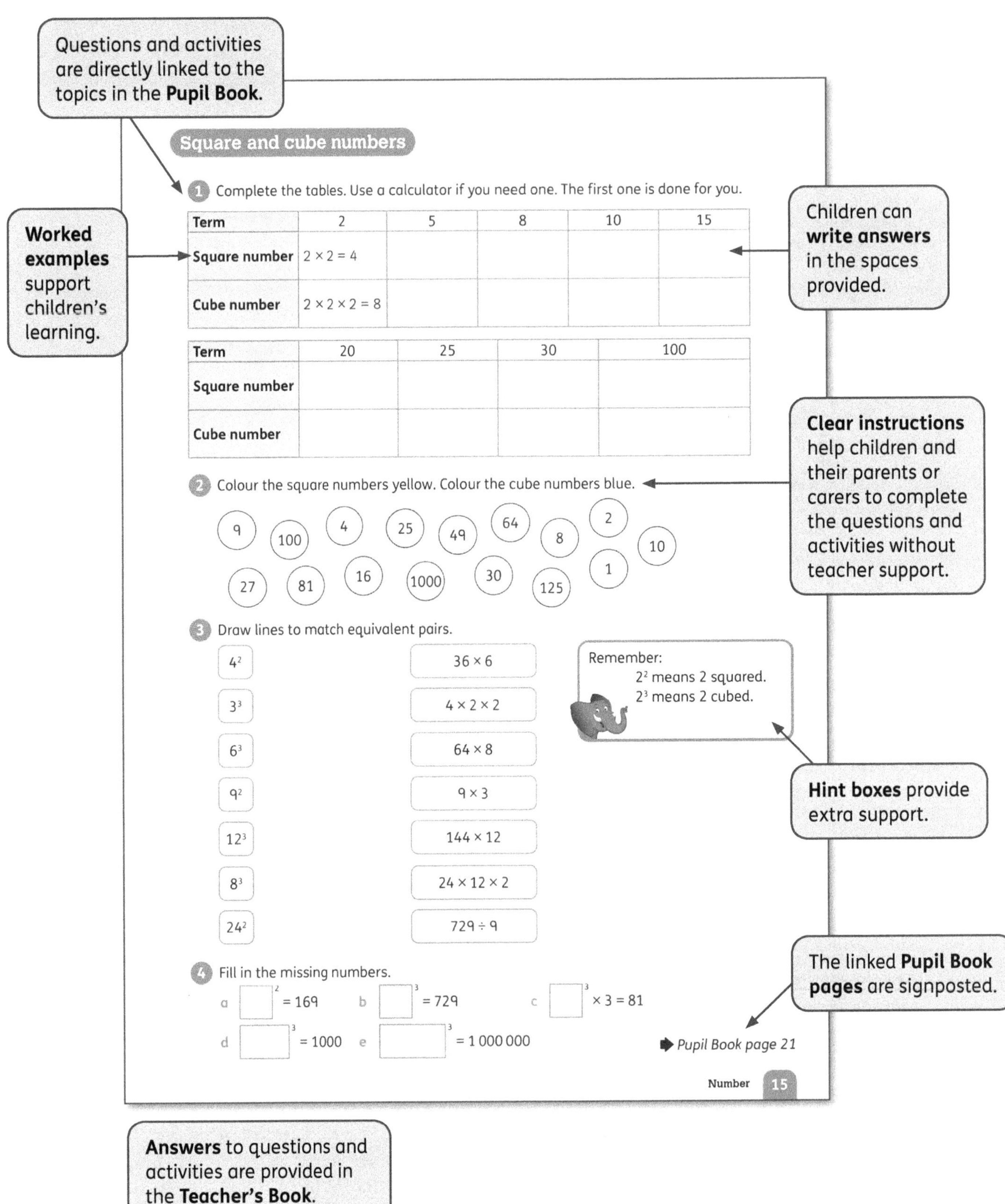

Using the Digital Content

This edition of *Nelson Maths* is supported by additional digital resources available online on **Oxford Owl**. These include:

- Digital versions of the Pupil Books, Workbooks and Teacher's Books to support planning and for front of class display
- A set of printable assessments with a progress tracking tool and mark scheme
- Guidance for you to share with the children's parents or carers about their child's learning, including support for homework and ideas for incorporating maths into everyday life
- Vocabulary support
- Curriculum correlation charts
- Planning support.

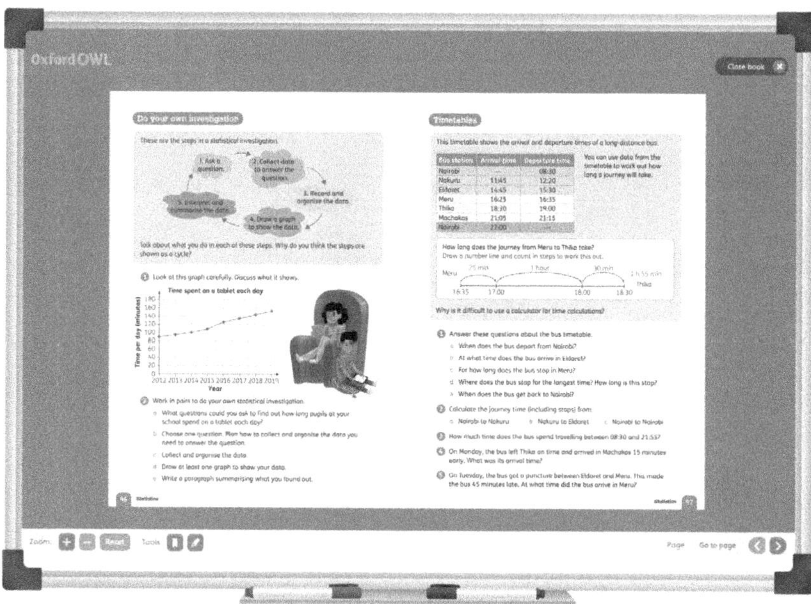

Using the Teacher's Book

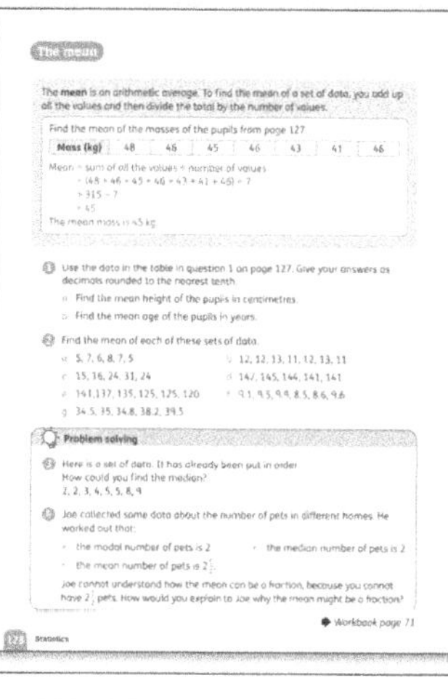

The **Teacher's Book** provides step-by-step lesson notes for each unit in the **Pupil Book** and **Workbook**, including learning objectives, answers to the questions activities and suggestions for additional challenge and support.

Before you start teaching the units, take some time to familiarise yourself with these useful sections at the start of the **Teacher's Book**. These four sections are designed to help you get the most out of the *Nelson Maths* teaching materials:

1. **Introduction** – find out more about the mathematical strands, skills and fundamental principles that underpin *Nelson Maths*. Get ideas and support for teaching maths in an inclusive way that acknowledges individual differences and human diversity.

2. **Teaching approach** – guidance for teaching *Nelson Maths* effectively, including:

   - how to develop robust learners with a growth mindset and an understanding of the importance of making mistakes
   - the importance of the do-talk-record model that underpins the lessons
   - using engaging teaching methods, from investigations and number talks, to exploring maths in real life contexts.

3. **An environment for exploration and play** – practical support on organising your classroom space; approaching whole-class, group and individual activities; and using manipulatives and games to deepen children's learning.

4. **Activity bank of warm-ups and mental maths** – a bank of engaging mental maths, support, and consolidation activities to use with your class.

Teacher's Book: using the unit pages

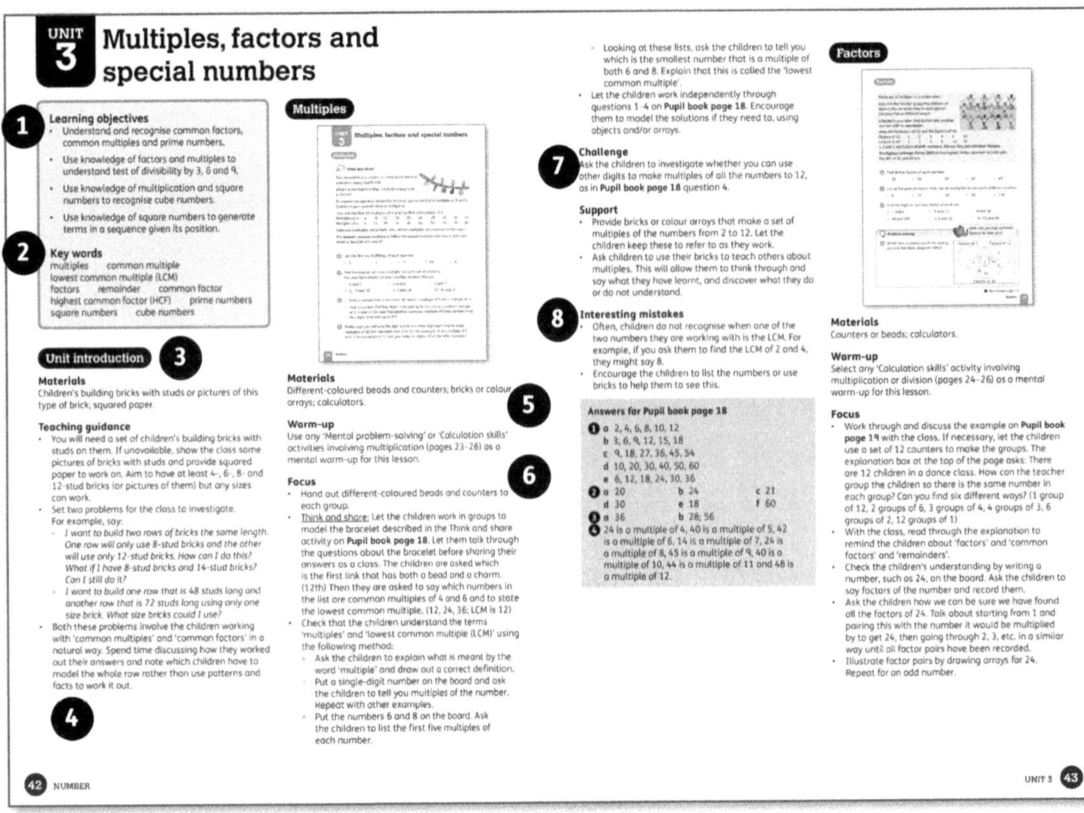

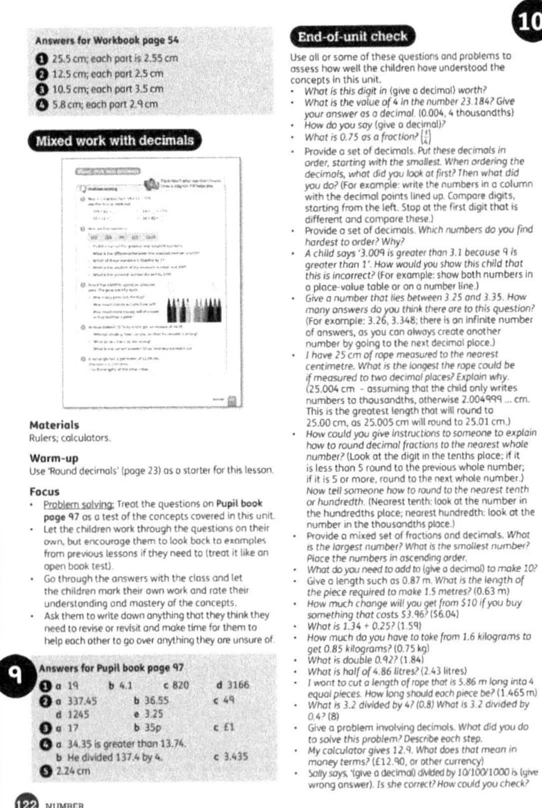

1 **Learning objectives** are listed at the beginning of each unit.

2 **Key words** match the important words highlighted in bold in the **Pupil Book**.

3 **Unit introduction** activities introduce the topics in the unit, engage children and find out what they already know.

4 The **linked Pupil Book and Workbook pages** are signposted.

5 **Materials** children need for the lesson are listed.

6 Most lessons are broken down into three stages:

- **Warm-up** – introduce the topic and unlock prior learning
- **Focus** – develop a firm understanding of concepts and skills by working through the **Pupil Book**
- **Follow-up** – recap the key learning points and extend the learning with further activity suggestions in the **Workbook**.

7 Where relevant, **Challenge** and **Support** sections give guidance on deepening the learning or making it more accessible, so that every child is catered for.

8 It is important for children to see mistakes as part of their learning journey. The **Interesting mistakes** section highlights common errors and misconceptions that children may have, along with what to look for and advice on how to support children.

9 **Answers** for **Pupil Book** and **Workbook** questions are provided to make marking easy.

10 An **End-of-unit check** allows you to assess how well children have understood the concepts in each unit, so you can monitor children's understanding throughout the course. A set of formative assessment questions or a closing activity are provided with answers.

Level 6 Scope and sequence

Unit	Unit title	Strand	Learning objectives	Key words
1	Think maths	Think maths	• Establish a classroom environment conducive to thinking and working mathematically • Set up positive norms • Establish key messages of growth mindset mathematics • Consolidate planning and problem-solving skills	strategy, critical thinking
2	Numbers and place value	Number	• Read, write, order and compare numbers up to 10 000 000 and determine the value of each digit • Compose, decompose and regroup numbers using standard and non-standard partitioning • Round any number as appropriate, including in context • Use negative numbers in context and calculate intervals across zero • Count on and count back in steps of constant size and extend beyond zero to include negative numbers • Solve number and practical problems that involve all the above	million, compose, decompose, ascending, descending, round, negative number, number sequence, term-to-term rule
3	Multiples, factors and special numbers	Number	• Understand and recognise common factors, common multiples and prime numbers • Use knowledge of factors and multiples to understand tests of divisibility by 3, 6 and 9 • Use knowledge of multiplication and square numbers to recognise cube numbers • Use knowledge of square numbers to generate terms in a sequence given its position	multiples, common multiple, lowest common multiple (LCM), factors, remainder, common factor, highest common factor (HCF), prime numbers, square numbers, cube numbers
4	Shapes. lines and angles	Geometry	• Classify, estimate, measure and draw angles • Know that the sum of angles in a triangle is 180° and use this to calculate unknown angles in a triangle • Recognise angles where they meet at a point, are on a straight line, or are vertically opposite, and find unknown angles • Draw 2D shapes using given dimensions and angles • Identify, describe, classify and sketch quadrilaterals, including reference to angles, symmetrical properties, parallel sides and diagonals • Compare and classify geometric shapes based on their properties and sizes and find unknown angles in any triangles, quadrilaterals and regular polygons • Illustrate and name parts of circles, including centre, radius, diameter and circumference, and know that the diameter is twice the radius • Recognise, describe and build simple 3D shapes, including making nets	protractor, right-angled, vertically opposite angles, two-dimensional (2D), polygons, regular polygon, equilateral, isosceles, scalene, acute, obtuse, equation, quadrilaterals, parallelogram, rectangle, rhombus, square, trapezium, kite, circle, circumference, centre, radius (r), radii, diameter (d), three-dimensional (3D), face, edge vertex, net, surface area
5	The four operations	Number	• Use knowledge of the order of operations to carry out calculations involving the four operations • Understand that brackets can be used to change the order of operations • Perform mental calculations, including with mixed operations and large numbers • Estimate, add and subtract integers • Solve addition and subtraction multi-step problems in context, deciding which operations to use and why • Multiply and divide by 10, 100 and 1000 (using place value) • Estimate and multiply whole numbers up to 10 000 by 1- and 2-digit numbers using formal written methods • Divide numbers of up to 4-digits by 2-digit whole numbers using the formal written method of long division and interpret remainders as whole number remainders, fractions or by rounding as appropriate for the context • Divide numbers of up to 4-digits by 2-digit whole numbers using the formal written method of short division where appropriate, interpreting remainders according to the context • Solve problems involving addition, subtraction, multiplication and division • Use estimation to check answers to calculations and determine, in the context of a problem, an appropriate degree of accuracy	brackets, inverse, long multiplication, perimeter, capacity, decimal fractions, long division

Unit	Unit title	Strand	Learning objectives	Key words
6	Fractions	Number	• Recognise when fractions can be simplified and use common factors to simplify fractions, use common multiples to express fractions with the same denominator • Compare and order fractions, including fractions greater than 1, using reasoning and choose between reasoning and writing fractions with a common denominator as a comparison strategy • Understand that a fraction can be represented as a division of the numerator by the denominator (proper and improper fraction) and calculate decimal fraction equivalents for a simple fraction • Add and subtract fractions with different denominators and mixed numbers, using the concept of equivalent fractions • Understand that proper and improper fractions can act as operators • Multiply and divide fractions by whole numbers • Multiply simple pairs of proper fractions, writing the answer in its simplest form	equivalent fraction, numerator, denominator, mixed number, improper fraction, simplify, cancelling, lowest terms, simplest form, common denominator
7	Position, direction and movement	Geometry	• Read and plot coordinates on the full coordinate grid (all four quadrants). • Draw and translate simple shapes on the coordinate grid and reflect them in the axes • Reflect 2D shapes in a given mirror line (vertical, horizontal and diagonal on square grids) • Rotate shapes clockwise and anti-clockwise around a vertex	coordinates, x-axis, y-axis, quadrants, reflection, translation, mirror line, horizontal vertical, (full) rotation
8	Decimals	Number	• Identify the value of each digit in numbers given to 3 decimal places • Multiply and divide numbers by 10, 100 and 1000, giving the answers to 3 decimal places • Compose, decompose and regroup decimals using standard and non-standard partitioning • Multiply 1-digit numbers with up to 2 decimal places by whole numbers • Use written division methods in cases where the answer has up to 2 decimal places • Round decimals to the nearest tenth or hundredth • Solve problems which require answers to be rounded to specified degrees of accuracy • Recall and use equivalences between simple fractions, and decimals, including in different contexts • Understand that a fraction can be represented as a division of the numerator by the denominator and calculate decimal fraction equivalents for a simple fraction	decimal point, thousandths
9	Percentages	Number	• Recognise percentages of shapes and numbers • Recall and use equivalences between simple fractions, decimals and percentages, including in different contexts • Solve problems involving the calculation of percentages and the use of percentages for comparison	percentage, discount
10	Measures and money	Measure	• Solve problems involving the calculation and conversion of units of measure, using decimal notation up to 3 decimal places where appropriate • Use read, write and convert between standard units, converting measurements of length, mass, volume and time from a smaller unit of measure to a larger unit and vice versa using decimal notation to up to 3 decimal places • Convert between miles and kilometres • Convert between time intervals expressed as a decimal and in mixed units	metric, convert, kilometre, line graph, mile
11	Data	Statistics	• Record, organise and represent data using a variety of tables, diagrams and graphs • Interpret and construct pie charts and line graphs and use these to solve problems • Calculate and interpret the mean, median, mode and range • Plan and carry out a statistical investigation	frequency table, grouped data, line graph, pie chart, sector, mode, range, median, mean

Unit	Unit title	Strand	Learning objectives	Key words
12	Ratio and proportion	Number	• Understand the relationship between quantities when they are in direct proportion • Use knowledge of equivalence to understand and use equivalent ratios • Solve problems involving the relative size of two quantities where missing values can be found by using integer multiplication and division facts • Solve problems involving similar shapes where the scale factor is known or can be found • Solve problems involving unequal sharing and grouping using knowledge of fractions and multiples	ratio, proportion, scale factor
13	Perimeter, area and volume	Measure	• Recognise that shapes with the same areas can have different perimeters and vice versa • Use simple formulae • Recognise when you can use a formula to find the area and volume of shapes • Use knowledge of the area of rectangles to estimate and calculate the area of right-angled triangles • Calculate the area of parallelograms and triangles • Understand the difference between volume and capacity • Calculate, estimate and compare the volume of cubes and cuboids using standard units, including cm^3 and m^3 and extending to other units (mm^3 and km^3)	perimeter, formula, area, composite shape, volume, cubic units, capacity
14	Algebra	Algebra	• Use simple formulae • Generate and describe linear number sequences • Find and use a position-to-term rule • Recognise the use of letters to represent quantities that vary in addition and subtraction calculations • Express unknown number problems algebraically • Find pairs of numbers that satisfy an equation with two unknowns • Enumerate possible combinations of two variables	number sequence, term-to-term rule, position-to-term rule, variable, formula, formulae, substitute
15	Probability	Statistics	• Use the language of probability and proportion to describe and compare possible outcomes • Identify when two events can happen at the same time and when they cannot, and know that the latter are mutually exclusive • Model probabilities through experiments using a large number of trials	probability, probability scale, mutually exclusive, experiment, outcome, possible outcomes

Section 1 Introduction

This Teacher's Book is designed to support the component parts of *Nelson Maths* Level 6.

Sections 1–4 of this Teacher's Book (pages 14–27) provide background and reference information, tips on classroom organisation, and a bank of activities and games to use for teaching practically. Section 5 provides lesson notes for the Pupil book and Workbook unit-by-unit and page-by-page.

Strands and skills

In Levels 1 to 6, the content is divided into strands:
- Number – numbers and the number system, and calculating
- Measure – money, length, mass and capacity, and time
- Geometry – shape, position and movement
- Statistics – organising, categorising and representing data
- Algebra – use of formulae, variables and equations

In this course, problem solving is not treated as a separate strand but is integrated into the materials at all levels.

Fundamental principles

This series makes the following assumptions about the teaching of mathematics:
- Children need concrete (practical, hands-on) experiences in order to acquire sound mathematical understanding. Like adults, children learn best when they investigate and make discoveries for themselves.
- Children refine their understanding and develop conceptual structures by talking about their own thinking and what they have done.
- Individual children develop at different rates. While some will find certain elements of mathematics difficult, others will understand them quickly.
- Children learn in a variety of ways, so mathematics teaching should provide a rich and wide variety of experiences. Children need plenty of opportunities to apply what they have learnt and to relate their mathematics work to other areas of the curriculum and their daily lives.
- Children will become more mathematically able if they are allowed to develop reliable personal ways of working. The formal recording used by mathematicians comes with time and experience, and we cannot expect young children to understand it or use it themselves before they are ready to do so. The conventions (formal methods) should be taught only once children are confident in their own knowledge, concepts and skills.
- Children learn mathematics most effectively when they enjoy what they are doing and when they can see the relevance of what they are learning.

This course reflects current thinking about the most effective ways of teaching and learning mathematics at primary level. It recognises the professionalism of the teacher, and acknowledges that teachers are the best judges of which experience(s) are most appropriate at different stages for the children in their classes. For this reason, this course does not impose a rigid, inflexible structure. Instead, it provides a wide variety of practical activities and games linked to clearly defined purposes and objectives. The teacher selects according to the needs of their classes, groups and individuals.

Individual differences and inclusivity

Everyone learns at their own pace, and in different ways. We also need to recognise that each child enters the classroom with their own experience, their own identity, and their own hopes, fears and curiosities.

This course recognises individual differences and aims to give children the chance to explore the world of mathematics and solve problems in their own way. To achieve this, we believe that each child needs to feel that maths is for them too. Therefore, examples and illustrations need to include a wide range of children of different genders and ethnic, cultural and linguistic backgrounds and should not promote stereotypes.

- When giving children additional maths problems to solve:
  - Use examples of girls doing things that are traditionally considered 'for boys'.
  - Use examples of carers (including dads as the primary carers) as well as working mothers and female leaders.
- Be aware of children's differences and acknowledge different bodies: not all children have ten fingers and ten toes, and some may not be able to walk or run as easily as others.
- Avoid comparisons of height or weight that some children may experience as body shaming.
- Celebrate slower workers in your classroom as much as faster workers. Praise reasoning and focused working rather than speed. Never focus on speed testing or 'who can find the answer the fastest'. Inclusivity means acknowledging that our slower, deeper thinkers are some of our best mathematical minds too. Maths is not about speed. It is about reasoning. Avoid any games that use countdown clocks or timers to turn mental maths into a race or speed test.

- Praise different ways of working. Some children need to move and fidget and play in order to solve problems. Include those who think visually or rely on practical manipulatives and drawings to solve their problems.
- Struggle and confusion are key parts of the process of maths learning. Never dismiss a child's need for additional support as evidence that they cannot do maths or that they are not mathematically minded.
- When you are planning classroom activities, bear in mind the abilities and needs of your group and change or adapt activities as necessary. For example, it may be more appropriate to count sets of ten counters, or use ten frames than to use fingers for counting. When activities expect children to identify, sort or match colours, adapt these activities for children who have colour-blindness, by focusing on pattern rather than colour.

Section 2 Teaching approach

A growth mindset

In recent years, scientists have made important discoveries about how the brain learns and grows. A key discovery is the notion of neuroplasticity or brain plasticity. Brain plasticity is the brain's ability to change connections and neural pathways. Put simply, this means that the brain can learn and grow throughout a person's life.

These discoveries have led many educators to talk about 'growth mindset' versus 'fixed mindset'. People with a fixed mindset believe their abilities are determined and fixed, for example by genetics or by factors outside their control. A fixed mindset can lead to fears of making mistakes. Children with fixed mindsets may believe that they are either 'good at' or 'bad at' maths, and may easily get frustrated when they find problems challenging. We need to help children to understand that our maths ability changes all the time as we learn and grow. By getting children to understand and believe that growth, development and improvement are always possible, we can foster a 'growth mindset', in which children are not afraid of tackling difficult problems.

Getting rid of maths myths

Every child has the potential to learn maths, irrespective of their existing level of mathematical ability. The ability to learn maths is also not related to other traits such as gender, race or ethnicity. There is no such thing as a 'maths person'. It is extremely important that teachers understand that all children are able to learn to think mathematically. Maths is not a special subject and does not require any special natural 'talent'. Some children may grasp concepts more quickly than others – just like in other subjects – but giving them the time to develop their understanding, allows all children to achieve highly in mathematics.

The importance of mistakes

Scientists have also studied the brains of people attempting to solve problems. They found something that might seem surprising – that more neural connections and pathways form when we make mistakes than when we get things right! Although getting correct answers is usually praised by teachers,

it is key that we also start understanding the importance of mistakes. Mistakes offer a valuable opportunity to explore why and how a concept makes sense. You can explore mistakes in a productive way with your class by:

- frequently reminding the children that making mistakes shows we are learning something new
- ensuring the children know that mistakes (and questions) are welcome in your classroom
- discussing mistakes with interest and curiosity
- instead of always asking a question that invites a correct answer, sometimes phrasing questions to ask the children to identify the *incorrect* answers and then explain why they don't work.

You can read more about growth mindset in books by Carol Dweck and Jo Boaler, or search for these online.

The do–talk–record model

The learning framework for this course can be summarised as: *do–talk–record*. In other words, the starting point is always to explore and investigate concepts. Talking, discussing and explaining follows. Finally, written work can record and represent the concepts that the child has already explored practically and in discussion.

Doing

The first step in developing and understanding concepts is to let children:

- handle and manipulate apparatus
- play games
- investigate patterns and rules using real objects
- model situations and problems using real objects.

Children need time to play and explore in this way before they are expected to communicate about their work.

Talking

Children can make sense of what they have been doing by discussing:

- what they have done
- why they have done it
- what they have found out.

This allows them to generalise concepts and ideas from their particular experiences. The teacher's role is to create situations for discussion and to ask open-ended but directed questions. Most of the activities in this Teacher's Book will help you to facilitate discussion and will encourage the children to listen to each other and experiment with different ways of thinking about and solving problems.

Recording

By Level 6, children are likely to have refined their skills and knowledge to some degree and they will have developed strategies that they find easy and useful for solving problems. They may still need to use informal and very personal methods (jottings) of recording steps in a process, or keeping track of what they have done. Jottings are an important step in moving towards non-standard methods of calculation (such as diagrams and jumps on a number line) that give the children a foundation for more concise standard written methods of recording.

It is very important that you allow, and in fact encourage, children to make use of jottings as they work. Here are some possible ways of doing this in the classroom:

- Do jottings of your own as you work out solutions. For example, if you are demonstrating how to calculate 144×5 you might jot the following on the board to show how you are thinking:

    1440
    720

- Talk through the jottings as you make them. For example: *144 times 10 is 1440, half of that is 720.* This modelling process helps children to see that jottings are important and useful.
- Make space for jottings in the children's exercise books. You can reinforce the importance of jottings as a means of showing your working by encouraging the children to jot as they work. If you only allow jotting on scrap paper, children may think it is not as important or valuable as their 'real' work in their book.
- Limit the use of prepared sheets with boxes for answers and no space for jotting down steps.
- Do activities where jotting is the point of the activity. For example, ask the children to represent $\frac{3}{4}$ visually in as many ways as possible, or ask them to work out problems where they will need to jot down interim steps to keep track of the process. Ask questions such as: *How many ways can you find of making one dollar using any combination of 50 cent and 10 cent coins?*
- Ask children to share their jottings and compare them to show that there are different methods of working. This can help the children to see that some strategies are more efficient than others and, in turn, refine their own thinking. In the 'make a dollar' task above you may find that some children draw coin combinations, others list them and those who are more confident may make a table and work more systematically. All of these methods may provide the correct answers, but obviously some will take longer than others.

In the early stages of using apparatus in a new way, recording may take the form of drawings or words and drawings. Some children will gradually find this time-consuming and will simplify their recording independently. Others may need your suggestions and encouragement. As a teacher, you will need to work out carefully when a child is ready to use a standard mathematical symbol or format, so that recording is based on full comprehension.

Although in Level 6 you will teach children some standard written methods for operations on larger numbers, it remains crucial that you do not force children into formal and standard methods of recording calculations before they have fully grasped the process and are confident in the methods.

Traditionally, primary mathematics has almost always tended towards short, directed tasks that result in 'right' or 'wrong' answers. The activities in this course provide a balance between short, fairly self-contained activities and open-ended investigations that may sometimes require the children to collect information, make their own observations, ask questions or complete activities at home. Most of the activities are designed to develop children's awareness of the range of mathematical possibilities open to them when tackling a mathematical task.

As much as possible, allow the children to take control, make decisions and explore the many avenues that can arise from a simple starting point. Always encourage children to ask *What if … ?* and *Why?* when investigating. These questions may lead to new challenges, fresh understanding and the development of new skills.

Many investigations have no final 'answer' or easily accessible generalisation for the children. Some have a simple pattern or rule that may be discovered and explained. However, many children will want to know why certain patterns repeat, and offer explanations about the rules that govern them. This is the first step towards generalisation, and you should encourage this by asking questions, for example: *Why is the same number added each time?* or *Can you guess what will happen next?*

The value in investigations is in children following them as far as they can, and in the new skills that they acquire on the way. For some children, the early, often concrete, experimentation is enough to give them confidence, and increase their enjoyment of using already acquired skills.

Many everyday objects can provide rich sources of investigative work. The 100 chart, addition square and multiplication square all contain many fascinating patterns. Children can also explore patterns in solid and flat shapes, such as the relationships between faces, edges and vertices of 3D shapes, and the relationships between sides, corners and angles of 2D shapes.

Use investigations to enrich the introduction of new concepts. For example, you can introduce number patterns through developing number chains and introduce geometric patterns through explorations of colour arrangements on geoboards. The children can explore the relationship between area and perimeter, and between volume and the dimensions of cuboids. As they develop an investigative approach, help children to become systematic in the way they work. This will help them to understand the structure and formal approaches of mathematical theory.

Number talks

Number talks are open-ended, meaning they present a question that has multiple possible answers, and multiple possible strategies for solving.

Children can use agreed hand signals as shown below, instead of calling out or putting their hands up. However, these may not be culturally appropriate in all countries, and you may want to come up with alternative signals that are suitable for your class.

- **Step 1: Present the problem and give time for independent thinking**
  First present the problem, then ask the children to think to themselves quietly how they could solve it. They can use hand signals (raising one finger means 'I have a way to solve it') to show when they have an idea.

I have an idea, and one way to solve it.

Note that when children have their fingers up with an idea, you can encourage them to think of more ways to solve it, and to indicate when they have more possible ways of doing it. They can add more fingers as shown.

I have an idea, and two ways to solve it.

- **Step 2: Share in pairs**
  In pairs, children share their suggestions for how to work a problem out. Note that explaining how they solved it is as important as (or more important than) the final answer. They can use a hand signal to indicate they are finished, such as a fist to the chest.

I have finished.

- **Step 3: Children share their strategies**
  Ask individual children to share their strategies with the class. Discuss each strategy as the child demonstrates for the class. Continue using hand signals (thumb and little finger out for 'I agree'; holding out both hands in front, palms down, and moving the hands in a criss-crossing motion for 'I disagree'; one more finger raised for 'I have another way to do it'), etc. so that children can engage with the discussion and take turns to share their own working.

I agree   I respectfully disagree

- **Step 4: Discussion**
  The discussion will take place as other children show that they agree or disagree, and offer alternatives strategies.

Principles of the number-talk strategy

The number-talk strategy aims to get all children to engage actively and creatively with mathematics. To achieve this engagement, it is important to remember the 'big ideas' behind the number talk:

- *Value the contributions of all children.* Throughout the discussion help them to articulate their thoughts and reach a clearer understanding of how they are working things out. The emphasis is on the process and understanding, not on reaching the right answer fast.
- Children often answer in a roundabout way, with a very general idea or 'sense' of what they want to say. Use questions to *clarify, refine and explore* the meaning behind each response.
- Encourage children to explore *conceptual explanations*, rather than procedures.
- Gently *explore interesting mistakes* because these provide opportunities for deeper understanding. We can often better understand how something does work by reaching a clearer understanding of how it doesn't work.
- Praise children for effort, for making sense of a problem and for *persistence in the face of difficulty*, rather than for efficiency. We like to tell children that if it is difficult, that tells us we are in the place of learning. If it is easy, that means we've already learnt it and it's time to find more challenge.
- Create a *safe community space* where children feel they can share their ideas without criticism or judgement.
- Encourage *sharing ideas*, rather than competition. Children should understand that learning comes from connecting with other people's ideas and building on them.

Finding out more about number talks

It is important to remember that number talk is a format or technique, not a specific lesson or set content.

- View number talks in action, shared by teachers, on the Internet.
- Find posters and strategies for number talks through an online image search.
- Books on number talks include: *Number Talks: Whole Number Computation, Grades K-5: A Multimedia Professional Learning Resource* by Sherry Parrish; *Classroom-Ready Number Talks for Kindergarten, First and Second Grade Teachers* by Nancy Hughes.

Numberless word problems

Many teachers fear problem solving because they present a problem, and then watch as children try to guess which operation to use, or which numbers to arrange into a number sentence. Numberless word problems offer a useful strategy for encouraging children to think and understand questions before looking for the 'maths answers'.

Here is an example. Say you want children to practise their addition. Instead of giving a problem in traditional form, we start with a numberless form of the same problem, and work very gradually towards presenting the numbers.

> *There are some children in the playground. Some more children come out to play. Now there are many children in the playground.*

The numberless word problem does two things:

- It removes the numbers.
- It removes the question.

This forces children to slow down and think about the problem before they can attempt to solve it.

To present a numberless word problem, follow these steps:

- **Step 1:** Present the word problem you want to use, but remove the numbers and question. So, reword it using words such as *some, more, joined, went away, took away, fewer* and so on. Ask the children to tell you what is happening in the story.
- **Step 2:** Ask children to think about what the question could be in this story. You will need to give plenty of thinking time, and invite the children to be 'question detectives', guessing from the clues what we are going to ask. They may have a variety of suggestions, such as: *How many were in the playground to start with? How many children came out? How many were there altogether?* Once they have identified the possible questions, you can reveal the question, but still without numbers.

> *There are some children in the playground. Some more children come out to play. Now there are many children in the playground.*
> *How many children came out to play?*

- **Step 3:** Have the children identify what information they need in order to solve the problem. First, ask the children if they can answer the question. They cannot – because they don't have the information they need. Let them suggest what they need to find out in order to solve it. Give them the first piece they need, by crossing out *some* and putting a number in:

> *There are ~~some~~ 127 children in the playground. Some more children come out to play. Now there are many children in the playground.*
> *How many children came out to play?*

Again, ask if they can answer the question. When they say no, let them work out what other information they need to solve it. Let them work out what else they need to know. Finally, cross out *many* and change it to a number (e.g. 251).

- **Step 4:** As with number talks, give the children time to solve the problem using their own strategies. Follow the steps of a number talk as you let them share and discuss their strategies.

Mathematics in real life

Some children may need additional support to understand the relevance of mathematics in their everyday lives. This course places great emphasis on making children aware of the relevance of mathematics in real life.

In this Teacher's Book, you will find ideas for using the child's own environment as a stimulus for mathematical activities. The Pupil book and Workbook frequently require the children to look at the mathematics in the classroom, the playground and their own homes. Each set of activities and problems requires new skills and fresh understanding. Many questions are open-ended or have no exact solution, and the children are asked to make predictions, generalisations and estimates, and to evaluate their own answers. Encourage these skills in all areas of the curriculum. Children use their understanding of mathematics at home and at school, in situations such as sorting toys or other items, telling the time, looking for and making patterns, helping to prepare food, and playing board and other games.

In school

In school, there are many opportunities for you to teach mathematics through familiar situations, so that the children understand why it is useful and appreciate the order and sense that mathematics gives to life. For example, the children can identify the date each day, as well as the time at various points throughout the lesson. Registration, dinner money, timetables, sorting and putting away equipment will provide a range of relevant experience in data work, measures, shape and space and geometry as well as number.

In play

Children of all ages should have opportunities to play both in and out of school. This offers them the freedom to explore new situations, to make discoveries for themselves and to be creative. Unfamiliar mathematics equipment should be introduced through play, with the children exploring the functions and possibilities of the materials. A good example of this is a geoboard – you can show the children how to use it, and then let them work out what different kinds of shapes they can make using the elastics and the board.

At home

Part of the teacher's role is to involve parents and guardians in the children's learning. Parents can do more than supervise their children's homework. Many activities can involve the parent in the child's learning and provoke mathematical discussion and language at home.

Encourage parents to extend their children's mathematical understanding through playing board and card games, and by encouraging them to help with normal home activities such as cooking, gardening, cleaning and organising the home, drawing up plans and measuring when redecorating, and estimating how many or how much when shopping.

Many children will voluntarily help and encourage younger siblings in games and getting organised. Family visits and holidays give children opportunities to see environments different from their own, and to experience time and distance. They are also likely to be budgeting pocket money, saving for special things and predicting how long it will take to afford treats.

Children may play with computer games that require a variety of mathematical skills. They might see and use a wide range of other electronic equipment at home, too, which also demands mathematical skills. Many children will be responsible for their own timekeeping and have a degree of responsibility for others.

Some homes will not actively encourage girls to use construction kits, computers or calculators, and some parents will not be confident of their own mathematical skills or understanding. As a teacher, you can help a great deal by making explicit the mathematical content of everyday experiences and activities.

You can find additional parent notes for each level on Oxford Owl. See page 9 for more details.

Section 3 An environment for exploration and reasoning

Organising the space

All teachers have their own preferences about how best to organise the available space. However, here are some useful guidelines for any classroom.

Storage

Always store equipment so that the children have easy access to it and you can check it periodically. Label all items clearly and encourage the children to make their own decisions about what they need. From the very beginning, insist that the children pack up and return equipment to the correct storage containers or shelves.

A mathematics centre

Some teachers may choose to have a dedicated area, maybe a corner, of the classroom where they keep maths materials such as counting frames and number frames, prepared number game boards, geometric and flat shapes. You can also display posters about shapes and numbers, patterns and other topics as these come out of the weekly lessons. Other teachers may prefer to integrate this into the rest of the classroom.

Recycled resources

Many household items can be repurposed to make mathematical equipment. Invite the children and their families to collect some of these materials at home in order to keep your classroom well resourced:

- bottle caps
- egg boxes
- clean glass jars
- buttons
- plastic caps from juice or milk cartons
- plastic yogurt pots
- beads
- natural objects such as shells, seed pods or stones for the children to arrange in size order.

Store the materials neatly in plastic containers or jars.

Whole class teaching

Often you will work with your whole class, especially at the beginning and end of a lesson. The course offers plenty of ideas for this kind of approach. The children may be seated together on chairs or on the floor, arranged in a circle for a discussion or clustered around you for a demonstration. If you are calling on individuals to show things to the group, make sure there is enough space at the front for them to come up, and make sure everyone can see and hear the volunteer child's answer.

Set up some classroom rules at the beginning of the year and go over these at the beginning of each term or even each week if necessary. Classroom rules might include listening when others are speaking and waiting turns. Many teachers find it useful to use signs for *agree*, *disagree* and *I know the answer* rather than having the children raise their hands or shout out. The traditional approach of raising hands can be intimidating for quieter children or those who take a little longer to think about their answers. Hand signals allow teachers to give more time to those in the group who need it. The illustrations on page 17 show some commonly used hand signals.

Group work

You can group the children in similar or mixed-ability groups, to suit the purpose of the work. This gives the children opportunities to collaborate and to discuss their work with each other and with you. It allows peer teaching to take place and for the work to be matched to their needs. It also allows you to work simultaneously with a number of children and this minimises the need for repeated explanations to individuals. Group teaching is an effective form of classroom organisation for both teacher and children.

Working individually or in pairs

Sometimes it may be appropriate for the children to work as individuals or in pairs, to provide extra help to children who need it, or to stimulate and challenge children who have grasped a concept quickly. Working individually gives the children the opportunity to concentrate on their own thinking, to develop this through investigations and problem solving, and to experiment with materials. Children working in pairs have the opportunity to develop collaborative skills, to play games together and to share ideas in an investigation.

Using manipulatives

Manipulatives give children an opportunity to experience the use of mathematics through concrete materials and objects. There are suggestions for different kinds of manipulatives in the unit-by-unit teaching guidance section of this book.

Counting apparatus
Interlocking cubes

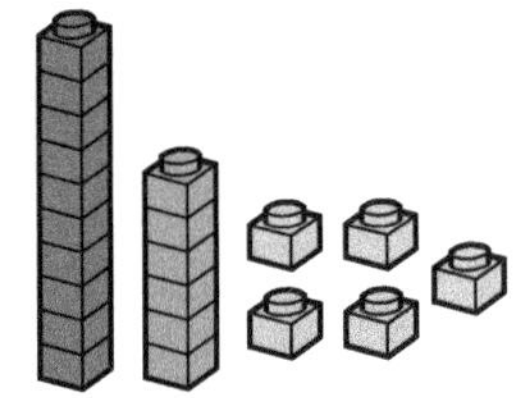

Interlocking or 'snap' cubes

Interlocking cubes are an invaluable resource for learning to count, especially as the children learn their number bonds up to 10, and then begin to work with tens and ones as they work with numbers up to 100.

Numicon (number frames)

Numicon

Numicon is another powerful and versatile concrete resource that supports children to visualise and work with numbers. Each hole in a shape represents a number.

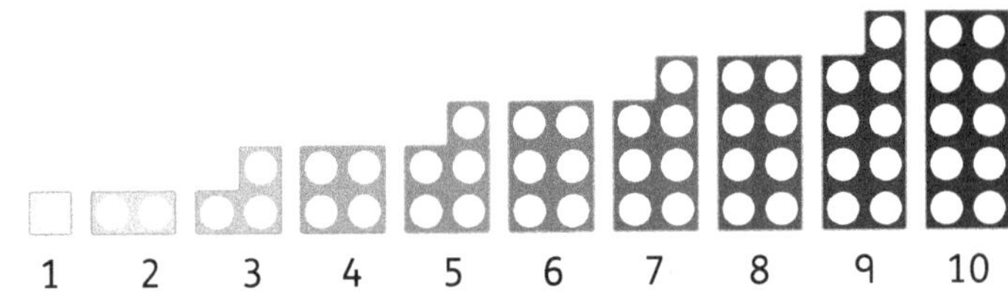

The children can use the Numicon shapes to count, represent, compare and calculate using fractions, decimals and ratios.

Base-ten blocks

Base-ten blocks are a concrete resource made up of blocks of varying sizes. Each block can represent a different value. For whole numbers, they represent:

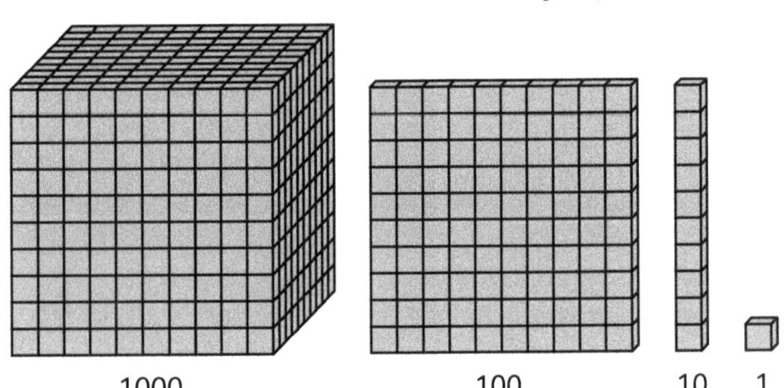

They can also be used to introduce decimals, representing ones, tenths, hundredths and thousandths, respectively.

Base-ten blocks help children to represent numbers, compare and calculate. They are particularly effective when used to support partitioning, for example into tens and ones.

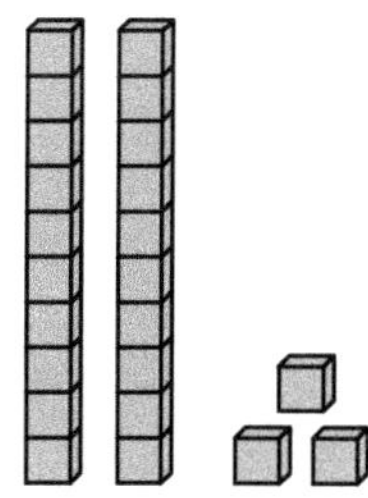

Representing 23 as two tens and three ones using two 10-rods and three ones cubes

Using base-ten blocks and place-value tables helps children to see how numbers change when multiplied and divided by powers of ten. Subtracting using base-ten blocks is a great, 'hands-on' way to introduce and support exchanging.

Bar models

Drawing bar models is excellent problem-solving strategy that children can use in both primary and secondary school. A problem that may otherwise seem very challenging can become more accessible when you 'draw' it.

Children should be encouraged to both interpret bar models as well as draw their own when solving word problems.

There are two main types of models: the part-part-whole model and the comparison model. Which you choose to use will depend on the problem you are trying to solve.

For example, Tyra has 32 marbles and she give 13 to Esme. How many marbles does she have left?

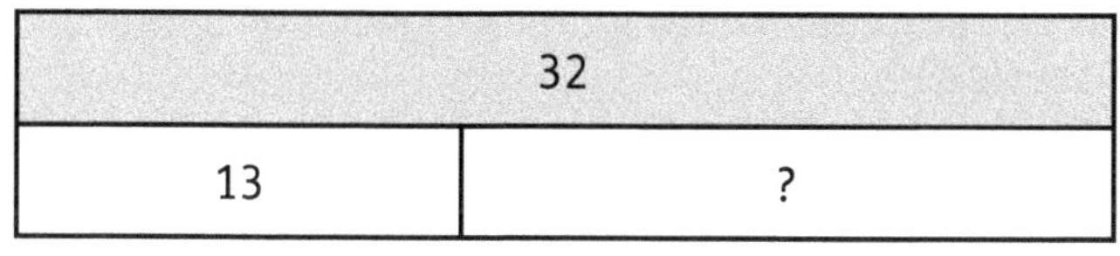

A part-part-whole bar model

Lewis completed 142 laps of the running track in one week. Joshua completed 52 laps in the same week.

How many fewer laps did Joshua complete than Lewis?

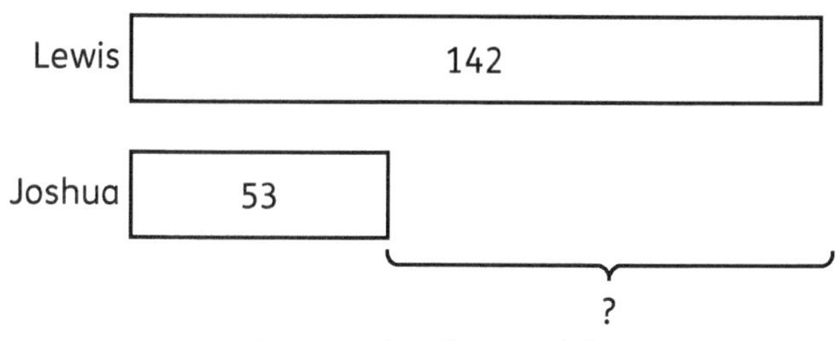

A comparison bar model

Place-value cards

Place-value cards for whole numbers have an 'arrow' or point on the right-hand side. Children can organise the cards horizontally or vertically to represent numbers in expanded notation. They can overlap cards and line up the arrows to form multi-digit numbers.

If any children have not previously worked with place-value cards, you will need to teach them how to use them. Begin by pointing out the arrows on the cards. Explain that these arrows always go on top of each other when you are making a number, for example:

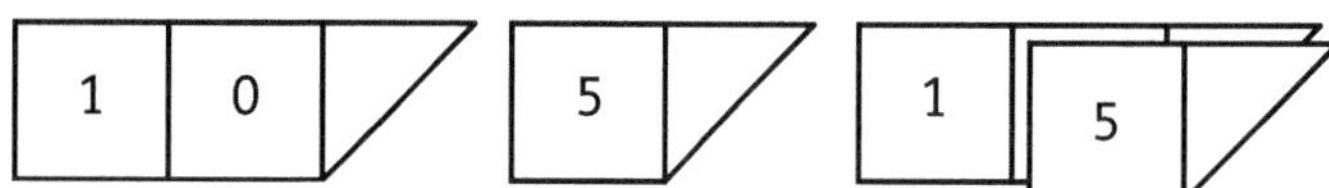

Place-value cards are an important teaching and learning resource and it would be useful to have a set available for each child. If possible, laminate the cards to make them more durable. (If you are making a set for each child, you may like to send the cards home for parents or carers to cut out.)

Place-value cards for decimals have an 'arrow' or point on the left-hand side, and a decimal point. For example:

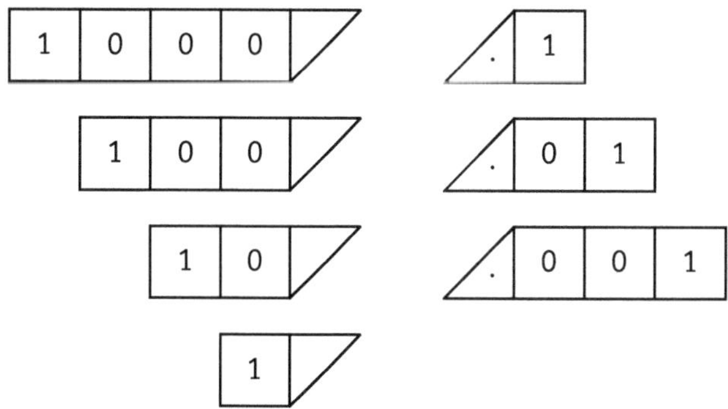

You can combine cards to make a decimal value, for example:

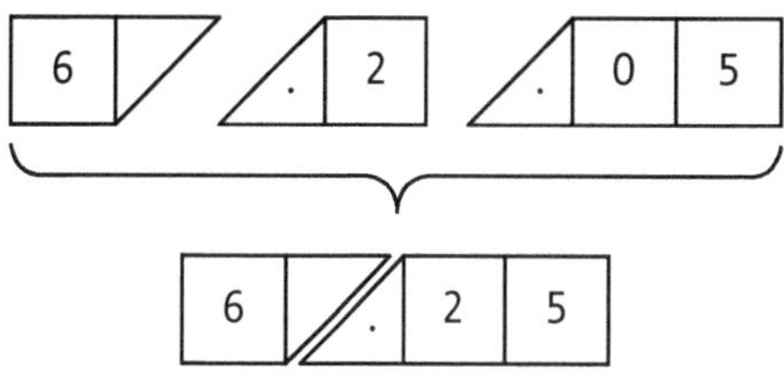

You can find printable place-value cards online.

Place-value tables

The children can use place-value tables to build numbers using counters or blocks. If possible, make place-value tables on card and laminate them to make them more durable.

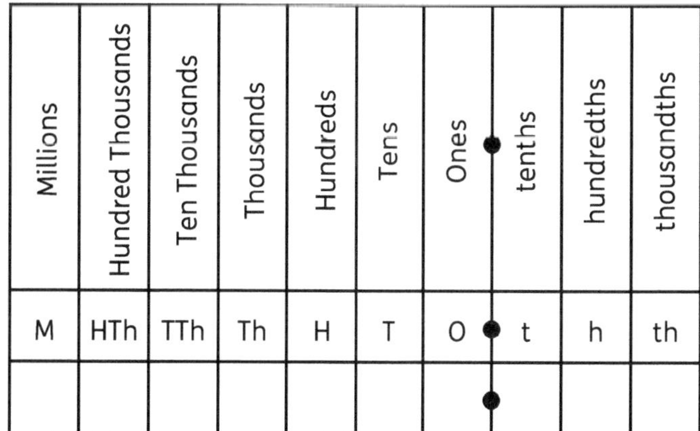

Millions	Hundred Thousands	Ten Thousands	Thousands	Hundreds	Tens	Ones	tenths	hundredths	thousandths
M	HTh	TTh	Th	H	T	O	t	h	th

Section 4 Activity bank of warm-ups and mental maths

This activity bank includes a range of mental maths activities, as well as some suggestions for support and consolidation activities. Most of the ideas are for number work and operations, although there are also some suggestions for the other strands. You can use this as a resource for activity ideas for your class. The unit-by-unit section in the next chapter refers back to specific activities in this section as unit introductions, suggestions for extra support and consolidation.

Try to include 10 minutes of mental maths activities each day. This section includes many examples that you can use as they are or adapt to suit your own classroom. It provides a range of different types of activities (factual recall, games, grids, tables, problem solving and puzzles) to show some of the ways in which you can approach the mental maths part of the lesson. However, this is not a definitive list and some activities will appeal more to some classes and teachers than others.

> Avoid any games or activities that use countdown clocks or timers to turn mental maths into a race or speed test. Speed testing is very damaging for children's mathematical development.

Place value and number sense

Identify the place value: Write any number on the board, for example 302 645.

- Ask the children to say the number aloud.
- Ask different children to say how many hundred thousands, ten thousands, tens, ones, etc. there are. (three hundred thousands, zero ten thousands, two thousands, six hundreds, four tens and five ones)
- Point to a digit and ask the children to say its value.
- Ask the children to reverse the digits and say the number. (five hundred and forty-six thousand, two hundred and three)

Make 6-digit numbers: Write six digits on the board, for example 5, 8, 2, 3, 1 and 4.

- Ask the children to make five different numbers using all of these digits.
- The children can exchange these and say each other's numbers aloud.
- Repeat the place-value questions using numbers the children have made.

Counting in given steps: The children work in pairs. Give them instructions such as:

- *Count forwards in ten thousands from 400 000 to 800 000.*
- *Count forwards in tens from 495 890 to 500 020.*
- *Count backwards in hundreds from 450 000 to 400 000.*

Make the greatest or smallest number: Give groups of children a set of five or six digits, each on a separate card. Ask the groups to make:

- the smallest possible number
- the greatest possible number.

Greater or smaller: Ask the children to write a 5- or 6-digit number on paper. Choose one child to come to the front of the class and display their number.

Ask: *Whose numbers are greater than this?* Let the children display their numbers. Repeat for 'smaller than'.

Choose one child at random. Invite them to come up and stand to the left or right of the other child (depending on whether their number is smaller or greater) and display their number. Choose other children to come up and position themselves appropriately between the 'numbers' already on display.

Number line: Write a list of positive and negative numbers on the board. For example:

12   –3   0   –5   –4   9   5   –7   –1   1   –9

Draw a blank number line on the board with only the start and finish marked.

Ask the children to find the smallest number. (–9) Write this on the left-hand side of the scale. Repeat for the greatest number, writing it on the right-hand side of the scale. (12)

Ask children to volunteer to come up to the board. Let them choose one of the numbers and position it on the scale as accurately as they can. Once all the numbers have been placed, discuss whether there are any inaccurate placements. Let the children decide and suggest how to move the numbers if necessary.

This activity can be adapted to work with decimals, fractions, mixed numbers and whole numbers.

Write all the possible numbers: Give the children some possible digits for each place value and ask them to work out how many numbers are possible with a given number of digits. Here is an example:

- *The hundreds place can have: 2, 3, 4 or 5.*
- *The tens place can have: 1, 2, 3, 4, 5 or 6.*
- *The ones place can have: 0, 1, 2, 3.*
- *How many 3-digit numbers can you make? (96)*

Make the number: Prepare a set of 40 cards with the digits 0–9 repeated four times. (If your class is larger than 36, you will need more cards.) Shuffle these and deal six cards at random to small groups of children

Ask the children to use all the cards to make:

* the greatest possible number
* the smallest possible number
* the number as close as possible to 1 000 000
* a number less than 500 000.

You can vary the task by changing the instructions and changing the number of digits. (For example, ask the children to make the largest possible 3-digit number.)

Fractions, decimals and percentages: Give the children a set of number cards with the digits from 0–9. They should choose two at random and write them as the numerator of a fraction with a denominator of 100. Repeat this five or six times. The children can then complete a table like this for each fraction they made. For example, with digit cards 6 and 0:

Fraction	Simplest form	Decimal	Percentage
$\frac{60}{100}$	$\frac{3}{5}$	0.6	60%

Rounding and estimating

Rounding mentally: Draw on the board a grid like the one below. If you are going to reinforce rounding to the nearest 10, make sure that all the numbers have a value other than 0 in the ones place.

456	1275	499	109
3245	6501	1295	1082
3509	8024	8019	876
103	562	901	1052

* Ask the children to copy the grid and to rewrite the numbers after rounding them to a given place value (for example, the nearest 10 or the nearest 100).
* Alternatively, you can work through the grid, pointing at each number in turn, and ask individual children to round it to the nearest 10.
* Repeat this for different place values.

Place 4-digit numbers on a number line: Draw an empty 0–10 000 number line with the thousand intervals marked but not labelled. Hand out a selection of 4-digit numbers to different children and invite them to come up and stick these in the appropriate positions on the number line. Allow the class to comment and discuss the placement of the numbers. Let the children decide whether any placements are inaccurate and move the numbers as necessary.

Real-life estimates: Display some large pictures in the classroom, for example photographs of a number of items, such as a tray of beans, stitches in a knitted jumper or bees in a hive, and ask the children to estimate how many there are. Ask them to explain how they find their answers.

Find the closest total: Write a set of six numbers on the board. For example, write a set of 3- and 4- digit numbers like these:

593   3199   5804   7083   908   427

Ask the class to find which two numbers should be added to get the total closest to a given number. Ask, for example: *Which two numbers should you add to get the total closest to 10 000?* (3199 and 7083)

Find the closest difference: You can adapt the previous activity for subtraction.

Make the target number: The children work in pairs. Ask them to choose any five digits and write them down. (You can vary this by stipulating that digits can or cannot be repeated.)

* Write a target 5-digit number on the board, for example 40 000. Ask the children to arrange their digits to make the number closest in value to 40 000.
* The children can compare their numbers and decide whose is closest.
* Repeat using different target numbers.

Wipe-out calculator game: The children work in pairs.

* One child enters a 4-, 5- or 6-digit number on a calculator. The other child should try to wipe out a specific digit without changing any other digits. For example, if the number is 345 234, the child needs to know to subtract 200 to wipe out the 2.
* Repeat, with the other child choosing the number.

Round decimals: Write a set of four digits on the board; two of these must be 5 and 9. For example, write: 6, 1, 5 and 9.

* Ask the children to make a decimal fraction that rounds to a suitable whole number, such as 60. The children can then make 59.6 or 59.61.
* Give the children some different target numbers. For example, ask them to make a number that rounds to 62. The children can then make 61.5, 61.9, 61.59 or 61.95.

Mental problem solving

Periods of time: Play a game with periods of time.

* Write a time period on the board (for example, 186 days). Then ask the children to estimate and write down how many weeks, months, hours, etc. this is.
* Repeat this for different units (for example, 4 weeks, 8 months).

Problem solving: Display this variation of the traditional rhyme 'As I was going to St Ives'.

As I was going to St John's
I met a man with seven sons
Every son had seven sacks
Every sack had seven cats
Every cat had seven kittens
Kittens, cats, sacks, sons
How many were going to St John's?

Ask the children work out the total number of kittens, sacks, cats and sons (although the answer to this riddle is 1 as it's the narrator going to St. John's). (7 sons, 49 sacks, 343 cats, 2401 kittens; multiply each number by 7) Ask them to explain how they worked it out. Change the activity by changing the numbers, for example: I met a man with 70 sons …

Number combination games: These should involve more than one operation, for example:

- *Use each digit 1, 3, 4, 6 and 9 once only to make a calculation that gives 3280.*

To find digits that work, do a calculation yourself. $91 \times 36 + 4 = 3280$. Then write them in size order for the children.

Goldbach's conjecture: Any even number can be written as the sum of two odd prime numbers, for example: $6 = 3 + 3$; $8 = 3 + 5$.

- Ask the children to express numbers in a given range as the sum of two odd primes. For example, write all the even numbers from 30 to 50 as the sum of two odd primes. (For example, $30 = 13 + 17$; $32 = 3 + 29$; $34 = 5 + 29$; $36 = 17 + 19$; $38 = 19 + 19$; $40 = 3 + 37$; $42 = 5 + 37$; $44 = 7 + 37$; $46 = 5 + 41$; $48 = 11 + 37$; $50 = 3 47$)

Letter-number code: Give each letter of the alphabet a numerical value. This can be A = \$1, B = \$2, C = \$3 and so on up to Z = \$26. Ask:

- *What is the value of your first name in this code?*
- *What is the value of your family name?*
- *Can you find a word that is worth \$50?*
- *Can you find a word that is worth more than \$75?*
- Can you make a word worth at least \$200?

Logic puzzles: These can be solved by mental calculations.

- *I have a 7-litre and a 5-litre container. How can I use these to accurately measure 4 litres of water?* (Fill the 7, pour it into the 5 leaving 2. Empty out the 5. Pour the 2 litres from the 7 into the 5. Fill the 7. Pour from the 7 to fill the 5. This will use 3 litres and leave 4 litres in the 7-litre container.)
- *How can you measure 1 litre with a 3-litre and a 5-litre container?* (Fill the 3 and pour the water into the 5. Fill the 3 again and pour the water into the 5 again. This will use 2 litres and leave 1 litre in the 3-litre container.)

Make a dollar: Ask the children to work out how many ways there are to make a dollar using any combination of 1¢, 2¢, 5¢, 10¢, 25¢ and 50¢ coins. This activity encourages the children to work systematically and to use jottings to keep track of their thinking. The most successful children will start either with the largest coins or the smallest like this:

```
50 + 50
50 + 25 + 25
50 + 25 + 10 + 10 + 5
50 + 25 + 10 + 5 + 5 + 5
```

You can make this problem easier by limiting the coin combinations.

Mathematical vocabulary: Give the children worded problems that can be solved mentally, for example:

- *Find the number that can be increased by 3 to make 21.* (18)
- *What is the product of 4 and a number 5 greater than 4?* (36)

- *What is the square of the difference between 16 and 7?* (81)
- *What number do you get if you halve the product of 9.8 and 100?* (490)
- *What is the sum of 345 and double 90?* (525)

Calculation skills

Multiplication practice:

Draw an empty 2×6 grid on the board. Ask the children to copy it into their books. Fill in the numbers from 1–12 on the grid in random order. For example:

6	1	10	11	8	4
2	5	9	7	3	12

Give the children a multiplier (for example $\times 7$, which suits work on converting weeks to days). They then fill in the product of multiplying each number on the grid by 7.

Mental addition and subtraction: The operations in these grids are given across and down. The children can work in any order to fill in the missing values.

$+7$ $\downarrow$	$+9 \rightarrow$				
	15	24			
	22	31			

-7 $\downarrow$	$-5 \rightarrow$				
	59	54			
	52	47			

- For subtraction, make sure you start with a large enough value to avoid negative numbers at too early a stage.
- You could prepare a variety of these grids on laminated cards.
- You can adapt this task to work with decimal values.

Doubling: Give one child a number, for example 7, and ask them to double it.

- Go round the class and see how far the children can get by doubling the number.
- Change the pattern by adding or subtracting a given number to get a different starting value. For example, once the children get to 448, say something like: *Next person, subtract 8.*

Halving: Repeat the previous task but ask the children to halve a given number. Remember that halving is more difficult than doubling as the children will get to fractional values as soon as they halve an odd number. When this happens, you can change the pattern by getting them to add a value such as $9\frac{1}{2}$.

Target number games: Give the class a target number.
- Either: ask the children to suggest different ways of getting to this number using whichever operations they like.
- Or: ask everyone to find ten different ways of getting to the number.

For example, if the target number is 86, the children could make the target by 80 + 6, 90 – 4, 8.6 × 10, 860 ÷ 10 and so on.

To make the game more challenging, you can ask the children to use particular operations. For example, make the target by adding three numbers, or make the target by doubling any number and then subtracting.

Decimal times-tables: Give the children the first four multiplications to establish the pattern involving decimals, for example:

    1 × 2.5
    2 × 2.5
    3 × 2.5
    4 × 2.5

Ask them to write the answers and to continue the pattern.

Find the sum: Ask the children to find the sum of all the numbers from 1 to 10, 1 to 25, 1 to 50 and so on using their own strategies. (1 to 10: 55; 1 to 15: 325; 1 to 50: 1275)

Fill in the gaps: Provide incomplete multiplication tables such as the ones below and ask the children to find the missing values. This requires them to think strategically and use inverse operations.

×			4
			12
			8
6	36	30	

×		8	
20	1800		320
		336	
		392	

- You can generate many of these tables yourself, but remember that you need at least one horizontal and vertical multiplier and at least four values for the children to find the missing values.
- You can extend these to a higher number range as the children become more competent at multiplication.

Odd one out: Provide a sheet of operations that form equivalent pairs with one odd one out. These could be fractions of amounts, equivalent fractions, decimal fractions paired with mixed numbers or percentages of amounts.

- Ask the children to find the odd one out and to explain how they decide. For example:

$\frac{3}{10}$ of 100	$\frac{9}{10}$ of 20	$\frac{1}{10}$ of 90
$\frac{4}{5}$ of 80	$\frac{1}{3}$ of 27	$\frac{1}{4}$ of 256
$\frac{1}{4}$ of 72	$\frac{1}{2}$ of 150	$\frac{1}{2}$ of 60

(75 as it's the only answer that does not make a matching pair.)

Same total: Make a set of cards with the digits 1 to 7 written on them. Display these on the board. Draw a grid like this one on the board.

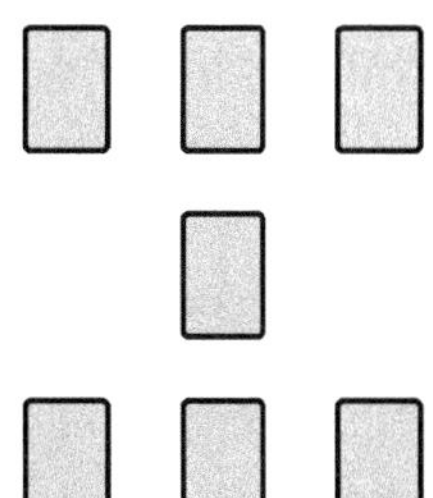

Ask the children to find ways of organising the digits so that the totals along any line (horizontal, vertical and diagonal) are the same. Encourage them to explain how they thought this out. (A possible answer is top row: 5, 1, 6; middle row: 4; bottom row: 2, 7, 3.)

Arrange the digits: Make a set of cards with the digits 1 to 6 written on them. Draw a grid like this on the board.

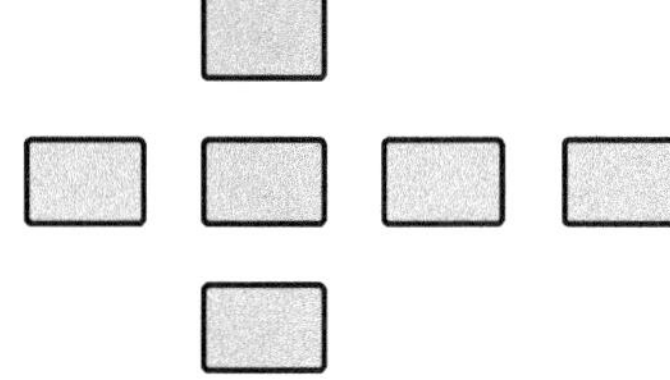

Ask the children to find ways of organising the digits so the sum of the vertical digits (the column) is equal to the sum of the horizontal ones (the row). (A possible answer is vertical digits: 2, 5, 6 ; horizontal digits: 3, 5, 4, 1)

You can make this more difficult by changing the instructions. For example, say:
- *Arrange the digits so that the sum of the horizontal row is 2 greater than the sum of the vertical column. (A possible answer is vertical column: 5, 1, 4; horizontal row: 2, 1, 3, 6)*
- *Arrange the digits so that the sum of the row is double the sum of the column. (A possible answer is vertical column: 4, 3, 1 horizontal row: 2, 3, 5, 6)*

Spinners: Give each pair of children a spinner with single-digit numbers on it.

The children take turns to spin the spinner and multiply the value they spin by a given multiple of 10 (for example × 40). They should take turns to spin five numbers and record their answers.

You could give the children these tasks:
- *Add up the products for the five calculations.*
- *Work out the difference between the highest and lowest scores.*

Minus the money: The children work in pairs. Each child starts with $10. They take turns to spin a 1–6 spinner and multiply the result by 10 (or any number you choose to reinforce). They must subtract that many cents from their total. They continue to play until one player has no money left.

Magic squares: Prepare a series of magic squares with one incorrect value. Ask the children to find the incorrect number and explain how they decided that it was incorrect. They should also suggest what the correct value is.

These examples use decimals with one place, but you can use decimals to hundredths, whole numbers, fractions and a mix of fractions and decimals to reinforce calculation in those areas.

2.8	3.5	3.0
3.3	3.1	2.9
3.2	2.6	3.4

7.1	7.3	6.9
6.6	6.8	7.0
6.7	7.2	6.5

9.3	9.2	9.7
9.8	9.4	8.9
9.1	9.6	9.5

The highlighted numbers are incorrect: this is for your information only. (The correct numbers are 2.7, 6.4 and 9.)

Equal sums: Show the class an arrangement of blank digit cards and give these instructions:

- *Arrange the digits from 1 to 6 so that the sum on each line is equal. (A possible answer is top row: 3; middle row: 5, 4; bottom row: 1, 6, 2)*

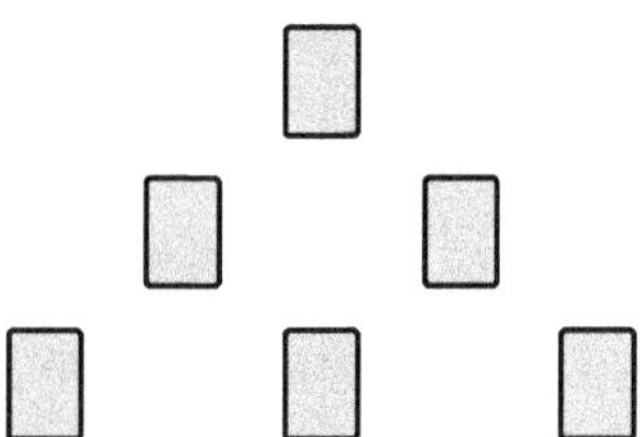

- *Arrange the digits 1 to 9 so that the sum of each line is equal. (A possible answer is top row: 6, 1, 8; middle row: 7, 5, 3; bottom row: 2, 9, 4)*

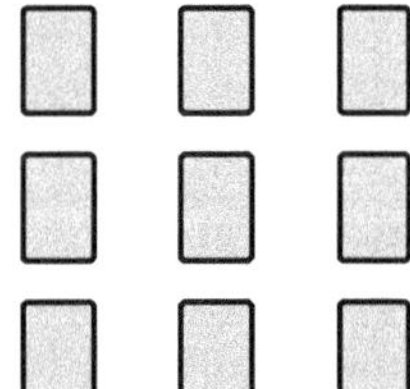

- *Arrange the digits 1 to 7 in these circles so that the sum on each line is equal. (Possible answer: 1 in the middle; clockwise from top: 5, 3, 2, 4, 6, 7)*

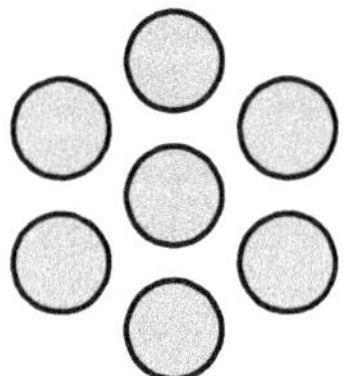

- *Use each digit from 1 to 7 once only to make a calculation with the answer 100. (A possible answer is $5 \times 6 \times 4 - 7 \times 3 + 2 - 1$)*

Multiples, factors and divisibility rules: Write a starting number on the board (for example, 64).

- Ask the children to jot down as many multiplications as possible to make this number.
- Repeat for other starting numbers including some prime numbers.
- You can also use decimal values once the children have had some practice in working with them.

Multiplication grid: Prepare decimal versions of the multiplication grid. Point to a square and invite individual children to give the product. For example:

×	0.1	0.2	0.5	0.8
3				
6				
8				
9				

Repeat this several times over a period of time, changing the values to reinforce multiplication facts and their translation to decimal values. Repeat for division as well.

Three or four in a row: This is a version of the game 'Noughts and crosses' that uses the ideas of equivalence between fractions, decimals and percentages. The children play in pairs. Give each pair a prepared grid with suitable families of fractions on it. For example:

$\frac{1}{10}$	$\frac{1}{5}$	$\frac{1}{2}$	$\frac{18}{40}$
$\frac{3}{50}$	$\frac{6}{20}$	$\frac{1}{25}$	$\frac{4}{25}$
$\frac{9}{100}$	$\frac{8}{10}$	$\frac{3}{5}$	$\frac{4}{10}$
$\frac{9}{20}$	$\frac{4}{5}$	$\frac{11}{20}$	$\frac{28}{100}$

- The children take turns to claim squares by offering an equivalent fraction, decimal or percentage. They can place counters on the squares they claim.
- The aim is to make a row of three (or four) while preventing their partner from making a row of three (or four).

Calendars and time

Calendar patterns: The dates for each month on a calendar offer a number of sequences. In a row, the dates count up in ones. The columns offer numbers with a difference of 7 and, on diagonals, the difference is either 6 or 8. The sum of three consecutive numbers will be 3 times the first number + 3 times the difference between them.

- Provide a blank calendar for a month with some dates filled in and let the children work on the missing ones (and only those). For example, fill in the first one and then let the children work out the dates going diagonally.

Finding times earlier and later than a given time:
Prepare a table like this one for display:

$\frac{1}{2}$ hour earlier	Time	15 minutes later	$\frac{3}{4}$ hour later	24-hour time	Time 4 hours ahead
	3.15 p.m.				
	2.30 a.m.				
	Half past five in the afternoon				
	10 to 6 in the morning				
	25 past nine at night				
	Midnight				

Point to cells in the table and invite different children give the missing values.

Geometry and measures

Making towers: Use a variety of solid objects (a few boxes, a can and ball) to build a small tower in the classroom. Ask the children questions such as these:
- *How many faces are there altogether in this structure? How did you work this out?*
- *How many corners (vertices) are there? How did you work this out?*

Angles on a straight line: Show the class a set of angles.
- Give the size of each angle in degrees and ask the children which angles can be put together to fit onto a straight line. (90° and 90°; 120° and 60°.; 150° and 30°; 100° and 80°; 135° and 45°)

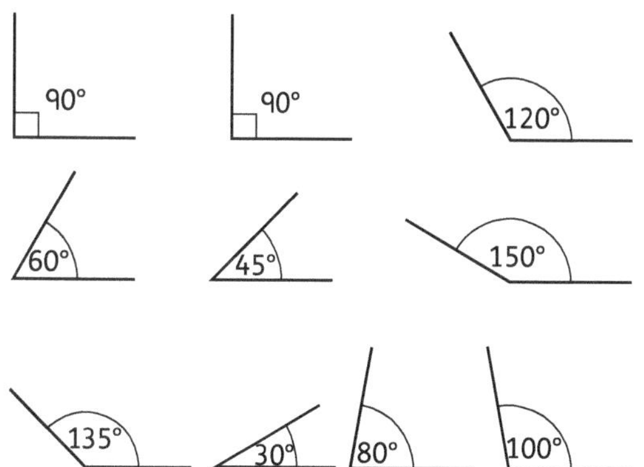

- To make it more difficult, include combinations such as 124° + 56° or 21° + 19° + 140°.

You can also do this activity using sectors of a circle. Ask the children which sectors or slices can fit together to make a circle (the angles sum to 360°).

Unknown angle game: Give the children two angles of a triangle and ask them to find the size of the third one (the value needed to make a total of 180°). For example:

$$90 + 30 + \square \ (60)$$
$$50 + 80 + \square \ (50)$$
$$34 + 56 + \square \ (90)$$

3D measures: Show the class a hard-cover book such as a dictionary. Give them dimensions for the book. For example, tell them that the covers are 5 mm thick and the pages are 3.5 cm thick. Ask the children to work out the height of a pile of 3, 5, 10, 25 and 100 books. Vary the activity by:
- adjusting the dimensions and the numbers of books in the pile
- telling the children that the books need to be packed on a shelf of a given length (for example, 1.2 m) then asking them to work out how many books can fit on the shelf.

Multiplying and dividing by 10: Give the children a variety of tables like this to complete:

	→ × 10 ← ÷ 10	
7.6		
9		
		156
		4

Multiplying and dividing by 100 and 1000: Give the children a variety of different tables like the one above that involve multiplying and dividing by 100 and 1000.

Converting metric units: The previous two activities can be adapted to give the children practice at converting between metric units of measurement.

Find the measurements: Draw several quadrilaterals on the board (use rectangles, squares, parallelograms, rhombuses and kites) and give their perimeter in different units. Ask the children to work out possible dimensions for the sides. To do this, they need to calculate but also apply the properties of each different shape.

Section 5 Lesson notes
Think maths

Learning objectives
- Establish a classroom environment conducive to thinking and working mathematically.
- Set up positive norms.
- Establish key messages of growth mindset mathematics.
- Consolidate planning and problem-solving skills.

Key words
strategy   critical thinking

Unit introduction

Teaching guidance

This short introductory unit aims to set up important norms in your class. It should be possible to complete this unit in two or three lessons.

As maths teachers, it is important to examine our own ideas about who can do maths, and how mathematical ability and learning work. It is also important to make sure the children understand that their own beliefs affect how they learn (or don't learn). From the earliest levels, this course aims to find real, experiential ways for the children to understand that:
- asking questions is the best way to learn
- making mistakes is a key part of brain growth and learning
- everyone's brain has the potential to learn maths. There is no such thing as a special 'maths brain'.

Just saying these things to the children will not make them believe us. We need to provide evidence of the ways that the brain grows and learns, and also to show them that maths can include everyone.

Think about representing numbers

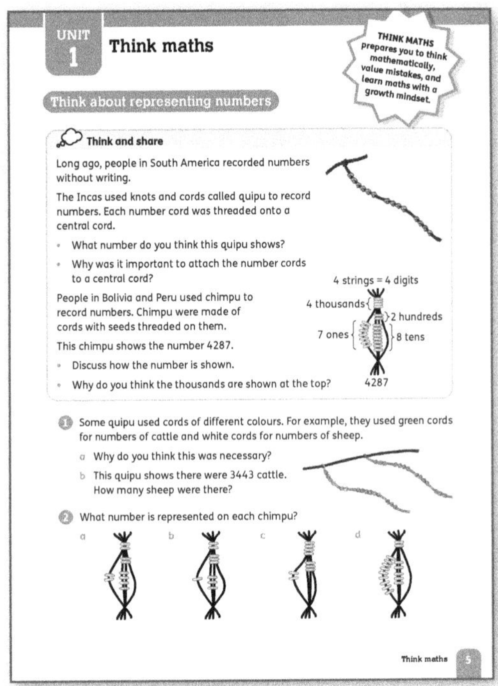

Materials
Wool of different colours; beads; counters; drawing equipment.

Warm-up
- Let the children work in groups.
- Ask them how they could show the numbers 523, 235 and 352 without writing anything. They might decide they could use counters, or draw diagrams. If so, ask: *How would you know what the place value of each counter is? Does a diagram count as writing?* Point out that any labels would count as writing. Ask the children to share their ideas with the class.
- Ask the children why they think people needed to invent numbers.
- Tell the class that the earliest counting systems used tally marks rather than numbers. For example, each morning some African goatherds placed one stone in a bag for each goat that left their farm. Without counting, they knew how many goats were out grazing. In the evening, when the goats returned to the farm, the goatherd removed one stone from the bag for each goat. Ask:
  - *What if some stones were left in the bag?* (Some goats had not returned.)
  - *What if there were not enough stones?* (Someone else's goats might have joined the herd.)

- Explain how other tally systems used notches on a stick. At the start of a count the person placed one notch per item down one side of the stick. To check the count again later, the person notched the other side.
- Spend time discussing the disadvantages of tally marks. Talk about the problems we might have today if we had no numbers for counting. Ensure the children understand that tally systems did not use place value and that many older tally systems had no numbers linked to amounts.
- Next, discuss how ancient civilisations kept numerical records. Talk about why they would do this. (It was mostly to keep track of a ruler's wealth or to tax cities and/or individuals.)

Focus

- <u>Think and share:</u> Look together at **Pupil book page 5**. Ask the children to work in groups to talk about the quipu and chimpu methods of showing numbers using knots. They could make their own, using coloured wool and beads. Note that both of these methods use place value.
- After they have looked at the number that the example chimpu shows (4287), the children are asked why the thousands are shown at the top. Spend some time getting answers from the different groups after their discussions. (The chimpu is important because we can attach other cords that express different things such as number of sheep, number of pots, etc. Suggestions about how the number is shown might be: the number is shown as four sets of seeds; each set represents a place value. The thousands are shown at the top of the cord.)
- The children can work on questions 1 and 2 in pairs.

Challenge

Ask: *How could you use a chimpu to record 439 people, 165 sheep, 432 cattle and 400 sacks of grain counted in one place?* Ask the children to draw their ideas and share their drawings with the class.

Answers for Pupil book page 5

1 a e.g. They needed different colour cords for different animals, so they could record the number of each type of animal.
   b 4263 sheep
2 a 2242   b 2331
   c 4302   d 2059

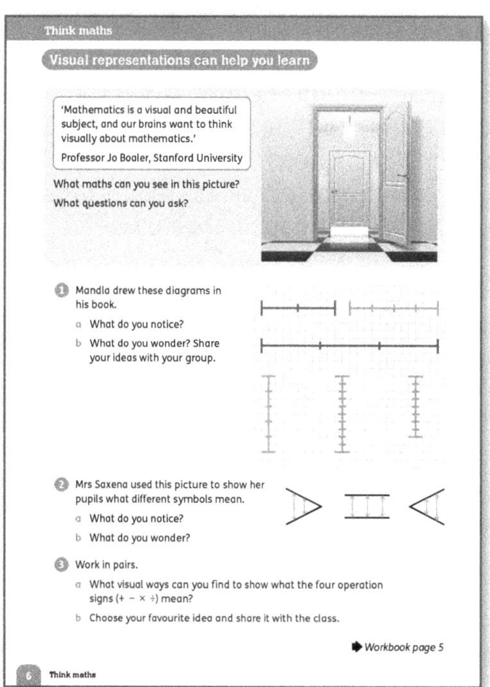

Materials

Drawing equipment; large sheets of paper.

Warm-up

Start by reading the quote by Professor Jo Boaler on **Pupil book page 6**. Explain that she is a professor of mathematics who believes that all people can learn maths to a high level, that maths learning is natural and that we can better understand things by using diagrams or other visual methods.

Focus

- Talk about the picture on **Pupil book page 6** as a class. The explanation box at the top of the page asks: What maths can you see in this picture? (*shapes: rectangles, a sphere, parallel lines, non-parallel lines*) What questions can you ask? (Examples are: *Where will the non-parallel lines going back into the picture meet? How many times larger is the door frame at the front of the picture than the door frame as the back of the picture?*)
- Let the children pose their questions. Let others suggest answers to these, or talk together about how you could find the answers.
- The questions in question 1 provide a powerful 'strategy' for developing growth mindsets, curiosity and creative and 'critical thinking'. It is important to give the children time to think before you take class feedback.
- Again, for question 2, let the children think about these images before taking class feedback.
- For question 3, let the children work in pairs and use large sheets of paper and drawing equipment to share and explore their ideas. The group should choose one to share with the class.

Follow-up

- Turn to **Workbook page 5**. Read through the poster with the class before asking them to think about when they've done each thing (perhaps in this lesson). Let the children add two points of their own to the poster. If they find this challenging, let them talk about it in groups or pairs.
- Ask the children to think about the group activities in this and the previous lesson and how they learnt from others in these activities. Then let them complete the mindmap on their own.

Challenge

Let the children work in pairs or groups to make a poster of images that they think are visually attractive and that have mathematical relevance. Many of the large photo libraries have online images the children can browse. Alternatively, the children could take their own photographs of mathematics in the environment and print these to make a class or school display.

Answers for Pupil book page 6

1 a e.g. The lines are different colours, lengths, positions and directions. Each line is divided into equal intervals. The numbers of intervals are different. The blue and green lines are the same length.

b e.g. Where do you use numbers like this in mathematics? Can you use any of the lines to help you find equivalent fractions? To round numbers? As a thermometer scale or probability scale.

2 a e.g. The symbols are > = and <. In > and < the distance between open ends of the lines is greater next to the greater number. In = the lines remain an equal distance apart.

b e.g. Would these diagrams help people to remember what the signs mean? Can you draw diagrams to help explain other symbols?

3 a, b Individual ideas about representations of the signs + - ÷ × that explain their meaning.

Answers for Workbook page 5

1 Individual answers.

2 Individual answers.

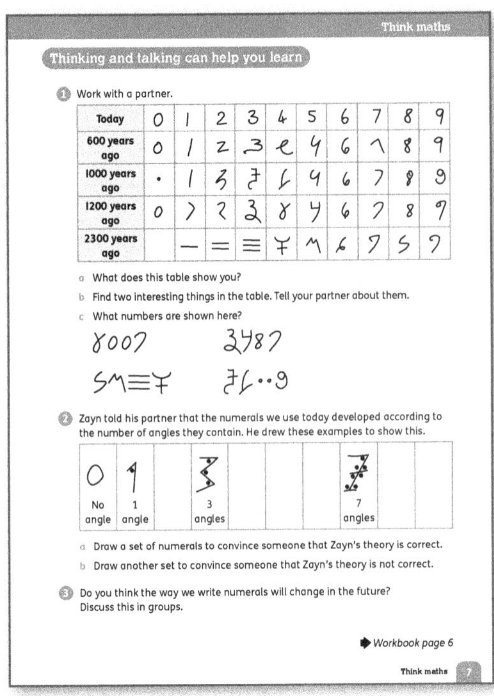

Materials

Rulers; drawing equipment.

Warm-up

Explain that the numbers we write today have developed over time from a number of different societies and systems of writing.

Focus

- Turn to the chart on **Pupil book page 7** question 1. Let the children work in pairs to discuss what is shown and then to say what is interesting. The children need to work out which time period the numbers in part c come from before they can work out what number is shown.
- For question 2, explain that: a theory is an idea that tries to explain something; a theory has to be proven before it is accepted as correct; if you can find an example that doesn't fit the theory, you can generally say that the theory is not correct. Let the children draw their own sets of numerals before asking them to convince someone else that the theory is correct, or not correct.
- The work in question 2 on the number of angles in each numeral is widely debated, but it makes for an interesting activity and discussion. You might like to extend this by looking at how all the digits from 0 to 9 are represented on digital displays by highlighting different parts of a seven-segment rectangular display. Display the blank '8' on the board and have the children show how they would represent each number on this LED display.
- Let the children discuss question 3 in groups and then have a whole class discussion about their ideas.

Follow-up

Use questions 1 and 2 on **Workbook page 6** to consolidate the idea that there is often more than one answer in a maths problem and that talking and thinking can help you see things differently and learn more about what you are doing. Let the children share their answers as a class and discuss any interesting ideas or disagreements.

Challenge

Let the children prepare their own sets of four numbers for a 'Which one doesn't belong?' activity. Let them spend some time talking about how these kinds of tasks work and how, very often, all four numbers can be said to not belong for different reasons. Display the children's sets in the classroom for others to think about, or use them as mental warm-ups in other lessons.

Answers for Pupil book page 7

1 a It shows the different notations of the numbers from 2300 years ago up to the present day.
  b e.g. There was no symbol for zero 2300 years ago. The symbols for 6 and 7 have not changed much over the years.
  c 4007, 3587, 8534 and 34 009.

2 a Individual drawings, e.g.
  b Individual drawings, e.g.

3 Individual ideas, e.g. Yes they may change to be more like the numbers we see on digital displays, or No, because people all over the world need to recognise numbers so we all need to use the same system.

Answers for Workbook page 6

1 e.g. 220 does not belong because it is the only even number. 11 does not belong because it is the only number with two digits.
143 does not belong because it is not a square number. 36 does not belong because it doesn't have the digit 1 in it.

2

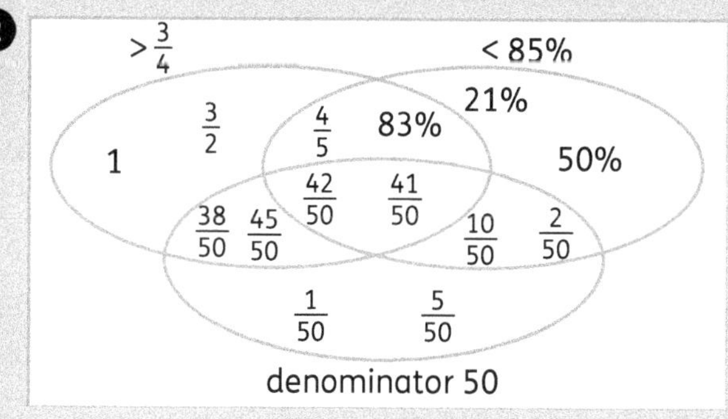

Interesting mistakes

- Many of us carry implicit beliefs and stereotypes about maths. These assumptions may come from our families or friends, from our earlier experiences with the subject, from other teachers, and even from popular culture and the media. Here are a few myths and stereotypes:
  - 'Maths is a difficult subject.'
  - 'Only some people are truly good at maths.'
  - 'If I get things wrong it means I'm useless at maths.'
  - 'Some people have a maths brain and others are better at creative/language subjects.'
  - 'Maths is a boys' subject.'
  - 'If I ask a question, everyone will know I don't understand, and they might think I'm stupid.'
  - 'Smart kids get everything right.'
- None of these myths are true, but they still persist – among both children and teachers. By focusing on how learning takes place, and the importance of mistakes and questions, we can constantly remind the children to question these messages.

End-of-unit check

Give children a table of fixed mindset statements and ask them to replace each one with a more positive growth mindset statement. You can find many examples of these online, but here is a set to start with:

Instead of this:	I can say this:
I can't do any better.	I can improve my understanding.
I just don't get this.	I can solve simple problems and then gradually understand more difficult problems.
This is too difficult for me.	If I keep trying, this will get easier.
I give up, I'll never be able to do this.	I will have another go.
I got some wrong, I'm not good at this.	I got some questions right, so I am beginning to understand this.
It's better to work on your own so that you can make sure you get everything right.	I can learn by sharing ideas.
I can't work out what to do with this problem.	A picture is worth a thousand words.

Numbers and place value

Learning objectives
- Read, write, order and compare numbers up to 10 000 000 and determine the value of each digit.
- Compose, decompose and regroup numbers using standard and non-standard partitioning.
- Round any number as appropriate, including in context.
- Use negative numbers in context and calculate intervals across zero.
- Count on and count back in steps of constant size and extend beyond zero to include negative numbers.
- Solve number and practical problems that involve all the above.

Key words
million   compose   decompose   ascending
descending   round   negative number
number sequence   term-to-term rule

Unit introduction

Teaching guidance
Have a number talk (pages 17–18) around the following problem to revise working with large numbers (to 'millions') place value and money amounts:
- Ask the class to discuss in groups whether it would be better to receive £1 million (use local currency) or 1p per day doubled each day for 30 days (in other words you get 1p on day 1, 2p on day 2 and so on).
- Give the children time to think about this and to talk about which is better (in terms of getting more money) before they do any working. Then let them share their ideas and thoughts. Let them use a calculator to work out how much money they'd get in 30 days, doubling 1p each day. The figures may surprise them.

1	2	4	8	16
32	64	128	256	512
1024	2048	4096	8192	16 384
32 768	65 536	131 072	262 144	524 288
1 048 576	2 097 152	4 194 304	8 388 608	16 777 216
33 554 432	67 108 864	134 217 728	268 435 456	536 870 912

- At the end of 30 days you would get 536 870 912 pence. Dividing by 100 to get pounds, you would have £5 368 709.12 – so you would have over 5 million as opposed to 1 million.

- Some children may argue that it is better to get the money up front and invest it (to earn interest, dividends or other returns) than to wait 30 days. This is a useful discussion point as you would need get over 500% return on the 1 million to equal the amount in pence. Given that really good investments would only get them around 20% (and they would risk losing money), it is still more profitable to take the pence option.

Revisit place value

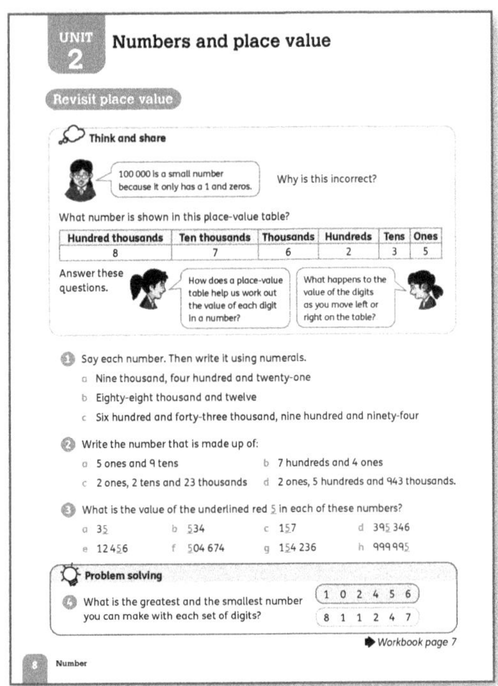

Materials
0–9 digit cards; place-value tables (page 21); place-value cards (page 21); drawing equipment.

Warm-up
- Use any of the 'Place value and number sense' activities (pages 22–23) as a mental starter for this lesson.
- Before moving to **Pupil book page 8**, ask the children to complete **Workbook page 7**. You can use their completed work to assess how well they know place value and numbers.

Focus
- <u>Think and share:</u> Turn to **Pupil book page 8**. Let the children work in small groups to read through the Think and share activity and to discuss their answers. The children are asked why it is incorrect to say that 100 is a small number because it only has 1 and zeros. (You need to look at the place value of the digits to decide whether a number is big or small. The 1 has place value one hundred thousand so it is not a small number. The place-value table shows the

number eight hundred and seventy-six thousand, two hundred and thirty-five.) Then the children are asked to explain how a place-value table helps us to work out the value of each digit in a number and what happens to the value of a digit as you move left or right in the table. (A place-value table helps us to see the value of a digit in a number. When you move to the left in a place-value table, the value of a digit becomes ten times greater; when you move to the right, the value of a digit becomes ten times smaller.)

- Next, ask different groups to share their ideas and to show the class how they represented the numbers. Try to get the children to tell you some of the advantages of the decimal place-value system we use today, for example:
  - It only uses ten digits and you don't have to tally.
  - It is a base-ten decimal system, so you can use place value to represent any number.
  - It has a zero that can be used as a place holder.
- Remind the children that these are not new ideas: thousands of years ago, the Babylonians and native South Americans used place value, the Chinese used 'negative numbers' and the Ancient Egyptians had systems of fractions.
- For question 1, ask the children to say each number aloud with the class, or in groups, before they write the numbers.
- If children need additional support with question 2, encourage them to draw simple place-value tables (page 21) and use dots to work out each number. Some children may also find place-value tables useful for question 3.
- <u>Problem solving:</u> For question 4, allow the children to use 0–9 digit cards and arrange them to help solve the problem.

Support

Give the children a set of place-value cards (page 21) or several sets of 0–9 digit cards. Say numbers less than 1 million and ask the children to lay out the cards to show each number.

Interesting mistakes

Children may need additional support to write and/or say numbers where zero is a place holder, for example: 4056 and 304 786. If they find this challenging, spend time asking the children to enter numbers into a place-value table and then say them in expanded form, emphasising where there is a zero in a column and what this means.

Answers for Pupil book 8

1 a 9421 b 88 012 c 643 994
2 a 95 b 704 c 23 022 d 943 502
3 a 5 b 500 c 50 d 5000
  e 50 f 500 000 g 50 000 h 5
4 654 210 and 112 478

Answers for Workbook page 7

1

	Hundred thousands	Ten thousands	Thousands	Hundreds	Tens	Ones
a				3	1	2
b			9	3	1	7
c		9	9	4	4	6
d	4	0	5	3	5	2
e	9	0	0	0	0	8
f		4	0	3	0	4
g		7	6	0	3	5
h	5	7	2	3	9	7
i	3	0	0	9	1	6
j	9	9	9	9	9	9

2

Group	What it means	Value	Name
10 groups of ten	10×10	100	one hundred
10 groups of 100	10×100	1000	one thousand
10 groups of 1000	10×1000	10 000	ten thousand
10 groups of 10 000	$10 \times 10 000$	100 000	one hundred thousand
10 groups of 100 000	$10 \times 100 000$	1 000 000	one million
10 groups of 1 000 000	$10 \times 1 000 000$	10 000 000	ten million

Millions

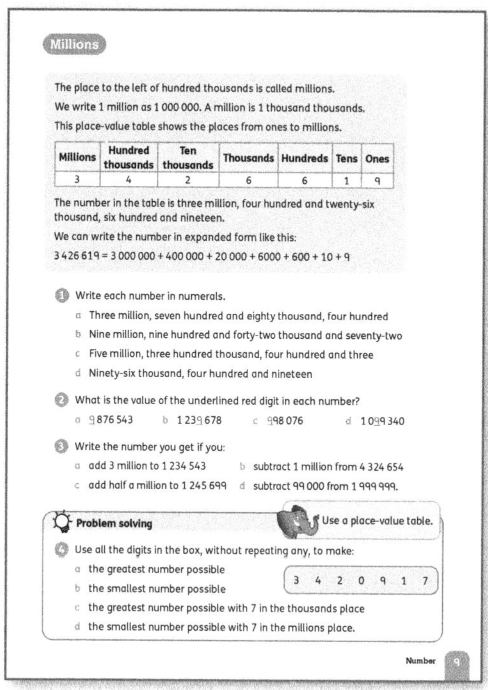

Materials

Large place-value table (page 21); counters and/or 0–9 digit cards.

Warm-up

- Use any of the 'Place value and number sense' activities (pages 22–23) as a mental warm-up for this lesson.

- Using the following examples, ask: *Which numbers will give 4 000 000 when you 'round' to the nearest million?* (3 589 143, 4 432 987, 4 199 999)

3 589 143	4 432 987	3 499 999
4 843 987	3 099 999	4 199 999

Focus

- Remind the children that the place-value table on **Pupil book page 9** is extended to the left to include higher and higher places, including millions (which they have met in Level 5). Make sure that the children understand the relationship between places in the table in terms of scaling the numbers by 10 or one-tenth. The children should be able to tell you that 1 000 000 is '10 times greater than 100 000' and that '100 000 is $\frac{1}{10}$ of 1 million' (in other words, it is ten times smaller).
- Read the number shown in the table and revise how to write numbers in expanded form.
- The children can work independently to complete questions 1–3.
- Check the children's answers before asking them to complete the problem-solving question 4.
- <u>Problem solving</u>: For question 4, provide a place-value table showing places from ones to millions or have the children draw up their own to keep and refer to as they work with larger numbers. The children will find this question easier to do if they model the numbers using 0–9 digit cards.

Challenge

- Let the children investigate binary numbers to see how numbers can be written using only the digits 1 and 0 in a different place-value system.
- You can help them understand that binary is a base two system by drawing up a place-value table for base two. Each place is 2 times as great as the one to its right.

16	8	4	2	1
1	1	0	1	0

- The binary number 11010$_{(base 2)}$ means $(1 \times 16) + (1 \times 8) + (0 \times 4) + (1 \times 2) + (0 \times 1) = 16 + 8 + 2 = 26$
- The only single digit numbers that are possible in this system are 0 and 1. As soon as you count beyond that, you move to the next place, so 2 is written as 10$_{(base 2)}$.
- The children could explore counting from one to ten in binary numbers and look at the patterns that develop.

Support

Draw a place-value table on the board. Put numbers into the table and use it to help the children say the numbers aloud. Say a number and ask children to come up and write the number in the place-value table.

Interesting mistakes

Children may need additional support to say large numbers correctly. Otherwise they may resort to simply saying the digits of the number in order. Do not allow them to do this, working with them to identify the value of each digit, saying the numbers in expanded form and then moving on to saying the numbers correctly.

Answers for Pupil book 9

1 a 3 780 400   b 9 942 072
   c 5 300 403   d 96 419

2 a nine million   b nine thousand
   c nine hundred thousand   d ninety thousand

3 a 4 234 543   b 3 324 654
   c 1 745 699   d 1 900 999

4 a 9 743 210   b 1 023 479
   c 9 437 210   d 7 0121 349

More millions

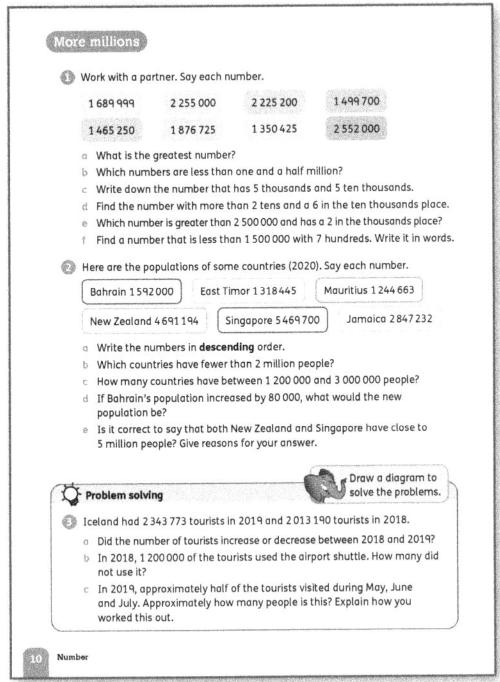

Materials

Atlas and/or world map; place-value tables (page 21); counters or 0–9 digit cards; calculators; drawing equipment.

Warm-up

Use any of the 'Place value and number sense' activities (pages 22–23) as a mental warm-up for this lesson.

Focus

- To reinforce how to say large numbers, let the children work orally, in pairs, through question 1 on **Pupil book page 10**. Then let them work on their own to write the answers.
- For question 2, use an atlas or world map to locate the countries. The children should realise that these are all fairly small countries and that other countries (possibly including your own) have much higher total populations.

- <u>Problem solving</u>: For question 3, remind the children that they can draw bar models (page 21) to show problems like these.

Interesting mistakes
- Some children find it difficult to read and write numbers given to them in words. They will need lots of practice to reinforce the idea that we read numbers in groups of three from left to right.

Answers for Pupil book page 10
1 a 2 552 000   b 1 499 700; 1 465 250; 1 350 425
   c 2 255 000   d 1 465 250   e 2 552 000
   f 1 499 700; 1 million, four hundred and ninety-nine thousand, seven hundred

2 a 5 469 700; 4 691 194; 2 847 232; 1 592 000; 1 318 445; 1 244 663
   b Bahrain; East Timor; Mauritius
   c 4
   d 1 672 000
   e Yes; both have a population of 5 million rounded to the nearest million.

3 a increase
   b 813 190
   c 1 171 887. Divide 2 343 773 by 2. It gives 1 171 886.5. Round it to the nearest whole number. As the tenths digit is 5, you round up.

Compose and decompose numbers

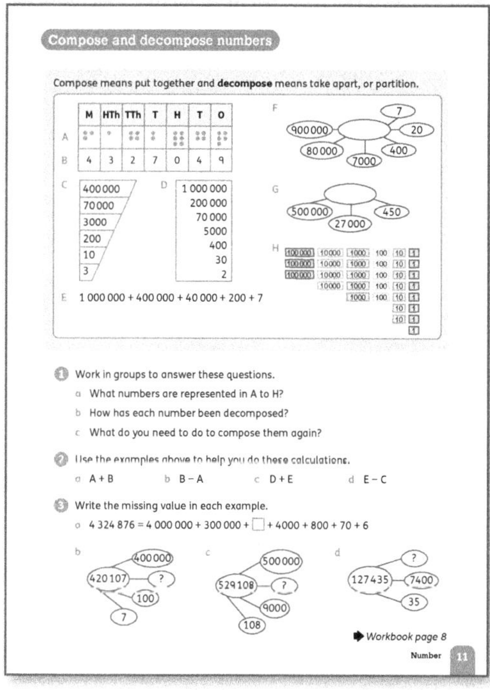

Warm-up
- Use any of the 'Place value and number sense' activities (pages 22–23) as a mental warm-up for this lesson.
- Write the words 'compose' and 'decompose' on the board. Talk about what they mean in everyday language (for example in: compose music; water is composed of hydrogen and oxygen; leaves and other organic matter decompose on the forest floor; when animals die, their bodies decompose).
- Elicit that compose generally means put things together, while decompose means to break things apart.
- Ask the children what these words mean in relation to numbers (they should remember this from earlier levels).

Focus
- Turn to **Pupil book page 11**. For question 1, let the children work in groups to discuss the pictures and answer the questions. Make sure they write the answer to part a. Let them discuss their answers and examples with the class.
- For question 2, you may need to check that the children have written the numbers correctly before they do the calculations.
- <u>Problem solving</u>: The children can use place-value cards to help them answer question 3.

Follow-up
Use **Workbook page 8** to consolidate composing and decomposing numbers and to assess whether the children can use these methods to help them calculate the sum of two numbers.

Challenge
- Draw a row of six boxes on the board. Ask the children to make a copy of this on paper.
- Explain to the children you are going to pick a card from a set of digit cards and then return it to the pack and shuffle. Each time, they must write the digit in one of the boxes. Once placed, the number cannot be moved.
- Explain that the aim is to create the largest number when all six boxes have been filled.
- Discuss with the children the strategies they used to make decisions about where to place numbers.
- Repeat, trying to make the smallest number.

Support
Children may need additional support when faced with non-traditional portioning because they may not initially see the value of breaking up numbers in different ways. It may help to use these representations in the context of comparing or finding the sum or difference between two numbers.

Materials
Place-value tables (page 21); counters; place-value cards (page 21).

Answers for Pupil book page 11

1 **a** A = 3 142 645; B = 4 327 049; C = 473 213;
D = 1 275 432; E = 1 440 207; F = 987 427;
G =527 450; H = 345 578

b They are decomposed using their place
values into millions, hundred thousands,
ten thousands, thousands, hundreds, tens
and ones.

c Add them together.

2 **a** 7 469 694   **b** 1 184 404
c 2 715 639   **d** 966 994

3 **a** 20 000   **b** 20 000
c 20 000   **d** 120 000

Answers for Workbook page 8

1 **a** 12 345 = 10 000 + 2000 + 300 + 40 + 5
b 329 410 = 300 000 + 20 000 + 9000 + 400 + 10
c 43 941   **d** 140 000
e 400 000 + 90 000 + 9000 + 10 + 9

2 **a** 884 438 + 391 902 = 1 276 340
b 1 413 279 + 946 043 = 2 359 322

3 e.g. Using a place-value table:

Thousands	Hundreds	Tens	Ones
4	3	1	4

Compare and order numbers

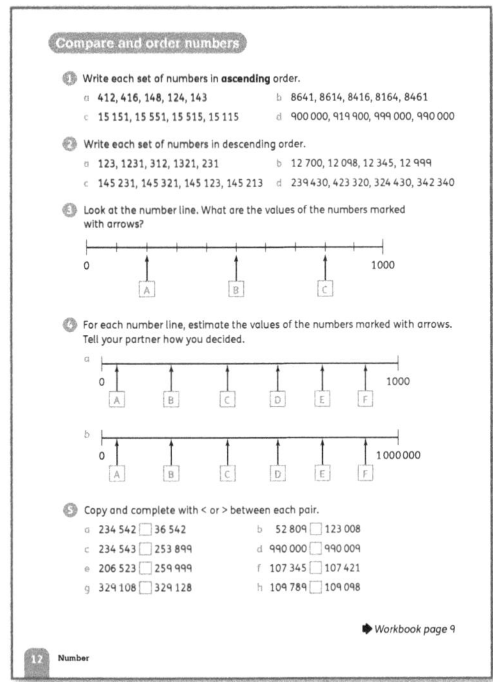

Materials
Number cards with numbers up to 10 000 000; blank
number lines; calculators.

Warm-up
Use any of the 'Place value and number sense' activities
(pages 22–23) as a mental warm-up for this lesson.

Focus
- Explain the meaning of 'ascending' and 'descending'
of numbers.
- Ask a group of children to stand up. Give each child a
card showing a number less than 10 000 000. Get the
rest of the class to instruct the group so they stand
showing the numbers in ascending or descending order.
- Give another child a different number card and ask
them to position themselves in the line.
- Ask the children to say a number that could lie
between two of the numbers in the line.
- Write several 5- and 6-digit numbers on the board.
Ask the children to order them in ascending order.
Encourage the children to explain how they made
their decisions about where to place numbers.
- The children can work on their own to complete
questions 1–5 on **Pupil book page 12**.

Follow-up
- Use **Workbook page 9** for additional practice
and to check that the children can compare and
order numbers.
- Let them complete the questions on their own and
then work in groups to check and compare answers.

Support
Children who find it difficult to order a set of numbers
can write them on cards and move them into order.

Interesting mistakes
- Although the children will have used number lines in
the earlier levels, they may find locating numbers on
an empty number line difficult.
- Spend some time looking at number lines with
different intervals to make sure the class can work out
what each interval represents and match the value of
a number to its correct position on the number line.
- In-between intervals can be worked out using the
children's knowledge of fractions. For example,
the number 2500 must lie halfway between 2000
and 3000.

Answers for Pupil book page 12

1 **a** 124, 143, 148, 412, 416
b 8164, 8416, 8461, 8614, 8641
c 15 115, 15 151, 15 515, 15 551
d 900 000, 919 900, 990 000, 999 000

2 **a** 1321, 1231, 312, 231, 123
b 12 999, 12 700, 12 345, 12 098
c 145 321, 145 231, 145 213, 145 123
d 423 320, 342 340, 324 430, 239 430

3 200; 500; 800

4 **a** 50; 250; 400; 550; 700; 850
b 50 000; 250 000; 400 000; 550 000; 700 000;
850 000

5 **a** 234 542 > 36 542   **b** 52 809 < 123 008
c 234 543 < 253 899   **d** 990 000 < 990 009
e 206 523 < 259 999   **f** 107 345 < 107 421
g 329 108 < 329 128   **h** 109 789 > 109 098

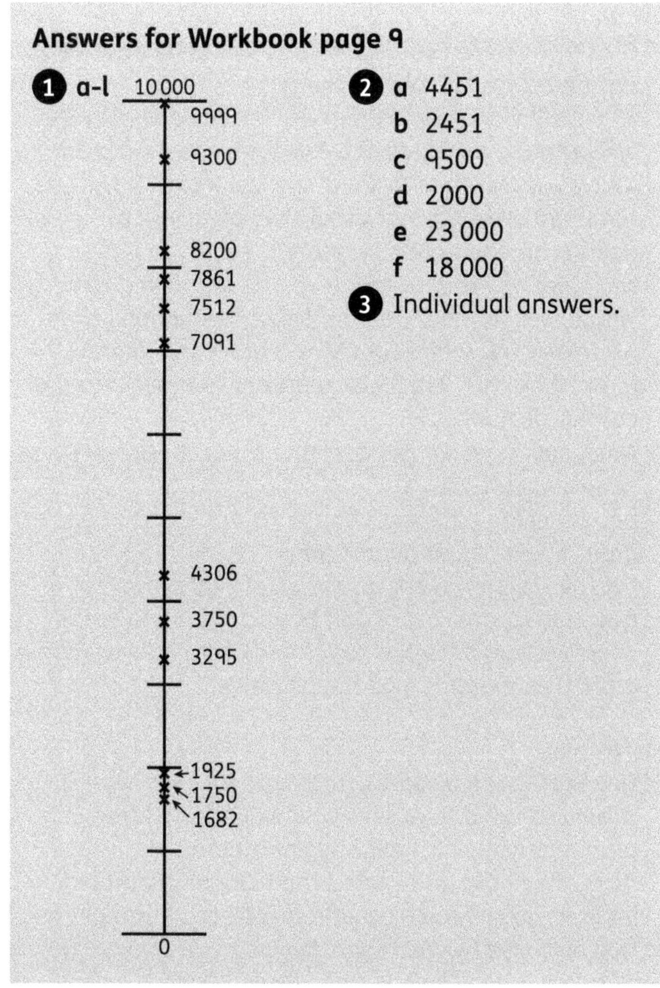

2 a 4451
   b 2451
   c 9500
   d 2000
   e 23 000
   f 18 000

3 Individual answers.

Round numbers to any place

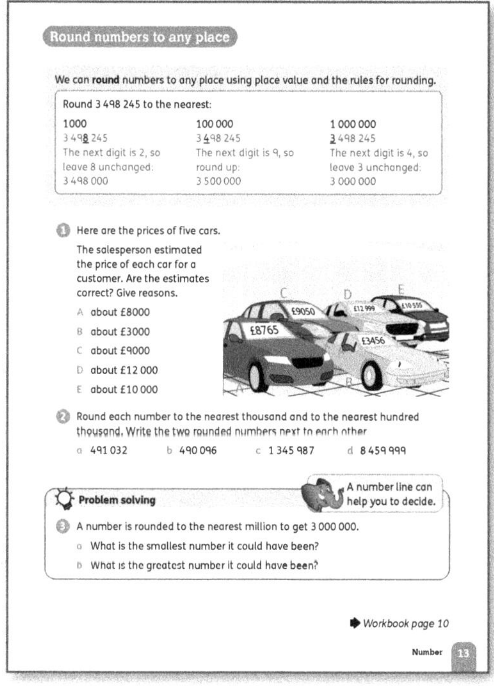

Warm-up

- Use any of the 'Rounding and estimating' activities (page 23) as a mental warm-up for this lesson.
- Write a multiple of 10 on the board. Tell the children you are thinking of a number that when rounded to the nearest 10 is the number on the board. Invite the children to tell you what the number could be. Also ask: *What numbers could it not be? Why?* Repeat with multiples of 100.
- Remind the children how they rounded numbers to the nearest hundred and thousand.
- Write 6-digit numbers on the board and ask the children to round them to the nearest thousand, then to the nearest hundred thousand. Highlight how we use exactly the same strategy as we did when rounding smaller numbers; that is, we look at the digit immediately to the right of the hundred thousand digit and use this to decide whether to round up or leave the digit the same.

Focus

- Let the children read through the information and examples on **Pupil book page 13** independently before they answer questions 1 and 2.
- Problem solving: For question 3, ask the children what number line they could draw to help them solve this problem. Try to elicit that it needs to show the numbers from 2 000 000 to 4 000 000 and that it's useful to indicate half millions as well.

Follow-up

- Use **Workbook page 10** to informally assess how well the children can round numbers to any given place.
- After they complete the questions on their own, you can check the answers with the class, allowing the children to give themselves a rating for understanding and accuracy.

Support

- Let the children do a survey at home to find out how and when people use rounded numbers and estimated values at home and at work. People who work with measurements are often very good at estimating, and a builder, for example, may be able to look at a load of bricks or tiles and estimate how many there are.
- The children can prepare a short talk or video for the class about how people use rounding and estimating.

Answers for Pupil book page 13

1 The prices for cars A, D and E are incorrect. Their correct prices are about £9000, £13 000 and £11 000, respectively. For car A, 7 In the hundreds place is greater than 5 so we round up to £9000. For car D, 9 in the hundreds place is greater than 5 so we round up to £13 000. For car E, 5 is the hundreds place so we round up to £11 000. A is incorrect; B is correct; C is correct; D is incorrect; E is incorrect.

2 a 491 000; 500 000   b 490 000; 500 000
   c 1 346 000; 1 300 000   d 8 460 000; 8 500 000

3 a 2 500 000   b 3 499 999

Answers for Workbook page 10

1 19 432: 19 430, 19 400, 20 000
21 376: 21 380, 21 400, 20 000
18 876: 18 880, 18 900, 20 000
27 654; 27 650, 27 700, 30 000
33 211: 33 210, 33 200, 30 000
29 876: 29 880, 29 900, 30 000
12 987: 12 990, 13 000, 13 000
14 897: 14 900, 14 900, 10 000
19 811: 19 810, 19 800, 20 000
122 456: 122 460, 122 500, 120 000
342 183: 342 180, 342 200, 340 000
665 581: 665 580, 665 600, 670 000
316 479: 316 480, 316 500, 320 000
1 223 718: 1 223 720, 1 223 700, 1 220 000
3 213 987: 3 213 990, 3 214 000, 3 210 000

2 **a** Round to 3 000 000 because population numbers are never exact and this is very close to 3 million.
b Round to 10 000 because it does not need to be very accurate.
c Round to £100 because it is easier to pay.
d Round to 120 000 because it is a number of people.
e Round to 45 000 000 because it is a number of people.

Negative numbers

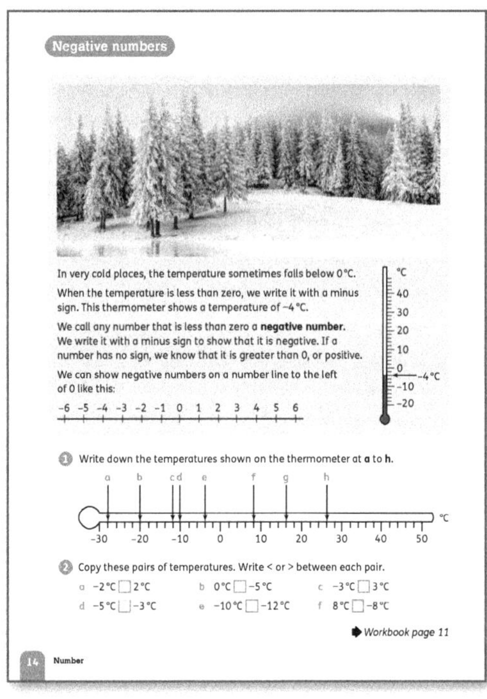

Materials

Cards showing temperatures (negative and positive); large number line extending below 0.

Warm-up

Use the large number line to do some skip counting backwards in intervals and bridging through 0 to remind the children that our number system uses negative numbers to represent amounts below zero.

Focus

- Give the children cards with positive and negative temperatures written on them.
- Ask one child to come up to the front and display the temperature on their card. Ask: *Who has a temperature that is warmer than this?* Let the children display their temperatures and check that they are correct.
- Repeat for colder temperatures.
- Next, get children to come up and stand in a row (to simulate a number line) and give instructions such as: *I need a temperature that is at least 5 degrees lower than this.* Ask for a volunteer to move into the correct position.
- Next, say: *I need a temperature that is between these two temperatures.*
- Repeat using different conditions and instructions to order five or six temperatures.
- Let the children work through the explanation and questions 1 and 2 on **Pupil book page 14** to revise negative numbers and to remind them how to read a scale that extends into the negative.

Follow-up

- Use **Workbook page 11** to consolidate the work in this lesson and to assess how well the children can read and work with negative numbers.
- After they have completed the work on their own, let them check each other's completed number lines and discuss any discrepancies.

Answers for Pupil book page 14

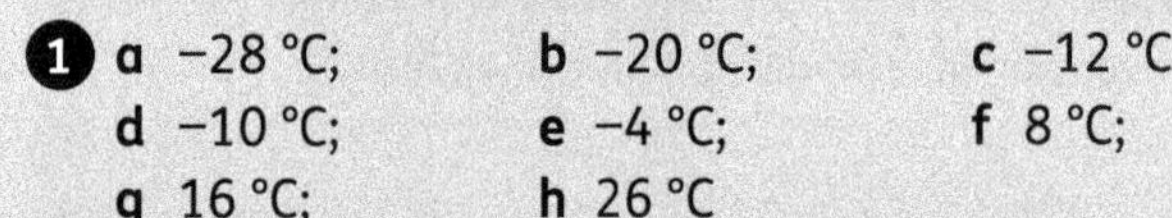

1 **a** −28 °C;   **b** −20 °C;   **c** −12 °C;
d −10 °C;   **e** −4 °C;   **f** 8 °C;
g 16 °C;   **h** 26 °C

2 **a** −2 °C < 2 °C   **b** 0 °C > −5 °C
c −3 °C < 3 °C   **d** −5 °C < −3 °C
e −10 °C > −12 °C   **f** 8 °C > −8 °C

Answers for Workbook page 11

1 **a**

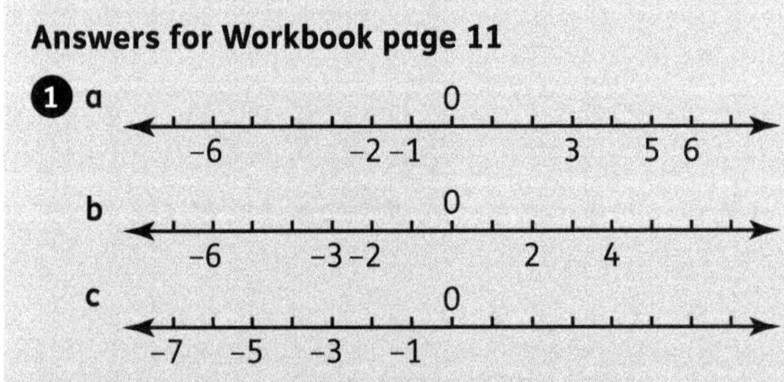

b

c

d

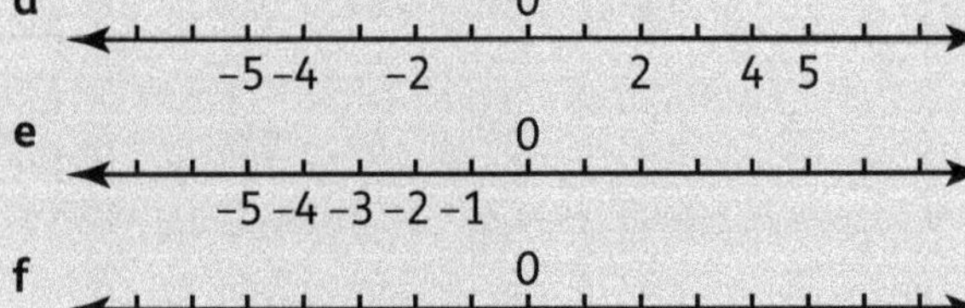

e

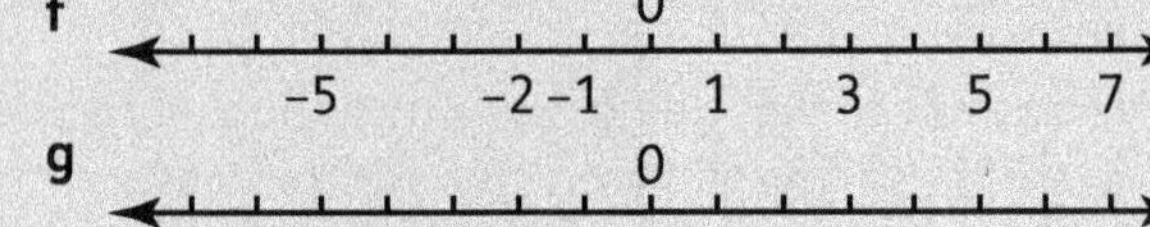

f

g

h

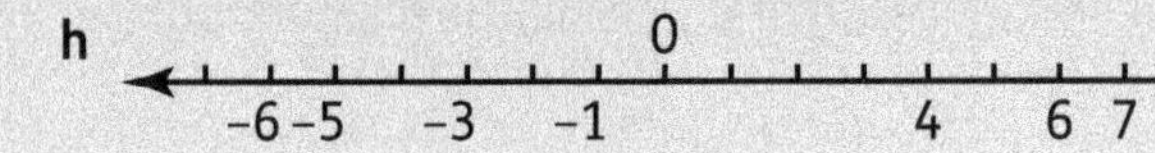

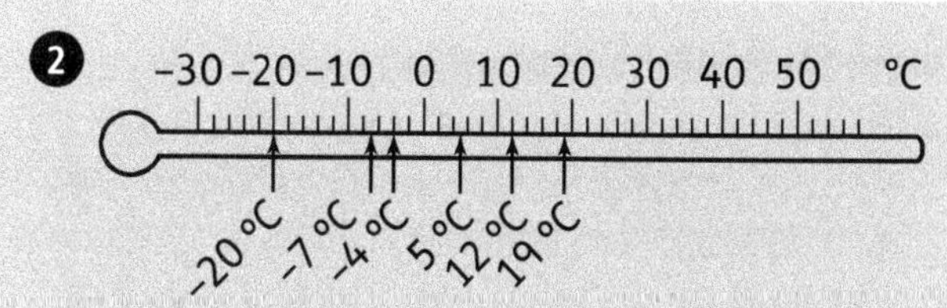

Work with negative numbers

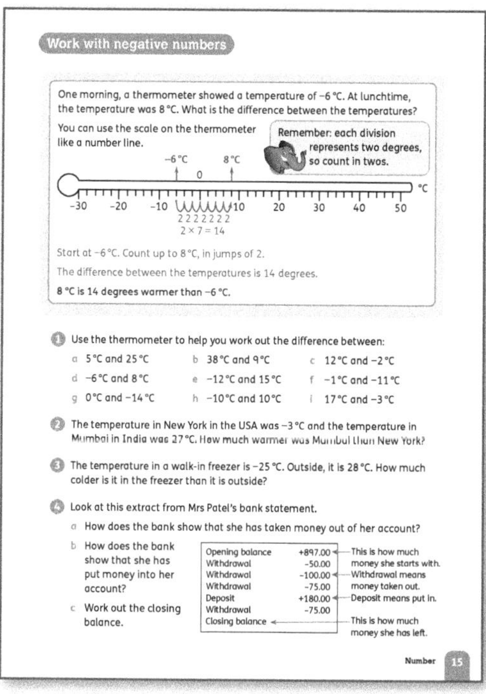

Materials
Thermometer; number lines.

Warm-up
- Show the class your thermometer and ask them how the scale can be seen as a number line.
- Let the children share their ideas. Try to elicit that you can count up (on) and down (back on the scale) to find the sum and/or difference between numbers.

Focus
- Read through the example on **Pupil book page 15** with the class before asking the children to work in pairs to complete questions 1–3.
- For question 4, spend some time discussing the bank statement and how it uses negative and positive numbers. Ask the children to suggest other contexts in which negative numbers are used and what the numbers mean. For example, they can be used to indicate:
  - movement in an elevator (up and down)
  - depth below sea level (negative measurements).
- Check the children's answers, and ask them to explain the strategies they used to answer the questions.

Support
- The children have already worked with negative numbers in the context of temperature. At this stage, they need to extend that work to use the thermometer scale like a number line to calculate differences.
- Encourage the children to use jottings and their own empty number lines to find solutions to the problems.

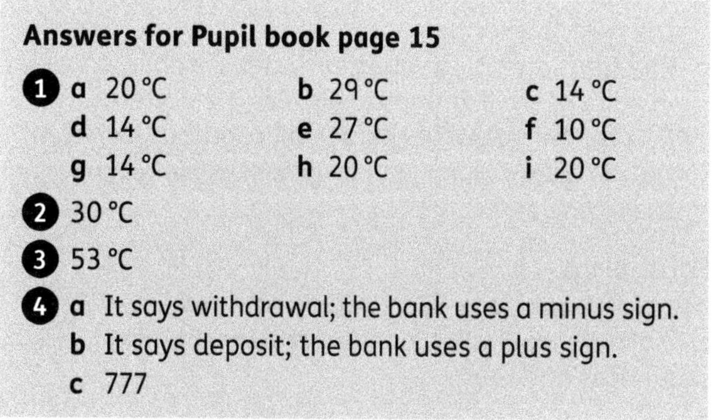

Answers for Pupil book page 15

1 a 20 °C   b 29 °C   c 14 °C
   d 14 °C   e 27 °C   f 10 °C
   g 14 °C   h 20 °C   i 20 °C

2 30 °C

3 53 °C

4 a It says withdrawal; the bank uses a minus sign.
   b It says deposit; the bank uses a plus sign.
   c 777

Count on and back in steps

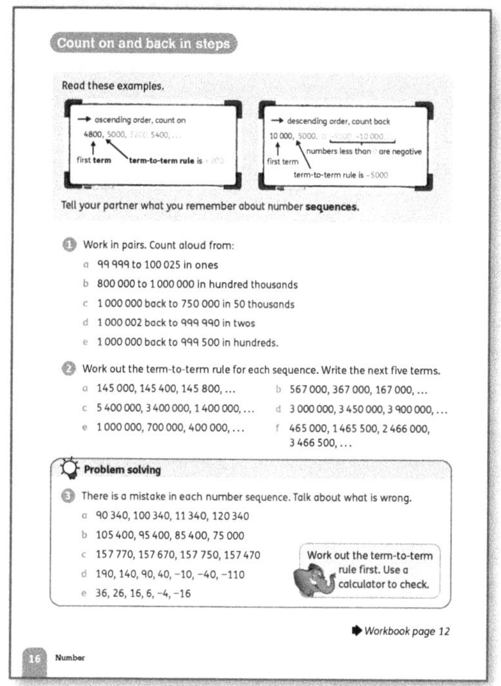

Materials
Calculators.

Warm-up
Do a number of different skip-counting activities as a mental warm-up for this lesson. Vary the counting interval and the starting numbers, and include counting forwards and backwards.

Focus

- Turn to **Pupil book page 16**. Let the children work through the examples and talk about what they remember about number 'sequences'.
- Make sure the children can tell you that the 'term-to-term' rule tells you what to do to get to the next term.
- Let the children complete question 1 orally in pairs.
- The children don't need to write down the rules for question 2, but they do need to work out how to find the next five terms.
- <u>Problem solving:</u> For question 3, let the children work in pairs or small groups to discuss the patterns and find the mistakes. Encourage all contributions from the children as long as they give sensible reasons for their choices.

Follow-up

Use **Workbook page 12** as additional questions to consolidate work with counting in steps, number sequences and rules.

Support

As the focus here is on patterns and the difference between terms, rather than calculation skills, allow the children to use a calculator so they can focus on the pattern rather than on trying to calculate.

Challenge

You can find a range of games involving patterns and missing numbers online. Search for number patterns and rules.

Interesting mistakes

- Children may need additional support when they have to bridge 0. Bridging across barriers such as 10 000 or 100 000 can cause difficulties in skip counting. For example, when asked to count backwards from 120 000 by thousands, some children will say: 120 000, 110 000, 190 000, 180 000, 170 000 … and then realise at some point that they have already said some of the numbers.
- Alternatively, children may omit the barrier number 100 000, saying 120 000, 110 000, 90 000, 80 000 and so on.
- Placing counters on number charts or showing hops on a number line can help children to see what happens when they reach the barrier number. Sustained practice will help them become more confident.

Answers for Pupil book page 16

1 a 99 999, 100 000, 100 001, 100 002, 100 003, 100 004, 100 005, 100 006, 100 007, 100 008, 100 009, 100 010, 100 011, 100 012, 100 013, 100 014, 100 015, 100 016, 100 017, 100 018, 100 019, 100 020, 100 021, 100 022, 100 023, 100 024, 100 025

   b 800 000, 900 000, 1 000 000

   c 1 000 000, 950 000, 900 000, 850 000, 800 000, 750 000

   d 1 000 002, 1 000 000, 999 998, 999 996, 999 994, 999 992, 999 990

   e 1 000 000, 999 900, 999 800, 999 700, 999 600, 999 500

2 a add 400   b subtract 200 000

   c subtract 2 000 000   d add 450 000

   e subtract 300 000   f add 1 000 500

3 a Term-to-term rule: + 10 000. The third term is missing a zero. It should be 110 340.

   b Term-to-term rule: − 10 000. The fourth term should be 75 400.

   c Term-to-term rule: − 100. The third term should be 157 570.

   d Term-to-term rule: − 50. The sixth term should be −60.

   e Term-to-term rule: − 10. The sixth term should be −14.

Answers for Workbook page 12

1 a 29 456, 32 456   rule: + 3000

   b 654 322, 504 322   rule: − 50 000

   c −30 542, −28 542   rule: + 2000

   d 388 699, 378 699, 368 699   rule: −10 000

   e 1 000 000, 250 000, 0, −250 000   rule: − 250 000

   f 5000, 55 000   rule: + 25 000

2

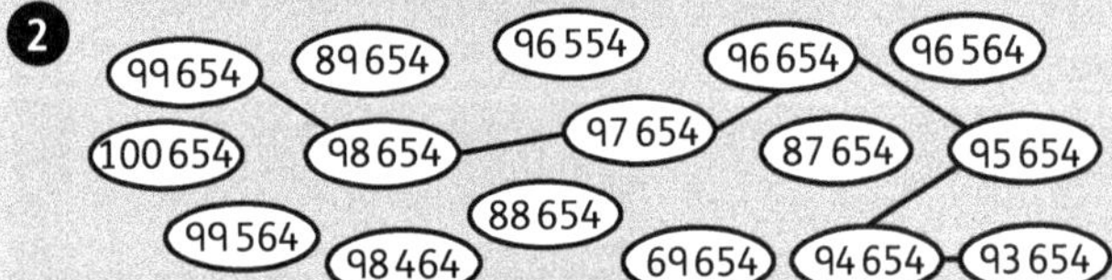

3 a 124 456   b −3 244 600

   c 1 000 876   d −19 456

4 a 113 456   b −3 255 600

   c 989 876   d −30 456

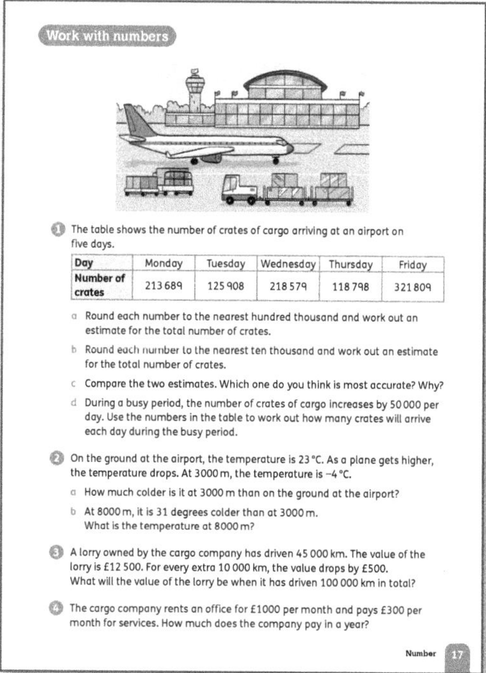

Materials
Calculators.

Warm-up
Choose any suitable 'Mental problem-solving' activities (pages 23–24) as a mental warm-up for this lesson.

Focus
- There are no new concepts in this lesson. The children need to apply what they have learnt to solve number problems. They are also expected to use basic operations.
- Let the children work in pairs or groups to discuss questions 1–4 and to find strategies for solving them.

Answers for Pupil book page 17

1 **a** 200 000; 100 000; 200 000; 100 000; 300 000
Total number of crates: 200 000 + 100 000 + 200 000 + 100 000 + 300 000 = 900 000

b 210 000; 130 000; 220 000; 120 000; 320 000.
Total number of crates: 210 000 + 130 000 + 220 000 + 120 000 + 320 000 = 1 000 000

c The second estimation is more accurate because each number was rounded to the nearest ten thousand and the first one was to the nearest hundred thousand.

d 263 689; 175 908; 268 579; 168 798; 371 809

2 **a** 27 °C   **b** −35 °C

3 £9750

4 £15 600

Use all or some of these questions and problems to check how well the children have understood the concepts in this unit.

- *What makes it possible for us to represent large numbers, using only the ten digits 0, 1, 2, 3, 4, 5, 6, 7, 8, 9?* (Place value: the position of a digit determines its value.)
- *Put the correct sign, < or >, between* (give a pair of numbers).
- *How do you order a set of numbers? Which part of each number do you look at to help you? Why?* (Look at the highest value place value first. If the place values are the same, look at the next highest place value and so on.)
- *What is the value of each digit in this number?* (Give examples of 5-, 6- and 7-digit numbers.)
- Write a number on the board. *How do you say this number?*
- *Draw a model to show 1 234 509.* (Part-whole model with whole 1 234 509 and parts 1 000 000, 200 000, 30 000, 4000, 500 and 9.)
- *Write (give a number) in expanded notation.*
- *What is wrong with this expanded notation: 3 445 098 = 3 000 000 + 400 000 + 50 000 + 90 + 8?* (40 000 is missing and 50 000 should be 5000.)
- Give a number sequence then ask:
  - *What is the term-to-term rule for this sequence?*
  - *Is there any other way of carrying on this sequence? How?*
  - *What comes next in this sequence? Why?*
- *Write these numbers in ascending order 1 230 954, 1 230 949, 1 230 994, 1 230 998, 1 230 940.* (1 230 940, 1 230 949, 1 230 954, 1 230 994, 1 230 998)
- *What is 100 more/100 less/1000 more/1000 less, 10 000 more/10 000 less than (give a number)?*
- *What is 32 375 rounded to the nearest 10?* (32 380)
- *What is 423 544 rounded to the nearest 10/100/1000?* (423 540/423 500/424 000)
- *How would you explain to someone how to round a number to the nearest 10/100/1000/10 000/100 000?* (Nearest 10: look at the digit in the ones column (place value to the right of the number being rounded to); if it is less than 5, round to the previous 10; if it is 5 or more, round to the next 10. Nearest 100/1000/10 000/100 000: look at the digit in the tens/thousands/ten thousands column.)
- *Which is colder, a temperature of 0 °C or a temperature of −2 °C?* (−2 °C)
- *What is the difference between a temperature of 2 °C and a temperature of −2 °C?* (4 °C)
- *Which of these numbers will give 7 000 000 when rounded to the nearest million?* (6 500 000, 6 789 123, 7 123 456)

6 498 299	7 603 000	6 500 000
7 500 000	6 789 123	7 123 456

Multiples, factors and special numbers

Learning objectives

- Understand and recognise common factors, common multiples and prime numbers.
- Use knowledge of factors and multiples to understand test of divisibility by 3, 6 and 9.
- Use knowledge of multiplication and square numbers to recognise cube numbers.
- Use knowledge of square numbers to generate terms in a sequence given its position.

Key words

multiples   common multiple
lowest common multiple (LCM)
factors   remainder   common factor
highest common factor (HCF)   prime numbers
square numbers   cube numbers

Unit introduction

Materials

Children's building bricks with studs or pictures of this type of brick; squared paper.

Teaching guidance

- You will need a set of children's building bricks with studs on them. If unavailable, show the class some pictures of bricks with studs and provide squared paper to work on. Aim to have at least 4-, 6-, 8- and 12-stud bricks (or pictures of them) but any sizes can work.
- Set two problems for the class to investigate. For example, say:
  - *I want to build two rows of bricks the same length. One row will only use 8-stud bricks and the other will use only 12-stud bricks. How can I do this? What if I have 8-stud bricks and 14-stud bricks? Can I still do it?*
  - *I want to build one row that is 48 studs long and another row that is 72 studs long using only one size brick. What size bricks could I use?*
- Both these problems involve the children working with 'common multiples' and 'common factors' in a natural way. Spend time discussing how they worked out their answers and note which children have to model the whole row rather than use patterns and facts to work it out.

Multiples

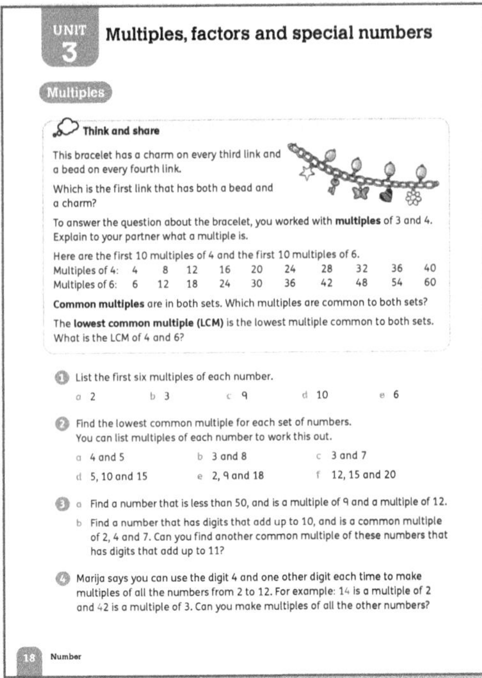

Materials

Different-coloured beads and counters; bricks or colour arrays; calculators.

Warm-up

Use any 'Mental problem-solving' or 'Calculation skills' activities involving multiplication (pages 23–26) as a mental warm-up for this lesson.

Focus

- Hand out different-coloured beads and counters to each group.
- <u>Think and share:</u> Let the children work in groups to model the bracelet described in the Think and share activity on **Pupil book page 18**. Let them talk through the questions about the bracelet before sharing their answers as a class. The children are asked which is the first link that has both a bead and a charm. (12th) Then they are asked to say which numbers in the list are common multiples of 4 and 6 and to state the lowest common multiple. (12, 24, 36; LCM is 12)
- Check that the children understand the terms 'multiples' and 'lowest common multiple (LCM)' using the following method:
  - Ask the children to explain what is meant by the word 'multiple' and draw out a correct definition.
  - Put a single-digit number on the board and ask the children to tell you multiples of the number. Repeat with other examples.
  - Put the numbers 6 and 8 on the board. Ask the children to list the first five multiples of each number.

○ Looking at these lists, ask the children to tell you which is the smallest number that is a multiple of both 6 and 8. Explain that this is called the 'lowest common multiple'.
- Let the children work independently through questions 1–4 on **Pupil book page 18**. Encourage them to model the solutions if they need to, using objects and/or arrays.

Challenge
Ask the children to investigate whether you can use other digits to make multiples of all the numbers to 12, as in **Pupil book page 18** question 4.

Support
- Provide bricks or colour arrays that make a set of multiples of the numbers from 2 to 12. Let the children keep these to refer to as they work.
- Ask children to use their bricks to teach others about multiples. This will allow them to think through and say what they have learnt, and discover what they do or do not understand.

Interesting mistakes
- Often, children do not recognise when one of the two numbers they are working with is the LCM. For example, if you ask them to find the LCM of 2 and 4, they might say 8.
- Encourage the children to list the numbers or use bricks to help them to see this.

Answers for Pupil book page 18
1 a 2, 4, 6, 8, 10, 12
  b 3, 6, 9, 12, 15, 18
  c 9, 18, 27, 36, 45, 54
  d 10, 20, 30, 40, 50, 60
  e 6, 12, 18, 24, 30, 36

2 a 20 b 24 c 21
  d 30 e 18 f 60

3 a 36 b 28; 56

4 24 is a multiple of 4, 40 is a multiple of 5, 42 is a multiple of 6, 14 is a multiple of 7, 24 is a multiple of 8, 45 is a multiple of 9, 40 is a multiple of 10, 44 is a multiple of 11 and 48 is a multiple of 12.

Factors

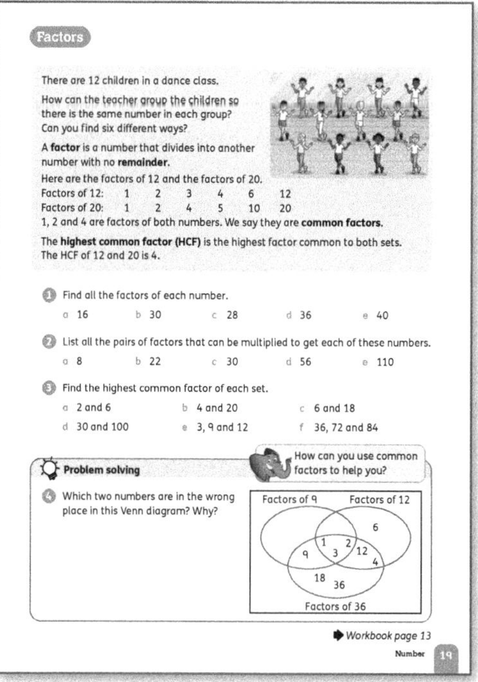

Materials
Counters or beads; calculators.

Warm-up
Select any 'Calculation skills' activity involving multiplication or division (pages 24–26) as a mental warm-up for this lesson.

Focus
- Work through and discuss the example on **Pupil book page 19** with the class. If necessary, let the children use a set of 12 counters to make the groups. The explanation box at the top of the page asks: There are 12 children in a dance class. How can the teacher group the children so there is the same number in each group? Can you find six different ways? (1 group of 12, 2 groups of 6, 3 groups of 4, 4 groups of 3, 6 groups of 2, 12 groups of 1)
- With the class, read through the explanation to remind the children about 'factors' and 'common factors' and 'remainders'.
- Check the children's understanding by writing a number, such as 24, on the board. Ask the children to say factors of the number and record them.
- Ask the children how we can be sure we have found all the factors of 24. Talk about starting from 1 and pairing this with the number it would be multiplied by to get 24, then going through 2, 3, etc. in a similar way until all factor pairs have been recorded.
- Illustrate factor pairs by drawing arrays for 24. Repeat for an odd number.

- Demonstrate how to find the 'highest common factor (HCF)' of two numbers:
  - Write 12 and 18 on the board.
  - Ask the children to list all the factors of each of these numbers.
  - Ask the children which are factors of both numbers. Looking at these, establish which of these common factors is the highest.
- Let the children work in groups to discuss and answer question 1 before taking feedback as a class. Introduce and explain the term 'prime number'.
- The children can work independently on question 2. Let them compare their answers to see if they have found all the pairs.
- The children can do question 3 with a partner.
- <u>Problem solving:</u> Work on question 4 as a number talk (pages 17–18).
- Discuss these questions:
  - *Which number is a factor of every other number? (1)*
  - *What number is a factor of every even number? (2) Why? (All even numbers are multiples of 2.)*
  - *What is the highest factor of any number? (the number itself)*

Follow-up

Let the children complete **Workbook page 13** independently to check that they can factorise numbers and identify the HCF from the factors.

Challenge

- Write a 2-digit number on a piece of paper without showing the children.
- The children must ask questions to which you can reply *Yes* or *No* in order to try and guess the number. Encourage them to ask questions relating to number properties, for example: *Does it have 2 as a factor? Is it a multiple of 3?* along with questions about, for example, the size of the number.
- Discuss with the children what your answer tells them and which numbers your answer eliminates.
- When the children guess the correct number, reveal the paper to confirm this.

Interesting mistakes

- There are a lot of key words in this unit that the children need to be familiar with and that they may find confusing, for example finding factors instead of multiples and vice versa.
- To overcome this, have short regular activities in which the children have to use the key words to describe numbers and their properties.

Answers for Pupil book page 19

1 a 1, 2, 4, 8, 16
  b 1, 2, 3, 5, 6, 10, 15, 30
  c 1, 2, 4, 7, 14, 28
  d 1, 2, 3, 4, 6, 9, 12, 18, 36
  e 1, 2, 4, 5, 8, 10, 20, 40

2 a 1 and 8; 2 and 4
  b 1 and 22; 2 and 11
  c 1 and 30; 2 and 15; 3 and 10; 5 and 6
  d 1 and 56; 2 and 28, 4 and 14; 7 and 8
  e 1 and 110; 2 and 55; 5 and 22; 10 and 11

3 a 2   b 4   c 6
  d 10   e 3   f 12

4 2 is not a factor of 9, and 6 is factor of 12 and 36.

Answers for Workbook page 13

1 a 8: 1 and 8; 2 and 4
  b 24: 1 and 12; 2 and 6; 3 and 4
  c 20: 1 and 20; 2 and 10; 4 and 5
  d 46: 1 and 46; 2 and 23
  e 48: 1 and 48; 2 and 24; 3 and 16; 4 and 12; 6 and 8
  f 100: 1 and 100; 2 and 50; 4 and 25; 5 and 20; 10 and 10
  g 60: 1 and 60; 2 and 30; 3 and 20; 4 and 15, 5 and 12; 6 and 10
  h 72: 1 and 72; 2 and 36; 3 and 24; 4 and 18; 6 and 12; 8 and 9
  i 40: 1 and 40; 2 and 20; 4 and 10; 5 and 8

2 a 1, 2, 3, 4, 6, 8, 12, 24   b 1, 2, 4, 5, 10, 20

3 a 50   b 56   c 300   d 96

Prime numbers

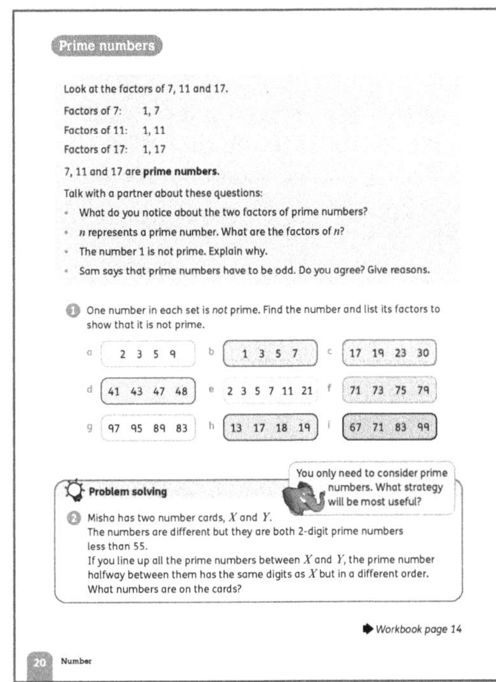

Warm-up

Use any 'Mental problem-solving' or 'Calculation skills' activity (pages 23–26) as a mental warm-up for this lesson.

Focus

- Turn to **Pupil book page 20**. Let the children read the examples and discuss the questions in pairs. The explanation box at the top of the page asks: What do you notice about the two factors of 'prime numbers'? (They are 1 and the number itself.) n represents a prime number. What are the factors of n? (1 and n) The number 1 is not prime. Explain why. (1 only has one factor. Prime numbers have exactly two factors.) Sam says that prime numbers have to be odd. Do you agree? Give reasons. (No, 2 is a prime number because it has two factors, 1 and itself. All other even numbers have 2 as a factor, as well as 1 and themselves, so they are not prime.)
- Discuss the answers as a class. Make sure the children are able to say that prime numbers have exactly 2 factors and that those factors are always 1 and the number itself. Ask them to make a general rule for this, for example: If X is a prime number then its factors are 1 and X.
- Turn to **Workbook page 14** and ask the children to complete question 1 to list the prime numbers to 100.
- Check their answers before moving on, so that they can refer back to the list when completing questions 1 and 2 on **Pupil book page 20**.
- <u>Problem solving:</u> The children can work in pairs on question 2, and then share their answers and strategies with the class.
- Let the children work independently to complete the questions on **Workbook page 14**.

Challenge

- Tell the class that any number can be written as the product of its prime factors. A prime factor is a factor that is also a prime number.
- Let the children investigate how to write numbers as the product of prime factors. Give them two examples like this:

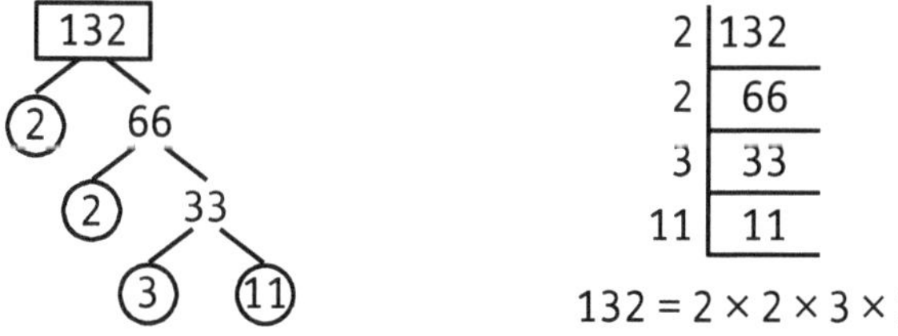

$$132 = 2 \times 2 \times 3 \times 11$$

- Let the children discuss these and work out how to find the prime factors of a number. If they are unsure, there are lots of suitable presentations available online that they could watch or read. Let them share what they find out.

Answers for Pupil book page 20

1   **a**   9; factors: 1, 3, 9
     b   1; factors: 1
     c   30; factors: 1, 2, 3, 5, 6, 10, 15, 30
     d   48; factors: 1, 2, 3, 4, 6, 8, 12, 16, 24, 48
     e   21; factors: 1, 3, 7, 21
     f   75; factors: 1, 3, 5, 15, 25, 75
     g   95; factors: 1, 5, 19, 95
     h   18; factors 1, 2, 3, 6, 9, 18
     i   99; factors 1, 3, 9, 11, 33, 99

2 X is 13 and Y is 53.

Answers for Workbook page 14

1   2, 3, 5, 7, 11, 13, 17, 19, 23, 29, 31, 37, 41, 43, 47, 53, 59, 61, 67, 71, 73, 79, 83, 89, 97

2   5th, 7th, 11th, 13th, 17th, 19th, 23rd, 29th, 31st

3   **a** $17 = 2 + 2 + 13$ **b** $21 = 2 + 2 + 17$
     c $35 = 2 + 2 + 31$ **d** $61 = 3 + 5 + 53$

Square and cube numbers

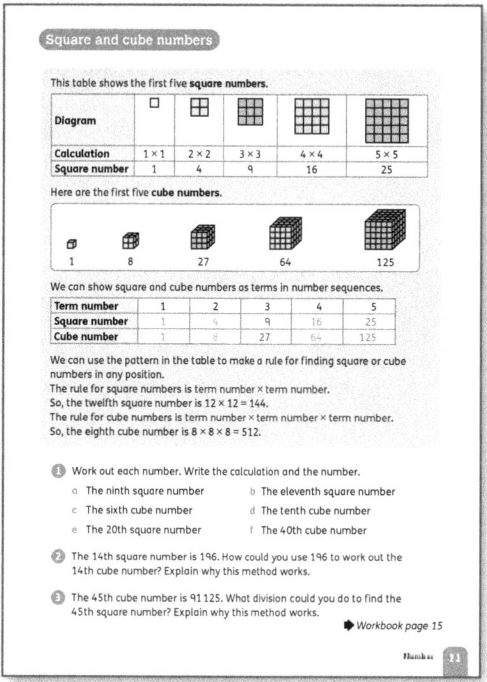

Materials

1000-cubes and 100-flats (page 20); calculators, if necessary.

Warm-up

Select any 'Calculation skills' activities involving multiplication or division (pages 24–26) as a starter for this lesson.

Focus

- The children have already met 'square numbers' in

Level 5, so let them work on their own or in pairs through the explanation and examples on **Pupil book page 21** and link these to 'cube numbers'.

- Ask questions to check understanding of the concepts:
  - Show the class a 100-flat. Ask them what number it represents (100) and how you can work this out (10×10).
  - Then look at the 1000-cube. Show the class that this is the same as $10 \times$ the 100-flat. In other words, it represents 10×100 which is 1000.
  - Ask how else you could find this by looking at the cube. (Each edge is 10 long, so you can calculate $10 \times 10 \times 10 = 1000$.)
- Let the children work on their own to complete questions 1–3 on **Pupil book page 21**. Allow them to use a calculator if necessary.

Follow-up

- Use **Workbook page 15** to informally assess how well the children understand the concept of square and cube numbers.
- Let them use a calculator as the focus is on their understanding of the concepts rather than their calculation skills.

Challenge

- Let the children explore how to find the square and cube of any number using the calculator functions (x^2 and x^y).
- They could also investigate how to find the square and cube root of a number using the $\sqrt{\ }$ and $\sqrt[3]{\ }$ functions.

Answers for Pupil book page 21

1   a $9 \times 9 = 81$ b $11 \times 11 = 121$
    c $6 \times 6 \times 6 = 216$ d $10 \times 10 \times 10 = 1000$
    e $20 \times 20 = 400$ f $40 \times 40 \times 40 = 64\,000$
2 $196 \times 14 = (14 \times 14) \times 14 = 2744$
3 $91\,125 \div 45 = (45 \times 45 \times 45) \div 45 = 45 \times 45$
    $= 2025$

Answers for Workbook page 15

1

Term	2	5	8	10	15	20	25	30	100
Square number	4	25	64	100	225	400	625	900	10 000
Cube number	8	125	512	1000	3375	8000	15 625	27 000	1 000 000

2   Square numbers: 1, 4, 9, 16, 25, 49, 64, 81, 100
    Cube numbers: 1, 8, 27, 125, 1000
3 4^2 to $4 \times 2 \times 2$ 3^3 to 9×3
    6^3 to 36×6 9^2 to $729 \div 9$
    12^3 to 144×12 8^3 to 64×8
    24^2 to $24 \times 12 \times 2$
4   a   13   b   9   c   3   d   10   e   100

Divisibility rules

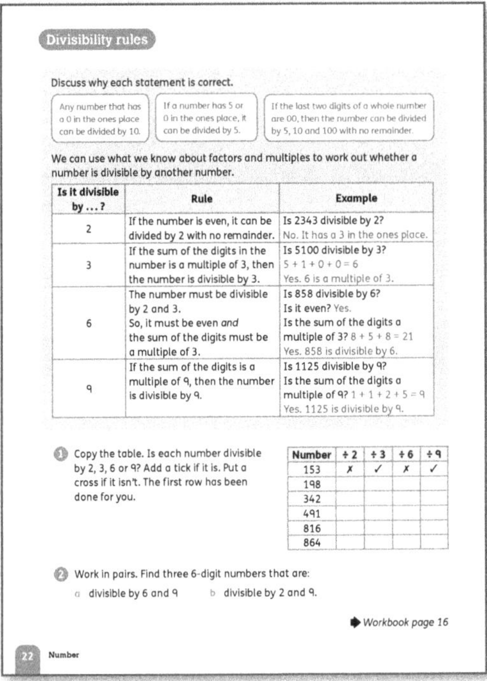

Discuss why each statement is correct.

> Any number that has a 0 in the ones place can be divided by 10.

> If a number has 5 or 0 in the ones place, it can be divided by 5.

> If the last two digits of a whole number are 00, then the number can be divided by 5, 10 and 100 with no remainder.

We can use what we know about factors and multiples to work out whether a number is divisible by another number.

Is it divisible by …?	Rule	Example
2	If the number is even, it can be divided by 2 with no remainder.	Is 2343 divisible by 2? No. It has a 3 in the ones place.
3	If the sum of the digits in the number is a multiple of 3, then the number is divisible by 3.	Is 5100 divisible by 3? $5 + 1 + 0 + 0 = 6$ Yes. 6 is a multiple of 3.
6	The number must be divisible by 2 and 3. So, it must be even *and* the sum of the digits must be a multiple of 3.	Is 858 divisible by 6? Is it even? Yes. Is the sum of the digits a multiple of 3? $8 + 5 + 8 = 21$ Yes. 858 is divisible by 6.
9	If the sum of the digits is a multiple of 9, then the number is divisible by 9.	Is 1125 divisible by 9? Is the sum of the digits a multiple of 9? $1 + 1 + 2 + 5 = 9$ Yes. 1125 is divisible by 9.

1. Copy the table. Is each number divisible by 2, 3, 6 or 9? Add a tick if it is. Put a cross if it isn't. The first row has been done for you.

Number	÷ 2	÷ 3	÷ 6	÷ 9
153	✗	✓	✗	✓
198				
342				
491				
816				
864				

2. Work in pairs. Find three 6-digit numbers that are:
    a   divisible by 6 and 9   b   divisible by 2 and 9.

➡ *Workbook page 16*

22   Number

Warm-up

Use any 'Calculation skills' activity on division (pages 24–26) as a mental warm-up for this lesson.

Focus

- Introduce and discuss with the whole class the rules of divisibility, starting with the statements on **Pupil book page 22**. This is quite an interesting topic for most children; they like checking numbers to see whether they obey the rules. The example box at the top of the page asks the children to discuss why three statements are correct. Any number that has a 0 in the ones place can be divided by 10. (Only the multiples of 10/numbers in the ten times table have a 0 in the ones place.) If a number has 5 or 0 in the ones place, it can be divided by 5. (Only the multiples of 5/numbers in the five times table have a 5 in the ones place. 10 can be divided by 5, so if a number can be divided by 10, it can also be divided by 5.) If the last two digits of a whole number are 00, then the number can be divided by 5, 10 and 100 with no remainder. (Only the multiples of 100/numbers in the 100 times table have last two digits 00. 100 can be divided by 10, and if a number can be divided by 10, it can also be divided by 5.)
- Let the children work through question 1 after you have discussed the rules.
- For question 2, many answers are possible. The children should work in pairs and use a calculator to check their work.

Follow-up

- The children can work in pairs on **Workbook page 16**.
- When they have finished, each pair can compare and check their answers with another pair.

Answers for Pupil book page 22

Number	÷ 2	÷ 3	÷ 6	÷ 9
153	✗	✓	✗	✓
198	✓	✓	✓	✓
342	✓	✓	✓	✓
491	✗	✗	✗	✗
816	✓	✓	✓	✗
864	✓	✓	✓	✓

2 **a**, **b** Individual answers.

Answers for Workbook page 16

1 534, 918, 456, 10 008, 2007, 9999

2 450, 621, 342, 8091, 4419

3 Yes, it is. As 6 = 2 × 3, if a number is divisible by 6 then it is divisible by 2 and by 3.
Yes, it is. As 9 = 3 × 3, if a number is divisible by 9 then it is divisible by 3.
No, they are not. For example, 27 is a multiple of 9 but it is not divisible by 6.

4 **a**

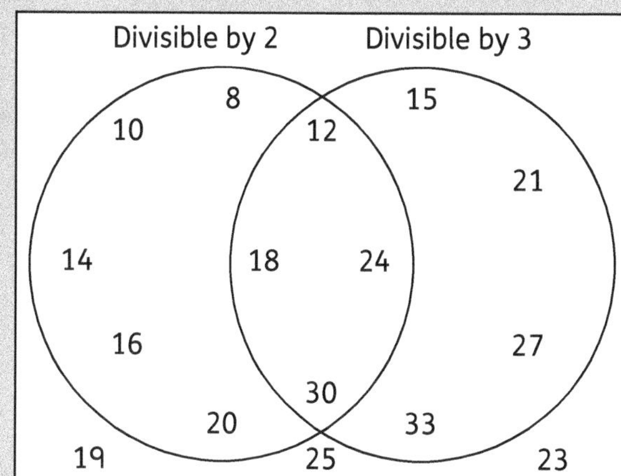

b The numbers that are not in circles are not divisible by 2 nor divisible by 3.

c 12, 18, 24 and 30. Where the two sets overlap, there are the numbers that are divisible by 2 and 3. So, they are divisible by 2 × 3 = 6.

Will it divide?

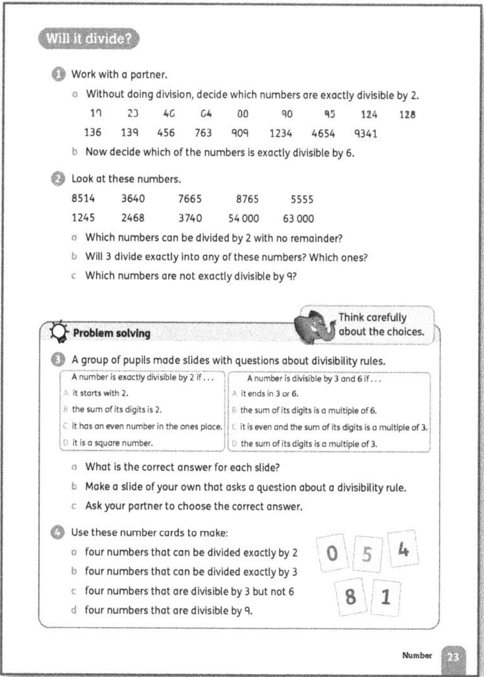

Materials

Calculators; 0–9 digit cards.

Warm-up

Select any suitable 'Calculation skills' activity using division (pages 24–26) as a mental warm-up for this lesson.

Focus

- Use this lesson to consolidate the rules of divisibility and to check that the children can apply them in different situations.
- Turn to **Pupil book page 23**. The children should work in pairs to complete question 1. They can do it orally, but they should list the correct numbers afterwards so that they can check their answers.
- The children can work independently to complete question 2. When they have finished, let them use a calculator to check their own answers.
- <u>Problem solving:</u> For question 3, after answering the questions the children work independently to make their own slide to try out on a partner.
- If you have time, you could get everyone to upload their slides and use them as a class task (in which everyone answers each slide) or as a mental starter for another lesson (the children can take turns to answer slides as you display them).
- <u>Problem solving:</u> For question 4, let the children use 0–9 digit cards to model numbers if necessary.

End-of-unit check

Use all or some of these questions and problems to check how well the children have understood the concepts in this unit.

- *What are all the factors of* (give a number)*?*
- *Give three multiples of* (give a number).
- Display a list of numbers, some of them prime. *Which of these numbers are prime numbers? How do you know?* (A prime number has exactly two factors: 1 and the number itself.)
- *What are the first ten prime numbers?* (2, 3, 5, 7, 11, 13, 17, 19, 23, 29)
- *What is* (give a number up to 12) *squared?*

- *4 squared is 16. How can you use this to find 4 cubed?* (Multiply 16 by 4.)
- *What is* (give a number up to 5) *cubed?*
- Write a 2- or 3-digit number on the board. *What are the properties of this number?* (for example, whether it is a square or prime number, its factors, some of its multiples, whether it is divisible by 2, 3, 6 or 9)
- Give children a challenging problem involving divisibility rules. For example: *Sort these numbers and write them in the correct parts of the Venn diagram:*

12	14	24	25	27	30	45	51	65	70	98	110
112	123	150	225	256	300	340	342	360	375	390	400

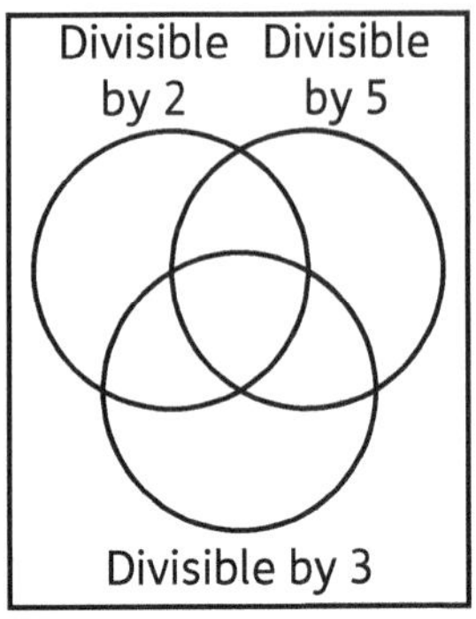

(only divisible by 2: 14, 98, 112; only divisible by 5: 25, 65; only divisible by 3: 51, 123; divisible by 2 and 5: 30, 70, 110, 340, 400; divisible by 2 and 3: 12, 24, 342 divisible by 3 and 5: 45, 225, 375; divisible by 2, 3 and 5: 150, 300, 360, 390)

You can find divisibility problems online.

UNIT 4 Shapes, lines and angles

Learning objectives

- Classify, estimate, measure and draw angles.
- Know that the sum of angles in a triangle is 180 and use this to calculate unknown angles in a triangle.
- Recognise angles where they meet at a point, are on a straight line, or are vertically opposite, and find unknown angles.
- Draw 2D shapes using given dimensions and angles.
- Identify, describe, classify and sketch quadrilaterals, including reference to angles, symmetrical properties, parallel sides and diagonals.
- Compare and classify geometric shapes based on their properties and sizes and find unknown angles in any triangles, quadrilaterals and regular polygons.

- Illustrate and name parts of circles, including centre, radius, diameter and circumference, and know that the diameter is twice the radius.
- Recognise, describe and build simple 3D shapes, including making nets.

Key words
protractor   right-angled
vertically opposite angles   two-dimensional (2D)
polygons   regular polygon   equilateral
isosceles   scalene   acute   obtuse
equation   quadrilaterals   parallelogram
rectangle   rhombus   square   trapezium
kite   circle   circumference   centre
radius (r)   radii   diameter (d)
three-dimensional (3D)   face   edge
vertex   net   surface area

Materials

Materials to make posters; camera; Internet images.

Teaching guidance

- Arrange the children into groups and explain that they are each going to produce a poster to show different shapes or angles in the environment.
- Decide which groups will focus on shapes (for example: triangles, rectangles, squares, other 'quadrilaterals', 'circles'); '3D shapes'; and angles.
- Let the children take photographs, make sketches or download and print images to make an informative poster on the topic they have been given. They should include headings and labels. For example, lines drawn over a photograph of a diamond-paned window can highlight 'right angles', 'acute' and 'obtuse' angles.

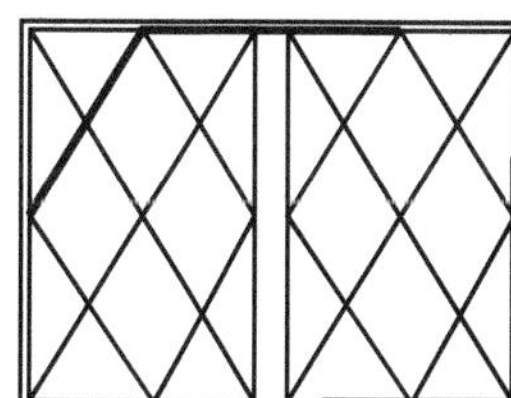

- Each group then presents their finished work to the class. Encourage questions and ask the children to say one thing they like and one thing they think could be improved in each presentation.

Measure and draw angles

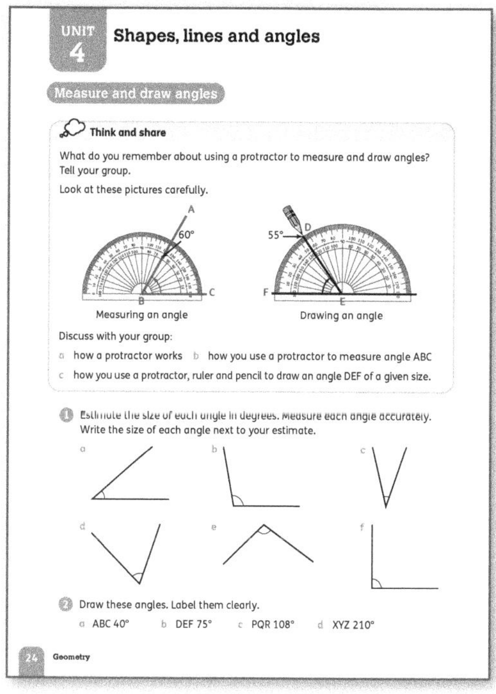

Materials

Protractors; rulers; sharp pencils; set of angle cards for loop game (see Follow up); sheet of angles with protractors on them.

Warm-up

Sketch angles on the board, or point out angles around the classroom, and ask children to classify them as acute (<90°), right angle (90°) or obtuse (between 90° and 180°).

Focus

- Think and share: The children used a 'protractor' in Level 5. Use the Think and share activity on **Pupil book page 24** to discuss what they remember.
- For question 1, make sure they estimate the angle sizes and record their estimates before measuring.
- For question 2, you may need to revise how to label angles, and stress that in angle ABC, B is the vertex. Let the children check each other's answers.

Follow-up

- You could round off the lesson with the class playing an angle loop game to check that the children can use a protractor to measure angles.
- Each set of cards should have an image of an angle drawn accurately and an angle that matches the number of degrees drawn on one of the other cards (see diagram). Make sure there is a card for each child.

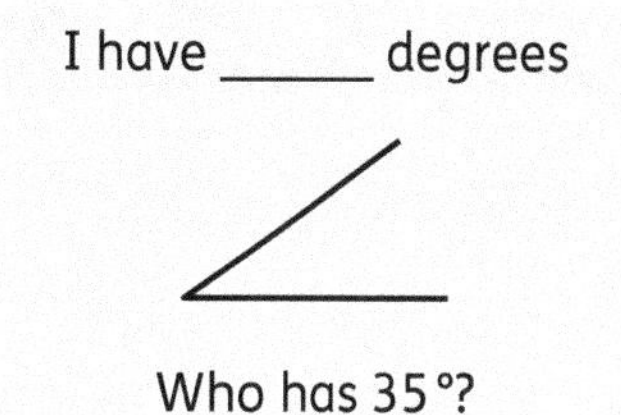

- To start, hand out a card to each child and ask them to measure the angle. They should record this in the space above the angle.
- Choose a child to start, encouraging them to read from their card: I have x degrees. 'Who has y degrees?' The child who has that number then says, 'I have y degrees. Who has z degrees?' Continue until all the cards are played.

Support

- If children need additional support to read and place the protractor, you could prepare a sheet of angles with protractors on them for them to read.
- For the first few, indicate with an arrow where to read the measure.

Interesting mistakes

- Many children need additional support to use a protractor correctly. Demonstrate how to line up a protractor and how to read the correct value.
- Encourage the children to think about whether their protractor reading makes sense; for example, if they have read 165° off their protractor but it looks like an acute angle, their reading must be wrong.

- The children may also read the protractor from the wrong side. Encourage them to use their finger to move from the zero line on the angle around the scale to the other arm. This helps them to use the correct scale.

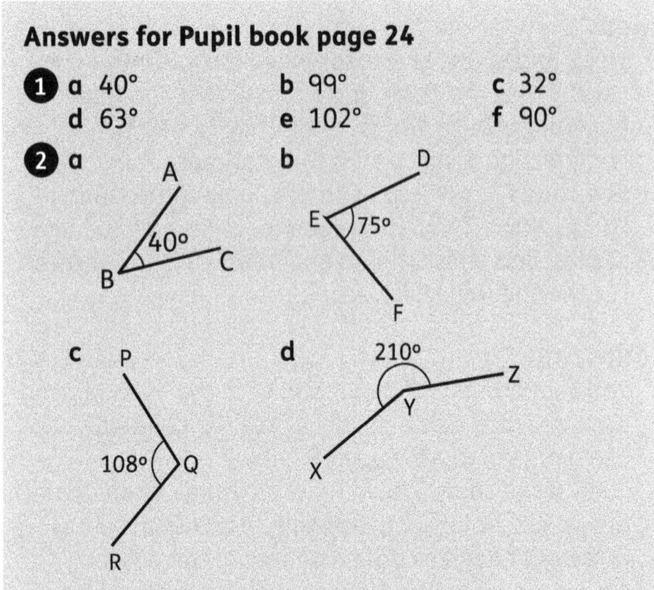

Answers for Pupil book page 24

1 a 40° b 99° c 32°
  d 63° e 102° f 90°

2 a

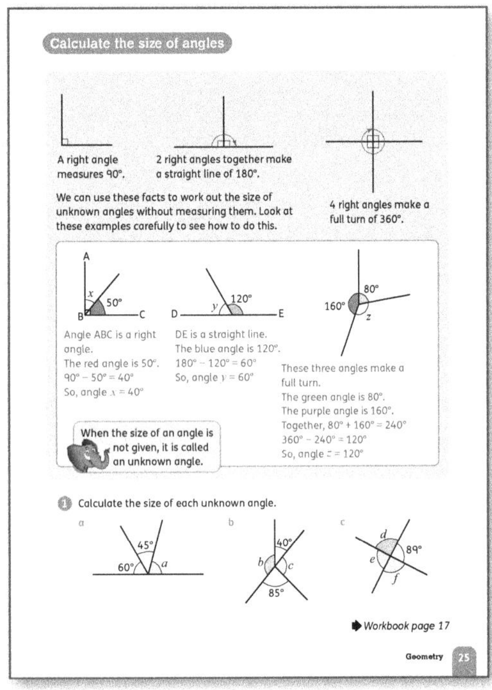

Calculate the size of angles

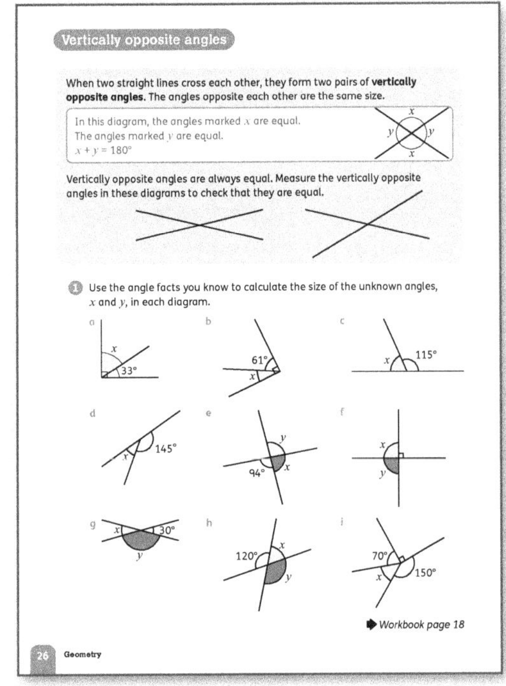

Materials

Protractors; set squares (optional); pencils.

Warm-up

Do some mental addition and subtraction activities using pairs of numbers that add up to 90, 180 or 360 as a starter for this lesson.

Focus

- Draw a four-point compass on the board and use it to remind the children that a full turn is 360°, a half turn or straight line is 180° and a quarter turn is 90°.
- Work through the explanation and examples on **Pupil book page 25** with the class. Emphasise that they are calculating the size of unknown angles and not measuring them.
- Be careful how you talk about unknown angles. It can be confusing for children if you say 'missing angles' because they are not missing (they appear on the diagram but without number labels).
- Let the children work independently on question 1.

Follow-up

Use **Workbook page 17** for additional practice and consolidation of the work on calculating angles.

Answers for Pupil book page 25

1 a $a = 75°$
  b $b = 140°$; $c = 105°$
  c $d = 91°$; $e = 89°$; $f = 91°$

Answers for Workbook page 17

1 a 138 ° [Provided as an example]
  b 107° c 113° d 93°
  e 132° f 147°

2 a $p = 45°$ and $q = 45°$
  b $p = 60°$ and $q = 30°$
  c $p = 126°$ and $q = 54°$

Vertically opposite angles

Materials

Two pencils, chopsticks or straws with a tack through them so they can rotate (see diagram); pair of scissors or garden shears; protractors; large sheets of paper; coloured pens.

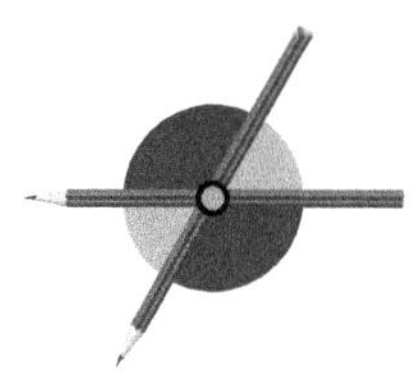

Warm-up

Do a different mental addition and subtraction activity using pairs of numbers that add up to 180 or 360 as a starter for this lesson.

Focus

- Start by showing the class a pair of scissors or shears. Open them slightly and point to the angles formed between the blades and the handles. Open the scissors more and more, asking the children what they notice about the size of both angles. They should observe that as the angle between the blades gets bigger, so does the angle between the handles.
- Use the rotating pencils to develop the idea further:
  - Start with four right angles. The children know that these are equal.
  - Move one pencil. Talk about what happens to the angles above the 'line' (one becomes less than 90°; the other becomes more than 90°).
  - Compare these with the angles on the other side of the line. The opposite happens.
  - Explain that these angles are called 'vertically opposite angles', and that they will always be the same size as each other.
- Turn to **Pupil book page 26**. Read through the information with the class to reinforce what you have just demonstrated.
- Let them measure the angles in the diagrams to check that they are equal.
- Next, ask the children to explain in their own words why these opposite angles will always be equal.
- For question 1, let the children work independently to calculate the unknown angles.

Support

Ask the children to work in groups to make a poster showing where you find vertically opposite angles in real life. They can either photograph items or draw them. Examples are: railway crossing signs; security gates and burglar bars with diamond patterning; ironing board stands; folding chairs; blades on ceiling fans.

Follow-up

- Use **Workbook page 18** to consolidate work on measuring and calculating angles.
- Hold a class discussion about why you cannot always measure to find the unknown angles (question 2).

2D shapes

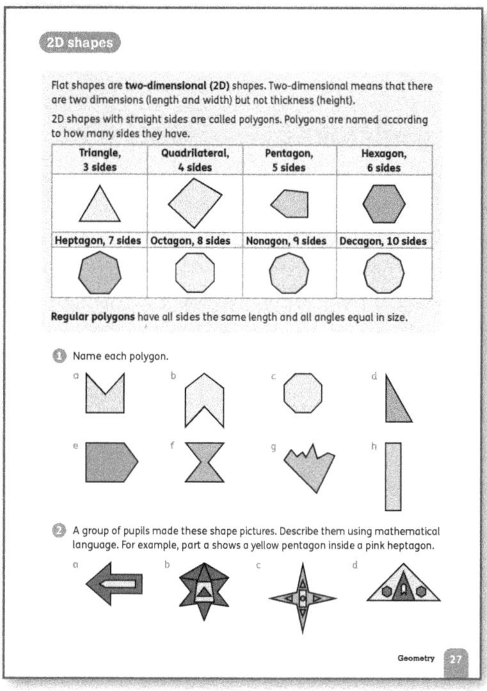

Materials

Large polygon; sheet of thick paper; shape vocabulary cards and matching diagram cards.

Warm-up

- Hide a large polygon behind a screen, such as a sheet of thick paper. Gradually move the shape to reveal part of it. Ask the children to suggest what the shape might be and why.
- Reveal more of the shape and ask the children to change their answers, for example to say what it cannot be: 'It can't be a triangle because there are already three vertices on show.' Continue until the shape is revealed. Use the term 'polygon' and explain 'regular polygon'.

Focus

- Invite a child to describe a shape by its properties (for example, number of sides, equal sides, equal angles) and ask other children to identify it.
- Turn to **Pupil book page 27** and read through the information to revise the names of polygons and the meaning of 'regular polygon'.
- The children can work on question 1 independently. Ensure they write their answers.
- Question 2 can be done orally. Give the children time to consider the designs and to decide how to describe each one. Then ask different children to give their descriptions. Decide whether each description is clear or whether it could be improved, and take suggestions for improvements.
- Ask: *Why are circles classified as 2D shapes but not as polygons?* (Circles do not have straight sides so they are not polygons.)

Support

- Make a set of shape vocabulary cards and a set of matching diagram cards for the children to match the vocabulary to the diagrams.
- Use these to play memory games, or card games such as 'Snap' where the children call 'Snap!' when the vocabulary and diagrams are played next to each other.

Interesting mistakes

Children may think that a polygon is regular if it has equal sides and ignore the equal angle condition. For example, they may say that a rhombus is regular. Use folding and/or measurement to match up angles so that they can see that these are not equal, so the shape cannot be regular.

Answers for Pupil book page 27

1 a pentagon   b hexagon   c octagon
   d triangle   e pentagon   f hexagon
   g decagon   h rectangle

2 a a yellow pentagon inside a pink heptagon [Provided as an example]
   b a pink triangle inside a yellow trapezium inside a green dodecagon
   c an orange circle inside a yellow rectangle inside a green octagon and 4 orange triangles inside the green octagon
   d a yellow hexagon inside a green triangle inside a yellow triangle and two pink hexagons inside the yellow triangle

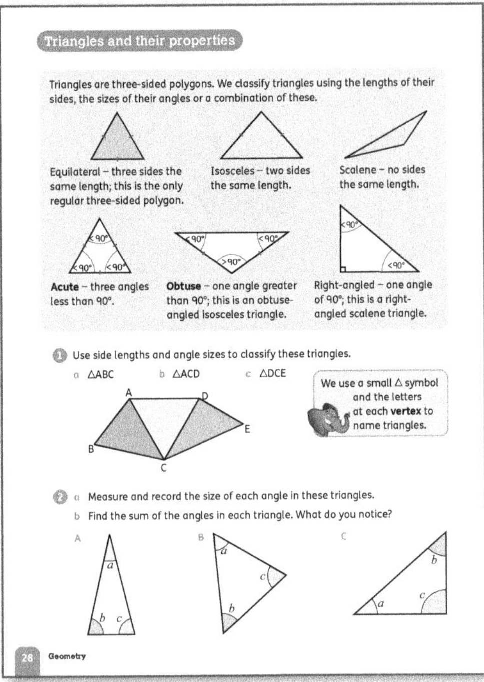

Materials

Flashcards, each showing a different type of triangle; drawing and colouring equipment.

Warm-up

Do some mental addition activities that involve adding sets of three 2-digit numbers.

Focus

- Hold up a flashcard showing a triangle. Ask the children how they know you are showing them a triangle. Ask them to identify acute, obtuse and right angles in the triangle.
- Explain the meaning of 'equilateral', 'scalene' and 'isoceles' and remind the children of the term 'right-angled'. Ask them to say what type of triangle you are showing. Encourage them to explain how they know, for example: 'It is an equilateral triangle because all the sides are equal.'
- Turn to **Pupil book page 28** to revise the classification of triangles using their side and angle properties. Make sure the children know the correct terminology and that they can name triangles using combined properties, for example 'obtuse-angled'. Introduce the notation △ABC used in question 1 and make sure the children can identify the triangles in the diagram.
- The children can work through question 1 and question 2 on their own. Discuss what they notice about the sum of the angles in a triangle. (180°)

Challenge

Write a criterion on the board, for example: Use 6 equilateral triangles and 4 isosceles triangles. Ask the children to construct a pattern that fits the criterion.

Support

Ask the children to draw a picture made completely out of different types of triangle. Ask them to colour all the equilateral triangles one colour, the isosceles another and so on.

Interesting mistakes

- Children may find it difficult to remember the names of the different types of triangle. Provide plenty of practice at identifying triangles and stating their specific properties.
- Children may have difficulties measuring the angles inside shapes correctly; they may need to be reminded about how to use a protractor properly.

Answers for Pupil book page 28

1 a isosceles and acute
   b equilateral and acute
   c scalene and right-angled
2 a For triangle A: $a = 27°$, $b = 78°$ and $c = 75°$.
     For triangle B: $a = 54°$, $b = 54°$ and $c = 72°$.
     For triangle C: $a = 41°$, $b = 49°$ and $c = 90°$.
   b The angles add to 180°.

Calculate angles in triangles

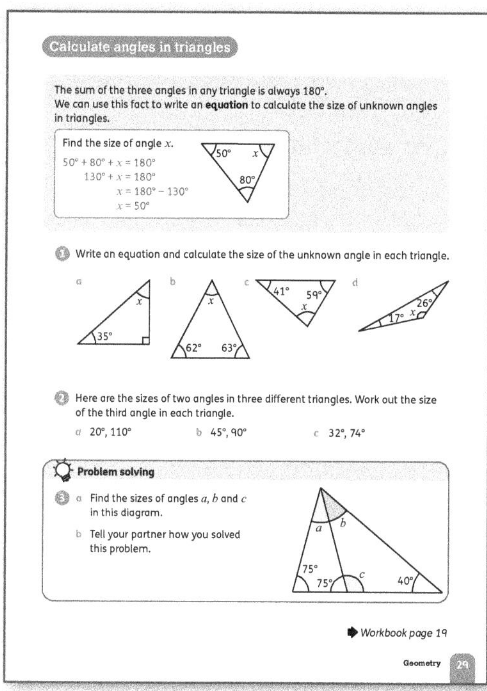

Materials

Rulers; scissors.

Warm-up

Use 'Angles on a straight line' (page 27) as a starter for this lesson.

Focus

- Ask the children what they found out about the angles in a triangle in the previous lesson, **Pupil book page 28** question 2. (Angles in a triangle add up to 180°.)
- It is helpful to demonstrate the rule by asking each child to draw and cut out a triangle and tear off the three corners (angles). By arranging these on a straight line or against a ruler (see diagram) they will see that the angles always total 180°.

- Turn to **Pupil book page 29** and work through the example with the class to show how to use this fact (together with triangle properties) to make 'equations' and calculate the size of unknown angles.
- Let the children work on their own or with a partner to complete question 1 and question 2.
- Problem solving: Let the children work on their own or with a partner to complete question 3.

Follow-up

- Use **Workbook page 19** as an informal assessment task to check that the children are able to write equations and solve them to find the size of unknown angles in triangles.
- Check their answers as a class and let the children rate their own work. Discuss any interesting mistakes and what the class can learn from them.

Answers for Pupil book page 29

1 a 55°   **b** 55°   **c** 80°   **d** 137°
2 a 50°   **b** 45°   **c** 74°
3 a $a = 30°$; $b = 35°$; $c = 105°$
   b Individual answers.

Answers for Workbook page 19

1 a $x = 77°$ **b** $x = 136°$
   c $x = 110°$ **d** $x = 53°$
   e $x = 69°$ **f** $x = 48°$
   g $x = 45°$; $y = 60°$ **h** $x = 147°$; $y = 32°$
   i $p = 50°$; $q = 60°$; $r = 70°$ **j** $x = 75°$

Quadrilaterals and their properties

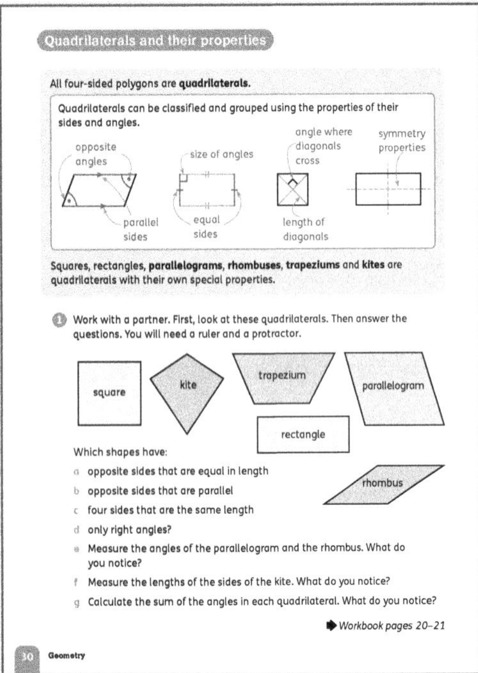

Materials
Rulers; protractors.

Warm-up
Play the 'Unknown angles' game (page 27) as a starter for this lesson.

Focus
- Explain to the class that this is an investigative lesson. Point out the shapes' names: 'parallelogram', 'rhombus', 'kite', 'trapezium', 'rectangle', 'square'.
- Tell the children they are to read through the information on **Pupil book page 30** and ask any questions they have before they start to work on the investigation (which involves the children measuring sides and angles in order to develop their own understanding of the properties of different quadrilaterals).
- The children can use **Workbook page 20** to help summarise and record what they have discovered about the shapes and their properties.
- Check their answers as a class; they will need to refer to accurate shape properties when classifying shapes in the next lesson.

Follow-up
- Use **Workbook page 21** to investigate the sum of angles in quadrilaterals and other polygons. The children will use the fact that the angles in a triangle add up to 180 to find the sum of angles in other shapes and to calculate angle sizes.
- Conclude with a class discussion about what they have learnt in this lesson, what they found interesting and/or surprising and what mistakes they made as they worked through the material. Discuss how any mistakes helped them develop their understanding.

Answers for Pupil book page 30

1
- **a** rectangle; parallelogram; square; trapezium; rhombus
- **b** rectangle; parallelogram; square; rhombus; trapezium (1 pair)
- **c** square; rhombus
- **d** rectangle; square
- **e** Opposite angles are the same size.
- **f** The top two sides are the same length and the bottom two sides are the same length.
- **g** The angles sum to 360°.

Answers for Workbook page 20

1
- **a** has four sides: square [Provided as an example]; rectangle; parallelogram; rhombus; trapezium; kite

  all sides equal in length: square; rhombus

  opposite sides equal in length: square; rectangle; parallelogram; rhombus

  adjacent sides equal in length: square; rhombus; kite

  both pairs of opposite sides parallel: square; rectangle [Provided as an example]; parallelogram; rhombus

  one pair of opposite sides parallel: trapezium

  no sides parallel: kite

  four right angles: square; rectangle

  opposite angles equal: square; rectangle; parallelogram; rhombus

2
- **a** quadrilateral
- **b** square; rectangle; parallelogram; rhombus
- **c** rhombus **d** rectangle **e** square
- **f** trapezium **g** kite

Answers for Workbook page 21

1
- **a** number of sides: 5, 6, 7, 8
  number of triangles: 3, 4, 5, 6
- **b** There are two fewer triangles than there are sides.
- **c** Add up the number of triangles and multiply by 180°.
- **d** pentagon: 108°; hexagon: 120°; octagon: 135°; decagon: 144°; dodecagon: 150°

2
- **a** 76° **b** 68° **c** Both angles are 140°.
- **d** 42° **e** 190° **f** All the angles are 140°.

Identify and classify quadrilaterals

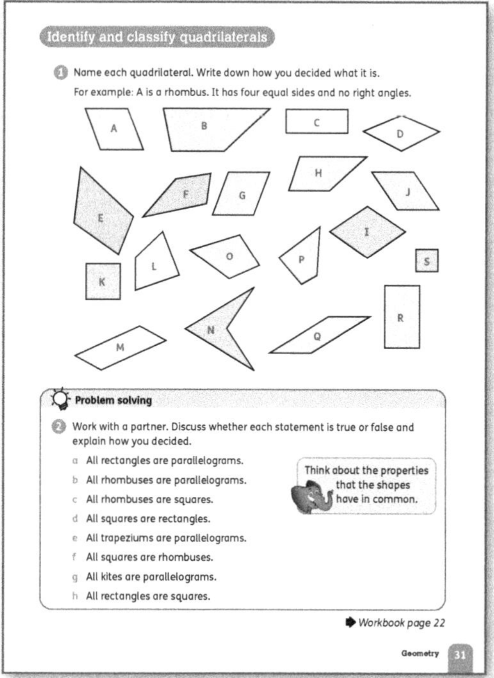

- As a class, discuss what makes one shape different from another. Turning them or looking at them from a different perspective does not make them different. Different dimensions make them different.

Support
- Encourage the children to draw shapes and use different colours to mark their properties. For example, they could use blue to show all the equal sides on a square.
- If they find another shape with all sides equal (such as a rhombus) they should use the same colour.
- They can also mark parallel sides, equal angles and right angles.
- This activity will help them to compare the properties of shapes. For example, the shapes with all four sides coloured blue will share that property.
- There is a lot of terminology and information for the children to remember in this unit. Help them to engage with it by writing statements on the board such as 'A parallelogram has four right angles' and asking the children to say whether the statements are true, false or sometimes true.

Interesting mistakes
Encourage the children to use correct terminology by modelling it yourself. Avoid referring to shapes as, for example, arrowheads or diamonds, because these are not mathematical terms and will confuse the children.

Materials
Geoboards and elastic bands; drawing equipment; rulers (see Follow up).

Warm-up
Use any 'Mental problem-solving' activity (pages 23–24) as a starter for this lesson.

Focus
- Turn to **Workbook page 20** and explain to the children that they can use the work they did here to help them with this lesson's activities on identifying shapes and making decisions about statements.
- Turn to **Pupil book page 31** and let the children work in pairs to complete the questions.
- The children can do question 1 orally, sharing their reasons for classifying shapes in different ways, but check that they write down the names of the shapes.
- For question 2, the children work in pairs to talk about each statement. Explain that they should be prepared to tell the class how they made their decisions.
- Once everyone has finished, hold a class discussion about the statements and their decisions.

Follow-up
- Turn to **Workbook page 22**. If possible, supply geoboards and elastic bands so the children can model the shapes before drawing them.
- Let the children work independently to draw the shapes accurately and then let them check each other's work.

Answers for Pupil book page 31

1 A, G, J, M and Q: parallelogram
B, F, H and P: trapezium
C and R: rectangle
D and I: rhombus
E, L, O and N: quadrilateral
K and S: square

2 a True. The opposite sides are parallel.
b True. The opposite sides are parallel.
c False. Rhombuses are squares only when their angles are all right angles.
d True. A rectangle with its length equal to its width is a square.
e False. Trapeziums have only one pair of parallel sides.
f True. Squares are rhombuses with all the angles equal to 90°.
g False. Kites did not have parallel sides.
h False. A rectangle does not always have length equal to its width.

Answers for Workbook page 22

1 Here are some possible answers.

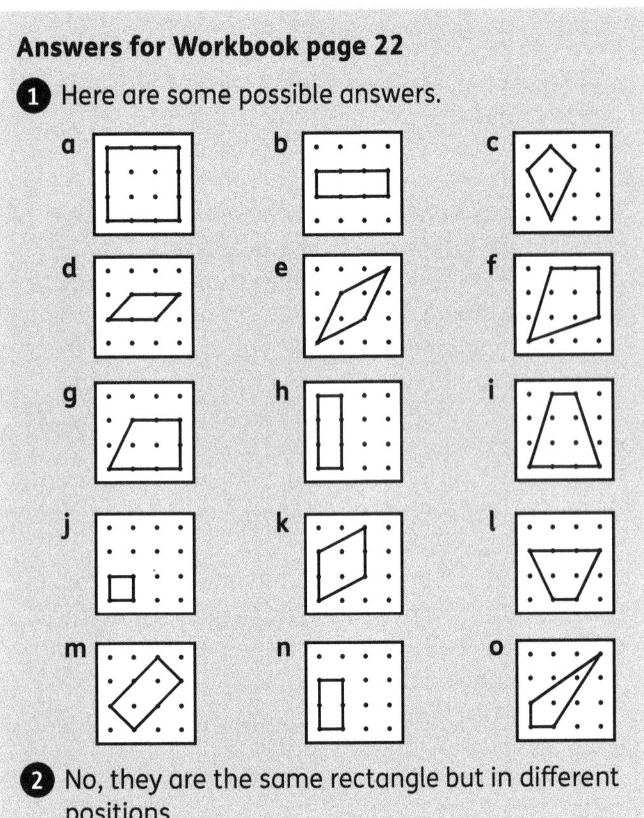

2 No, they are the same rectangle but in different positions.

Special parallelograms

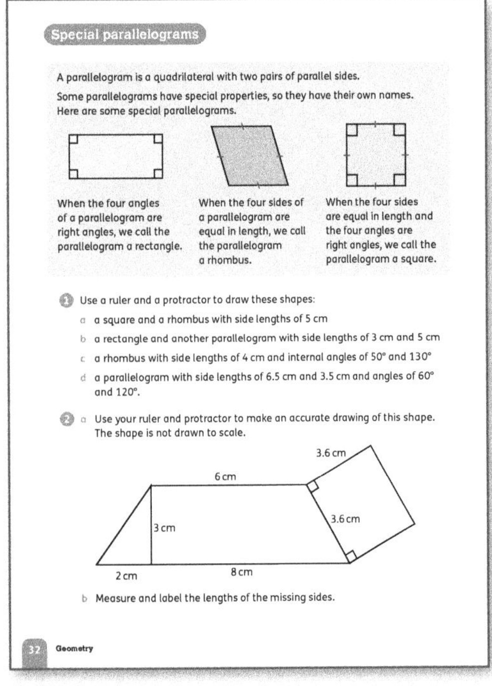

Materials

Rulers; protractors; squared paper.

Warm-up

- Reuse the angle loop game from 'Measure and draw angles' on (page 50) as a warm-up for this lesson.
- The children need to revise measuring angles and using a protractor before they draw accurate shapes.

Focus

- Work through the explanation on **Pupil book page 32** with the class. Make sure they understand that all shapes with two pairs of opposite and equal parallel sides are parallelograms, so squares, rectangles and rhombuses are parallelograms as well.
- For question 1, let the children work independently to draw the shapes accurately. Ask them to check each other's work.
- For question 2, explain that the children need to use the given dimensions and what they know about shapes and angles to make an accurate copy. Again, let them check each other's work.

Challenge

Let the children draw a shape of their own that combines at least three different polygons without writing any dimensions on their shape. Then they swap shapes with a partner to measure the sides and make an accurate copy.

Support

- Check that the children can accurately measure and draw lines of different lengths, including measurements given as decimals (for example, 2.5 cm). Revise these skills as necessary.
- You could provide centimetre or half-centimetre squared grids for the children to draw on, to help them check measurements and angle sizes.

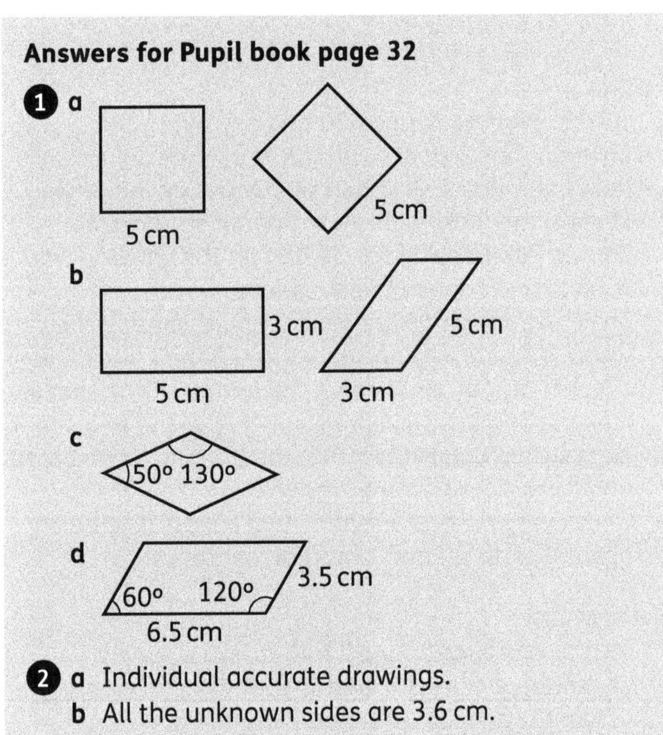

Measure and draw shapes

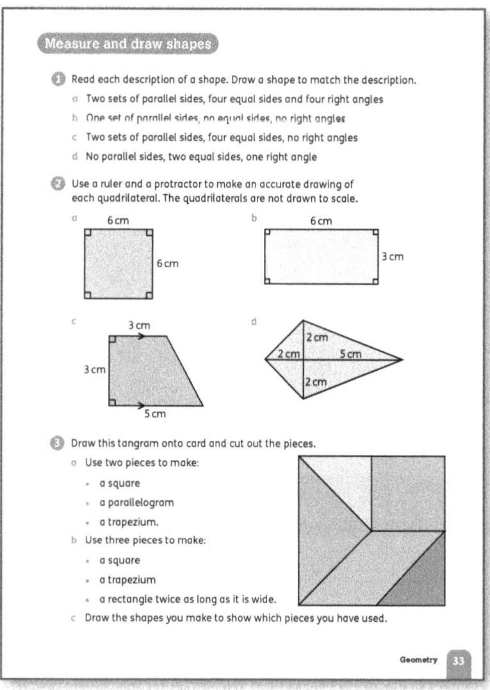

Materials
Drawing equipment; rulers; protractors; card; scissors.

Warm-up
- Play a quick vocabulary game with the class as a starter for this lesson. Display the name of a polygon (including types of quadrilaterals) and ask the children to sketch a rough diagram of what you have named.
- Also, give definitions of shapes and let the children jot down what they could be.

Focus
- This is a practical activity lesson. Use questions 1–3 on **Pupil book page 33** to informally assess that children can read and interpret instructions correctly and that they are able to use a ruler and protractor to draw accurate shapes. Observe them as they work and assist anyone who needs additional support.
- For question 3, the children should make their own tangrams and cut out the pieces, but they can work in pairs or groups to make the shapes. However, ensure they sketch (rough sketches are fine) the shapes they make in their own books.

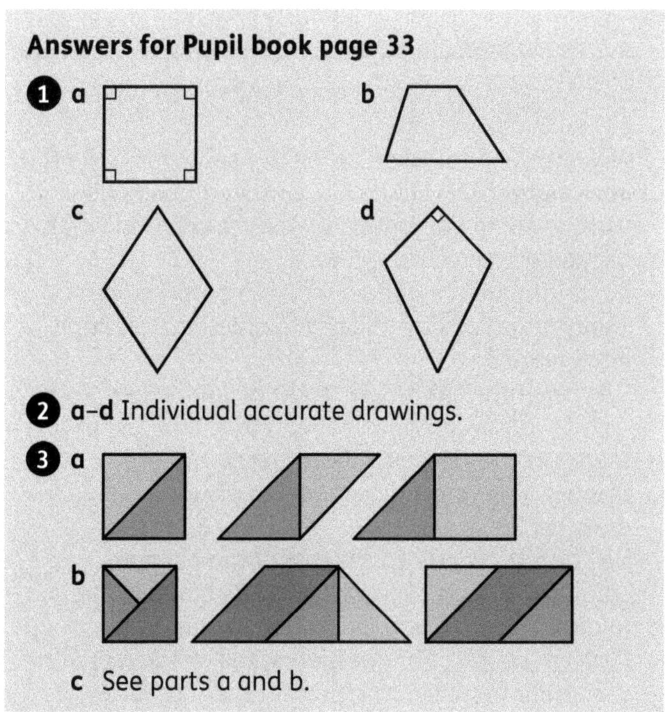

Circles

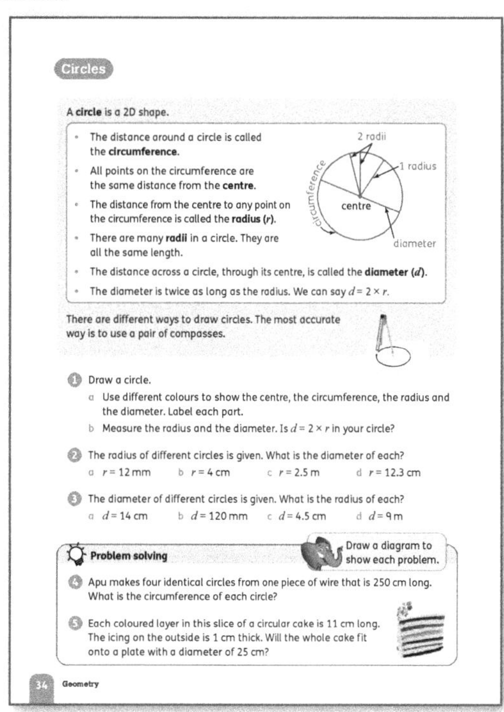

Materials
Rope or string; pairs of compasses; rulers; stencils; printed circles with the parts labelled; scissors; glue.

Warm-up
The children will be using the fact that the diameter of a circle is twice its radius, so warm-up with 'Doubling' and 'Halving' (page 24).

Focus

- Ask the children to point out circles in the classroom, and name other circular objects they can think of, for example cake, hoop.
- Ask one child to stand still holding one end of a rope, and another child to hold the other end and begin to walk, keeping the rope taut. Ask the children what shape path the child will walk.
- Read through the definitions of 'circumference', 'centre', 'radius r', 'radii' and 'diameter d' on **Pupil book page 34**.
- Demonstrate how to use a pair of compasses to draw a circle, or use an instructional video to show the class. Let the class practise drawing circles.
- Discuss how you can use the pair of compasses to draw circles of a particular radius or diameter.
- Use questions 1–3 to check that the children can draw a circle and label its parts and use the relationship between radius and diameter.
- <u>Problem solving:</u> For question 4, the children could use string to model this; prompt them to draw diagrams to help them solve question 5.

Support

Print a circle with the parts labelled for the children to cut out and stick in their exercise books for reference.

Challenge

- Let the children investigate the relationship between the circumference and diameter of a circle. The formula for finding the circumference is $C = \pi \times d$, so $C \div d = \pi$. π (pi) is a constant ratio (and a decimal with an infinite number of places).
- Ask the children to find or draw five circles of different sizes.
- For each circle, ask them to use string and a ruler to measure the circumference as accurately as possible and a ruler to find the diameter. They should write the measurements in a table like this:

Circle	Circumference, C	Diameter, d	$C \div d$
1			
2			
3			
4			
5			

- Next, they use a calculator to work out $C \div d$ to find the ratio of the circumference to the diameter. (They should get a constant ratio that rounds to 3.142, but it's likely they won't have measured the circumferences accurately.)
- Ask them to discuss with a partner:
  - *What do you notice?*
  - *What does this tell you about circles?*
  - *Pi (π) is a special number. Pi rounds to 3.142. How could you use this to find the circumference of a circle if you know its diameter?* (Multiply the diameter by π.)
- Let them share their ideas with the class.

1 a

b Yes, $d = 2 \times$ radius

2 a 24 mm b 8 cm
   c 5 m d 24.6 cm

3 a 7 cm b 60 mm
   c 2.25 cm d 4.5 m

4 62.5 cm

5 Yes; $d = 2 \times 11 + 1 + 1 = 24$ cm

Revisit 3D shapes

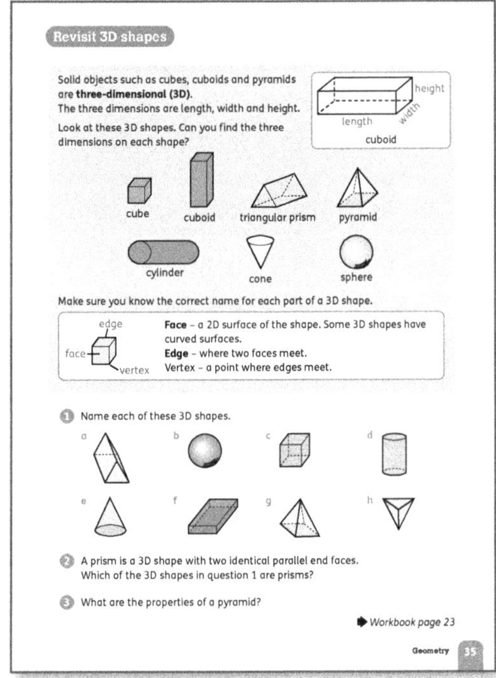

Materials

Various 3D shapes, such as food containers; flashcards, each with the name of a different shape; screen; model prisms and pyramids.

Warm-up

Use any suitable 'Mental problem-solving' activity (pages 23–24) as a starter for this lesson.

Focus

- Remind the class that they have worked with 3D shapes before. Display a variety of 3D shapes for the class. Ask the children to name the shapes they recognise.
- Put flashcards showing shape names on a table and ask the children to place the shapes next to the appropriate card.
- Ask the children if they can think of any common objects that could be added to each set.

- Hold a shape out for the children to see. Ask the children to point to a 'face', an 'edge' and a 'vertex'. Ask questions such as: *How many faces has this shape got? What shape is this face?*
- Hide a shape behind a screen. Describe how many faces it has and what shape the faces are. Can the children guess what the hidden shape is?
- Again, hide a shape behind a screen. This time encourage the children to ask questions about the shape to which you can give 'Yes' or 'No' answers, for example: 'Does it have eight faces?' After each question, the children can guess what the shape is.
- Turn to **Pupil book page 35**. The explanation box at the top of the page asks: Can you find the three dimensions on each shape? (*The two horizontal dimensions, at the widest part of the shape, are length and width, the vertical dimension is height*). Give the children time to read through the information before asking them to complete questions 1–3.

Follow-up

- Use **Workbook page 23** to check that the children are able to name 3D shapes and identify their edges, faces and vertices.
- Check their answers as a class, letting them mark their own work.

Support

- To help the children to name, sort, describe and classify shapes, show them a selection of model prisms and pyramids.
- Ask the children to sort the shapes into two categories and name each one.
- Ask the children to describe the general features of a prism, for example: two opposite, identical faces; and then of a pyramid.

Interesting mistakes

- Some children may have difficulty with the terminology of properties of 3D shapes.
- Regularly use the correct terminology by playing games that involve the children in describing shapes or listening to descriptions to identify shapes.

Answers for Pupil book page 35

1 a triangular prism b sphere
c cube d cylinder
e cone f cuboid
g square-based pyramid
h triangular-based pyramid/tetrahedron

2 a, c and f

3 A pyramid has a base that is a polygon and all its faces are triangles that meet at a vertex (a point).

Answers for Workbook page 23

1 a cone b cylinder c square-based pyramid
d cuboid e sphere f cube
g triangular-based pyramid/tetrahedron
h hexagonal-based prism i triangular prism

2

Shape	Number of faces	Number of vertices	Number of edges
cube	6	8	12
cuboid	6	8	12
cone	2	1	1
cylinder	3	0	2
sphere	1	0	0
triangular pyramid	4	4	6
square-based pyramid	5	5	8
triangular prism	5	6	9
hexagonal prism	8	12	18

3 a They have the same number of faces, vertices and edges.
b A cube must have square faces and a cuboid must have some rectangular faces.
c Yes, due to the answer in **b** and the fact that a square is a special type of rectangle.
d No; due to the answer in **b** and the fact that a square is a special type of rectangle.

3D shapes and their nets

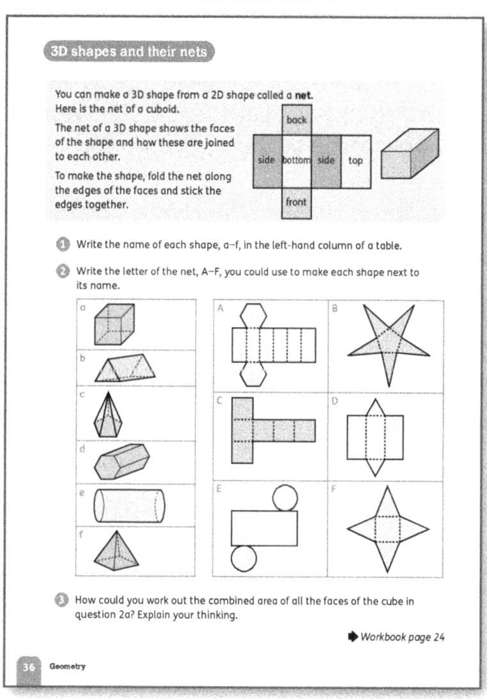

Materials

Variety of solid objects; selection of cardboard boxes; scissors; squared paper; drawing equipment.

Warm-up

Use 'Making towers' (page 27) as a starter for this lesson and to revise the terms used to describe 3D shapes.

Focus

- Spend some time deconstructing cardboard boxes. Let the children work in pairs or small groups to take a box apart in a way that leaves one 2D shape. Let them draw the 'pattern' or 'net' of the box they have created. Then ask them to fold it up again to remake the box.
- Encourage the children to look at the different possible nets for a cuboid box.
- Ask the children what is meant by the net of a shape. Develop their answers until there is a clear definition.
- Show the class a container with a more unusual shape, such as a pencil box shaped like a triangular prism or a box of chocolates with a pyramid shape.
- Discuss what the net of each box will look like. Discuss the shapes and number of faces and ask the children to sketch the nets before you flatten the box.
- Look together at nets of different shapes, such as those on **Pupil book page 36**. Ask the children to name the shapes each net will make. Encourage the children to explain how they know what shape will be created. They can then work independently to draw and complete the table in questions 1 and 2.
- For question 3, give the children time to think about the question before holding a class discussion to agree on the answer. Explain that the area of all the faces is called the 'surface area' of the 3D shape.

Follow-up

- Use **Workbook page 24** to consolidate the children's understanding of nets. Let the children work in pairs to complete the questions, then check the children's answers as a class.
- If you have time, let the children design and build their own gift boxes using nets.

Support

When drawing nets, some children may find it difficult to do this accurately. Support the children in using a ruler, cutting out, etc. where required.

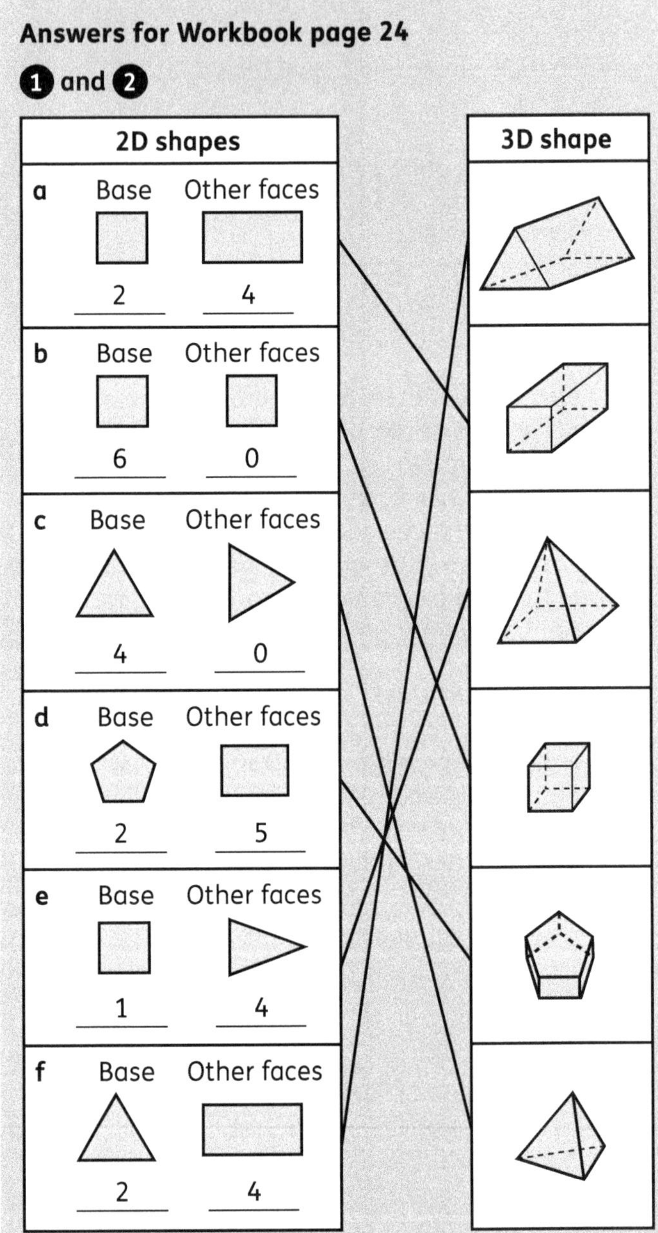

Investigate nets

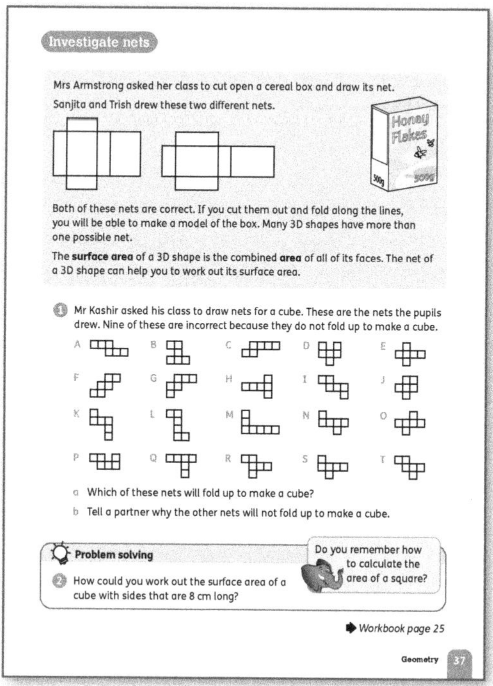

Materials
Boxes from the previous lesson; colouring equipment; (see Follow up): squared paper; scissors; tracing paper.

Warm-up
Use any suitable 'Mental problem-solving' activity (pages 23–24) as a starter for this lesson.

Focus
- Remind the children that a net is a 'plan' of the shape. Show them the cut open boxes from the previous lesson to remind them what the net looks like.
- Let the children work in small groups to complete **Pupil book page 37**. Provide squared paper to any groups that need to cut out and model the nets of cubes. Discuss the answers to questions 1 and 2.

Follow-up
- When you are sure the children understand the concept of a net and the fact that different nets can be used to produce the same 3D shape, let them work in pairs to complete **Workbook page 25**.
- Let the children use squared paper to cut out and model nets as necessary.
- Question 2 is challenging and if the children need additional support with spatial reasoning you may wish to let them take the problem home to consider before trying to solve it in class (remember: we learn from being challenged).
- Allow children who need additional support with this task to trace and cut out the nets so that they can physically manipulate them.

Challenge
You can extend question 2 on **Workbook page 25** by giving the children other shapes with an area of 60 cm² to fill with the shapes, but be aware that many of the solutions will involve flipping some of the shapes.

Answers for Pupil book page 37

1. **a** A; E; F; H; I; L; N; O; R; S; T
   b Individual answers, for example, some of the faces will overlap when folded..
2. First, multiply 8 by itself to get the area of the square. Then multiply by 6 to get the surface area of the cube.

Answers for Workbook page 25

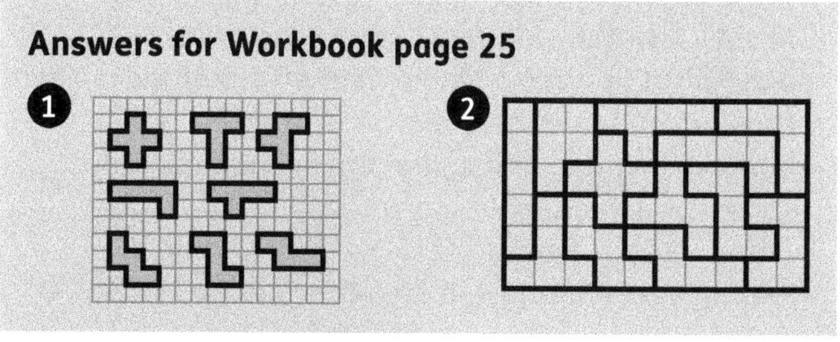

End-of-unit check

Use all or some of these questions and problems to assess how well children have understood the concepts in this unit.

- Write statements on the board such as: *A triangle has three equal sides.* Ask: *Is the statement 'true' 'false' or 'sometimes true'? Explain your answer.* (A triangle has three equal sides: sometimes true; an equilateral triangle has three equal sides.)
- Show a group of triangles. *How can you sort these triangles into four groups? Name each group and describe the properties of the triangles in the group.* (right-angled triangle: one right angle; acute-angled isosceles triangle: two equal sides, two equal angles, all angles acute; obtuse-angled isosceles triangle: two equal sides, two equal angles, one angle obtuse; equilateral triangle: all sides and all angles equal)
- *What facts do you know about parallelograms?* (for example: four sides, two sets of parallel sides, opposite sides equal, opposite angles equal)
- *Is it possible for a quadrilateral to have only three right angles? Why?* (No. The only quadrilaterals with three rights angles are squares and rectangles, and they have four right angles.)
- *What can you say about the diagonals of a parallelogram?* (Apart from in squares and rectangles, they are different lengths and do not intersect (cross) at right angles.)
- *What is meant by the term 'regular polygon'?* (a 2D shape with straight sides and all sides and angles equal) *Give an example of a regular polygon.* (for example: equilateral triangle, square)
- Show different angles. *How can you estimate the size of this angle?* (Use whether the angle is less than or greater than 90°/180° and whether it is close to a multiple of 90° or close to, for example, half of 90°.)
- *What important tips would you give someone to help them use a protractor properly?* (Make sure that the protractor is lined up correctly, so the line on the protractor between 0° and 180° is along one of the lines of the angle. Read the size of the angle starting

from the 0° that lies on the line of the angle. Think about whether the reading makes sense, for example is the value read for an acute angle less than or greater than 90°?)

- Display a set of angles. *Label each of these angles as acute, obtuse, right angle or reflex.* (Acute angles are less than 90°, right angles are equal to 90°, obtuse angles are greater than 90° and less than 180°, and reflex angles are greater than 180°.)

- *Is the statement 'The angles of a triangle add up to 180°' always true? How do you know?* (Yes. For example: Cut out a triangle, tear off the three corners (angles) and arrange them on a straight line.)

- Show a diagram with some angles marked and some unmarked. *How can you find this unknown angle without measuring? How did you do it?* (For example, in a triangle, the angles add up to 180°. In

a parallelogram and a rhombus, opposite angles are equal. Vertically opposite angles are equal. Angles on a straight line add up to 180°.)

- Give children a group of shapes. *How could you group these shapes into sets? What are the properties for your sets?* (for example: number of sides, regular and irregular)

- Show different 3D shapes. *How many faces/edges/ vertices does this shape have?*

- Show a 2D or 3D shape. *How can you describe this shape?* (For example, 2D shapes: number of sides, number of angles, number of equal sides/angles, regular/irregular; 3D shapes: number of sides, number of faces, number of vertices, shapes of faces, number of equal sides)

- *Draw a net for a triangular prism/pyramid.* (See the nets on **Pupil book page 36.**)

UNIT 5 The four operations

Learning objectives

- Use knowledge of the order of operations to carry out calculations involving the four operations.

- Understand that brackets can be used to change the order of operations.

- Perform mental calculations, including with mixed operations and large numbers.

- Estimate, add and subtract integers.

- Solve addition and subtraction multi-step problems in context, deciding which operations to use and why.

- Multiply and divide by 10, 100 and 1000 (using place value).

- Estimate and multiply whole numbers up to 10 000 by 1- and 2-digit numbers using formal written methods.

- Divide numbers of up to 4-digits by 2-digit whole numbers using the formal written method of long division and interpret remainders as whole number remainders, fractions or by rounding as appropriate for the context.

- Divide numbers of up to 4-digits by 2-digit whole numbers using the formal written method of short division where appropriate, interpreting remainders according to the context.

- Solve problems involving addition, subtraction, multiplication and division.

- Use estimation to check answers to calculations and determine, in the context of a problem, an appropriate degree of accuracy.

Key words

brackets   inverse   long multiplication   perimeter
capacity   decimal fractions   long division

Unit introduction

Teaching guidance

- This unit does not contain any new concepts. The children already know how to calculate with whole numbers, and they can apply what they know to work with larger numbers.

- The children will learn how to set out and use 'long division', but this builds on what they already know about using written methods of division.

- Have a number talk (pages 17–18). First, divide the class into groups and ask them to think about which of the four operations (+, −, × and ÷) they find the easiest. They should try to explain why they find one operation easier than the others.
- Take feedback as a class, and then pose the following questions to the groups: *A mathematician said that all operations can be seen and done as addition. What do you think this means? And how is this possible?*
- Give the groups time to talk about this and encourage them to jot down ideas and to use examples to show what they mean. The children should be able to demonstrate using what they know about 'inverse' operations, counting on to subtract, repeated addition instead of multiplication, and counting up or adding multiples to divide.
- The activity is useful as it shows the children that even the operations they sometimes find challenging can be reduced to addition (and addition is likely to be the operation that most of the children chose as the easiest).

The order of operations

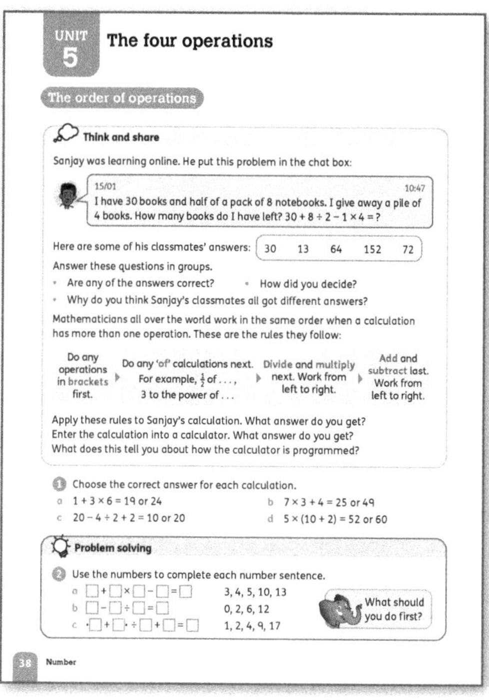

Materials
Calculators that apply order of operations rules; card; sicissors.

Warm-up
Choose any suitable 'Calculation skills' activity (pages 24–26) as a mental warm-up for this lesson.

Focus
- <u>Think and share:</u> The children have worked with the order of operations in previous years, so use the Think and share activity on **Pupil book page 38** to discuss the rules and to remind the children of the order in which they need to work. Include revision of 'brackets'. The children can discuss

the questions in the same groups they worked in for the Unit introduction. (The answer to Sanjay's problem is 30 books. The children can check this on a calculator. Sanjay's classmates may have carried out the operations in the wrong order to get the other answers, for example, working out 30 + 8, then dividing by 2, then subtracting 1 and finally multiplying by 4 gives 72.)
- For question 1, let the children work in pairs to discuss the calculations and decide which is correct. Then they should write the answers in their books.
- <u>Problem solving:</u> For question 2, the children may find it helpful to make small number cards and move them around to find the solutions.

Interesting mistakes
- If you teach the rule as *Brackets first, then 'of', then 'multiplication and division', and lastly 'addition and subtraction'*, children may assume that they always have to multiply before they divide and always add before they subtract. This is not the case, and they need to work from left to right through both pairs of operations. When you discuss the rules and the order, always try to say: *Multiply and divide from left to right. Add and subtract from left to right.*

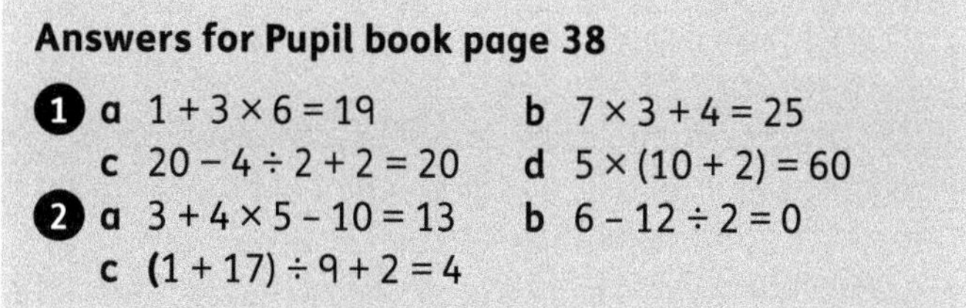

Answers for Pupil book page 38

1 a $1 + 3 \times 6 = 19$ b $7 \times 3 + 4 = 25$
c $20 - 4 \div 2 + 2 = 20$ d $5 \times (10 + 2) = 60$
2 a $3 + 4 \times 5 - 10 = 13$ b $6 - 12 \div 2 = 0$
c $(1 + 17) \div 9 + 2 = 4$

Brackets and the order of operations

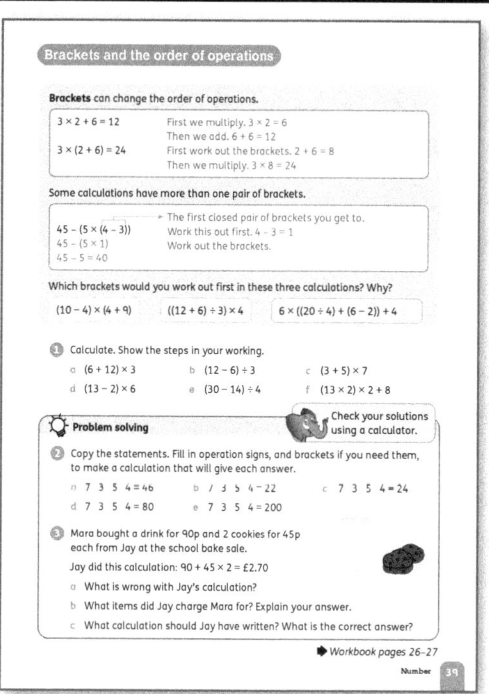

Materials
Calculators that apply order of operations rules.

Warm-up

Use any suitable 'Calculation skills activity (pages 24–26) as a mental warm-up for this lesson.

Focus

- Tell the children: *There are 4 children. Each child has 2 red sweets and 5 green sweets. How can you work out how many sweets there are altogether?* Confirm that you add 2 and 5, then multiply by 4.
- Write $2 + 5 \times 4$ on the board and ask: *Will this give the correct answer?* (No, the answer to this is 22.)
- Explain to the class that sometimes to solve a problem you might need to add or subtract before you do anything else. In examples like these, mathematicians put anything that needs to be done 'first' in brackets. Write the brackets to give the calculation $(2 + 5) \times 4$ and confirm this gives the correct answer to the sweets problem (28).
- Work through the explanation and examples on **Pupil book page 39** with the class.
- The example box at the top of the page asks: Which brackets would you work out first in these three calculations? Why?
$(10 - 4) \times (4 + 9)$
$((10 - 4)$ the first closed pair of brackets you get to)
$((12 + 6) \div 3) \times 4$
$((12 + 6)$ the first closed pair of brackets you get to)
$6 \times ((20 \div 4) + (6 - 2)) + 4$
$((20 \div 4)$ the first closed pair of brackets you get to)
- Ask the children to identify which of these solutions is correct and to explain why.
a $50 - (10 \times (4 - 3)) = 13$ **b** $50 - (10 \times (4 - 3)) = 40$
(**b** is correct: $50 - (10 \times (4 - 3) = 50 - (10 \times 1) = 50 - 10 = 40))$
- The children can work on question 1 independently.
- <u>Problem solving:</u> The children can work in pairs on question 2 and question 3, and then discuss their answers.
- For question 3, make sure the children use a calculator that allows them to enter brackets as necessary.

Follow-up

- Use **Workbook page 26** to consolidate the children's understanding of the order of operations and to informally assess whether the children can apply them.
- Let the children work on their own to complete the questions and then let them work in pairs to check each other's work.
- Let the children work on their own to complete question 1 and question 2 on **Workbook page 27** and check their answers as a class.
- For question 3, put the children into groups to draw the flow diagrams. Display these in the classroom.

Support

Children may find it difficult to work out what operations to use and where to insert brackets even when they are confidently following the rules of operations. Treat these questions as interesting problems to think about and let the children talk with each other and share their strategies for working out the answers.

Challenge

Say: *Choose digits of your own. Use any of the operations and brackets to make calculations. Try to get as many different answers as possible.*

Answers for Pupil book page 39

1 **a** $(6 + 12) \times 3 = 18 \times 3 = 54$
b $(12 - 6) \div 3 = 6 \div 3 = 2$
c $(3 + 5) \times 7 = 8 \times 7 = 56$
d $(13 - 2) \times 6 = 11 \times 6 = 66$
e $(30 - 14) \div 4 = 16 \div 4 = 4$
f $(13 \times 2) \times 2 + 8 = 26 \times 2 + 8 = 52 + 8 = 60$
2 **a** $(7 + 3) \times 5 - 4 = 46$
b $7 \times 3 + 5 - 4 = 22$
c $(7 - 3) \times 5 + 4 = 24$
d $(7 - 3) \times 5 \times 4 = 80$
e $(7 + 3) \times 5 \times 4 = 200$
3 **a** He added 90 to 45 first then multiplied by 2.
b It should be $90 + 45 \times 2 = 90 + 90 = 180$p $= £1.80.$

Answers for Workbook page 26

1 **a** $2 + 3 - 4 = 1$ **b** $2 \times 3 - 4 = 2$
c $2 - 3 + 4 = 3$ **d** $2 \times 3 + 4 = 10$
e $8 \times 3 - 2 = 22$ **f** $8 \times (3 + 2) = 40$
g $8 \times 3 \div 2 = 12$ **h** $8 - (3 \times 2) = 2$
i $(8 - 3) \times 2 = 10$ **j** $8 + 3 - 2 = 9$
2 **a** $5 \times (7 - 3) = 20$
b $28 - (13 - 6) = 21$
c $(6 - 5) \times 12 = 12$
d $38 - (23 + 17 - 12) = 10$
e $48 \div (12 + 6 \times 6) = 1$
f $(23 - 18 - 5) \times 7 = 0$
g $5 \times 6 \div (13 - 10) = 10$
h $8 \times (7 - 2) = 40$
3 **a** $(8 + 7) \times 5 > 8 + 7 \times 5$
b $7 + 2 \times 4 < 5 \times 2 + 6$
c $5 \times 9 \div 3 < 4 \times 8$
d $23 - 12 \div 6 > 2 + 6 \times 3$
e $5 \times 5 \div 5 = 5 \div 5 \times 5$
f $36 \div 12 \div 3 = 24 \div 12 \div 2$

Answers for Workbook page 27

1
a $9 \times (2 + 3) = 45$ b $(16 - 7) \times 3 = 27$
c $20 \div (4 + 1) = 4$ d $(20 \div 4) + 1 = 6$
e $(18 + 9) \div 3 = 9$ f $(64 \div 8) - 6 = 2$
g $(6 + 6) \times 3 = 36$ h $(10 - 4) \times 5 = 30$
i $5 + 2 \times (3 + 7) = 25$ j $12 + 6 \div (7 - 4) = 14$
k $(7 + 10 - 5) \div 2 = 6$ l $(6 - 3) \times 2 = 6$
m $10 \times (6 + 4) = 100$ n $27 - (14 - 8) = 21$

2
a true   b true
c false   d false
e true   f true

3 e.g.

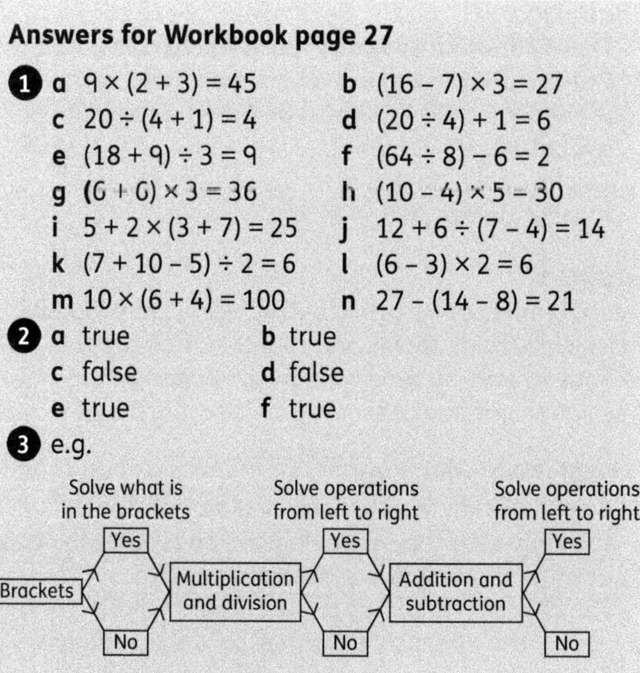

Number relationships

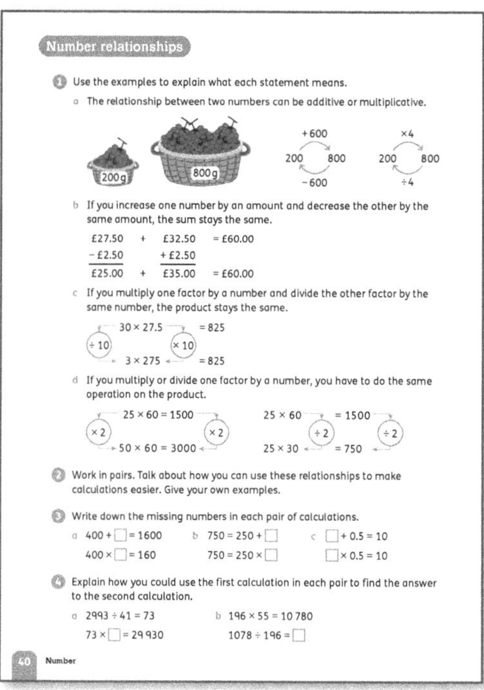

Materials

Calculators; large sheets of paper; marker pens.

Warm-up

Use any suitable 'Mental problem-solving' activity (pages 23–24) as a starter for this lesson.

Focus

- This lesson looks at some important number relationships that the children might already know but not have really thought about or verbalised before.
- For **Pupil book page 40** question 1, put the children into groups or pairs to read the statements and work out what each means.
- Ask different groups to share their explanations.
- For question 2, the children should work together to talk about how the relationships can be used to make calculations easier. Let them write their examples on sheets of paper to show to the rest of the class.
- For question 3, the children should work together to complete the missing number statements using the relationships they've talked about.
- If necessary, for question 4, ask questions to check that the children understand that multiplication and division are inverse operations. Use basic facts to remind them.

Interesting mistakes

- Some children may have fairly good recall of multiplication facts but need additional support with division facts. A child may know $6 \times 7 = 42$ but not use this to work out that $42 \div 7 = 6$. This is an additional problem when they have to scale-up facts using a combination of place value and inverse operations. Use place-value tables (page 21) and counters to model the operations and to show how the digits move when you scale them up or down.
- Let the children explain what is happening and what they are doing to help consolidate understanding and to see if and where they are confused.
- Continue to revise times table facts and fact families, and include scaling.

Answers for Pupil book page 40

1
a Given two numbers, we can get from one to the other by adding/subtracting or by multiplying/dividing.
b Adding and subtracting the same number to a sum does not change it.
c Multiplying and dividing a product by the same number does not change it.
d If we multiply or divide one factor by a number we must multiply or divide the product.

2 Individual answers.

3
a $400 + \underline{1200} = 1600$ $400 \times \underline{0.4} = 160$
b $750 = 250 + \underline{500}$ $750 = 250 \times \underline{3}$
c $\underline{9.5} + 0.5 = 10$ $\underline{20} \times 0.5 = 10$

4
a $73 \times \underline{410} = 29\,930$ We multiply by 10 times 41.
b $1078 \div 196 = \underline{5.5}$ We divide one tenth of $10\,780$ by 196.

Addition

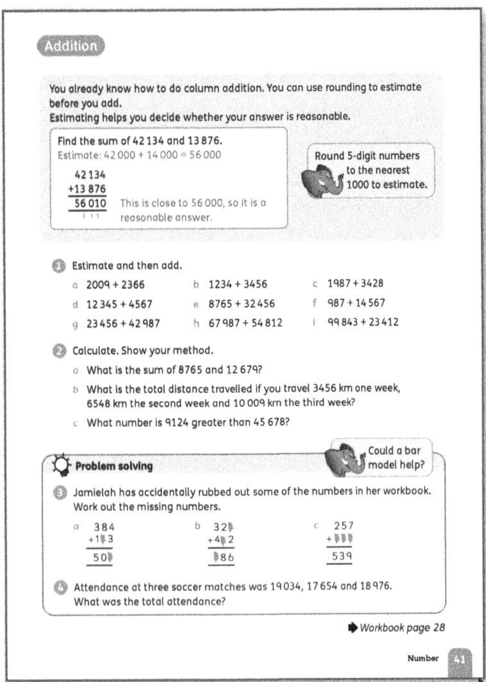

Materials

Place-value tables (page 21); counters; blank number lines.

Warm-up

Use any suitable 'Mental problem-solving' activity (pages 23–24) on addition as a starter for this lesson.

Focus

- Discuss the example on **Pupil book page 41** with the class. Explain that compact methods with carrying are the most efficient for recording and working out when you have to add large numbers. However, stress that the children should use the method they feel most comfortable with if they get stuck.
- Write 2367 + 4109 + 1863 on the board. Below the calculation write 10 339, 8339 and 5339. Ask the children to guess which is the right answer. Ask them to explain how they made their choice.
- Talk about how to use rounding to estimate and check answers.
- For question 1, check that the children are writing their estimates before they calculate.
- For question 2, check that the children are using the correct operations.
- Problem solving: The children can work in pairs on questions 3 and 4.

Follow-up

- Use **Workbook page 28** for additional practice and to consolidate modelling and using different methods to add large numbers.
- Check the answers as a class and discuss any interesting mistakes.

Challenge

- The children could work in pairs with a calculator.
- One child sets a multi-digit addition question and the second child estimates and then works out the answer.
- The first child then checks the answer with the calculator.

Support

The children could practise adding using blank number lines, place-value tables and counters. This could be done in small groups with one child devising a calculation and the other children carrying it out.

Interesting mistakes

- Some children tend to use calculators without thinking what they are doing and consequently write down answers that are not sensible.
- The children should do a rough approximation of the sum in their heads before they do the calculation. For example, the sum 3186 + 4612 is approximately 3000 + 4500, so they should expect an answer around 7500. If the answer obtained is very different, they have made a mistake on the calculator.

Answers for Pupil book page 41

1. a 4000; 4475   b 4000; 4690
   c 5000; 5415   d 17 000; 16 912
   e 41 000; 41 221   f 16 000; 15 554
   g 66 000; 66 443   h 123 000; 122 799
   i 123 000; 123 255
2. a 21 444
   b 3456 + 6548 + 10 009 = 20 013 km
   c 45 678 + 9124 = 54 802
3. a 384 + 123 = 507
   b 324 + 462 = 786
   c 257 + 282 = 539
4. 55 664

Answers for Workbook page 28

1. e.g. 44 791 + 520 457 = 565 248
   44 791 + 209 364 = 254 155
   520 457 + 209 364 = 729 821
   44 791 + 520 457 + 209 364 = 774 612
2. 201 + 349; 160 + 415; 589 + 111
   223 + 377; 109 + 516; 97 + 703
3. a 150 + 270 = 420
   b 3936 + 5610 = 9546
   c 504 + 647 = 1151

Subtraction

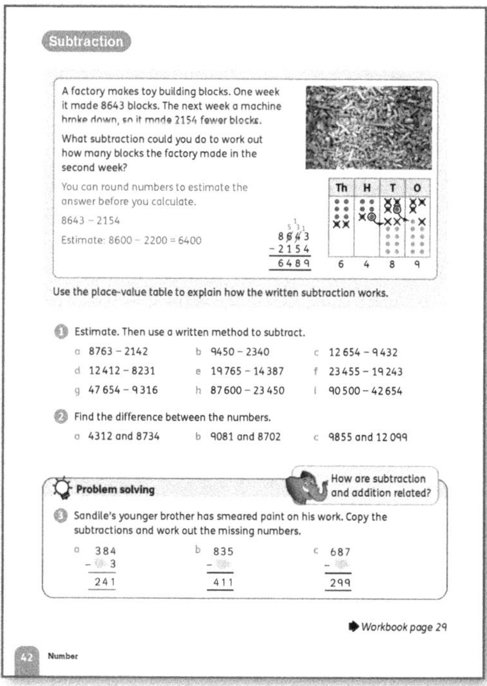

Subtraction

A factory makes toy building blocks. One week it made 8643 blocks. The next week a machine broke down, so it made 2154 fewer blocks.

What subtraction could you do to work out how many blocks the factory made in the second week?

You can round numbers to estimate the answer before you calculate.

8643 − 2154

Estimate: 8600 − 2200 = 6400

Th	H	T	O
6	4	8	9

$$\begin{array}{r} 8\,6\,4\,3 \\ -\ 2\,1\,5\,4 \\ \hline 6\,4\,8\,9 \end{array}$$

Use the place-value table to explain how the written subtraction works.

1 Estimate. Then use a written method to subtract.

a 8763 − 2142 b 9450 − 2340 c 12 654 − 9432
d 12 412 − 8231 e 19 765 − 14 387 f 23 455 − 19 243
g 47 654 − 9316 h 87 600 − 23 450 i 90 500 − 42 654

2 Find the difference between the numbers.

a 4312 and 8734 b 9081 and 8702 c 9855 and 12 099

Problem solving

How are subtraction and addition related?

3 Sandile's younger brother has smeared paint on his work. Copy the subtractions and work out the missing numbers.

a
$$\begin{array}{r} 3\,8\,4 \\ -\ \ \ 3 \\ \hline 2\,4\,1 \end{array}$$
b
$$\begin{array}{r} 8\,3\,5 \\ -\ \ \\ \hline 4\,1\,1 \end{array}$$
c
$$\begin{array}{r} 6\,8\,7 \\ -\ \ \\ \hline 2\,9\,9 \end{array}$$

➡ Workbook page 29

42 Number

Materials

Grids on laminated cards (page 24); place-value tables (page 21); counters; squared paper; calculators.

Warm-up

Use the 'Mental addition and subtraction' activity (page 24) as a starter for this lesson.

Focus

- Start with the problem on **Pupil book page 42**. Let the children read through the question and think about what subtraction they could do. Give the children time to read through the worked example and then ask them to share their ideas about how the place-value table and column subtraction show the calculation. The example box at the top of the page says: Use the place-value table to explain how the written subtraction works. (When you cannot work out a subtraction in one column, for example 3 − 4, you can decompose the number in the column to the left, for example exchange 1 in the tens column for 10 ones, and put these in the ones column. Then you can subtract 13 − 4.)
- Let the children work on their own to complete the calculations in question 1 and question 2. Remind them that they can use different methods, and encourage estimation and approximation.
- Problem solving: For question 3, talk about how they could find the missing values in the calculations and allow them to suggest and try out different methods.

Follow-up

Use **Workbook page 29** to consolidate work on addition and subtraction. Observe the children as they complete the questions and assist anyone who needs help or seems confused.

Support

- Encourage the children to see the value of estimating answers by doing some calculations with them, for example:
  - Write a subtraction and ask the children to round and estimate the answer. Enter the calculation into the calculator, sometimes making an error.
  - For example, omit one digit, or press + instead of −.
  - Read out your answer and let the children use their estimates to decide whether your answer is reasonable. If it is not, let them suggest what you might have done wrong.
  - Work out the correct answer as a class.

Challenge

- Encourage the children to research a set of numbers they can use to make up subtraction problems (or combined addition and subtraction problems). Some higher number sets that work well include:
  - areas of different countries
  - population figures
  - areas covered by forests on each continent
  - distances between cities
  - values of imports/exports
  - passenger numbers on rail or air transport
  - crowds attending sport or cultural events.

Interesting mistakes

- When subtracting, children sometimes subtract the smaller number from the larger one digit by digit. Children may also need additional support to decompose tens into ones (and so on).
- Modelling the subtractions on a place-value grid with counters can help them to see how subtraction works and how they have to subtract the 'bottom' number from the top even when that digit is greater than the one above.
- Avoid saying you cannot subtract a larger number from a smaller one because mathematically it is possible, for example: 5 − 7 = −2. Point out how we cannot have negative digits in a number, so we do not subtract negative numbers using the column method.

Answers for Pupil book page 42

1 a 6700; 6621 b 7100; 7110
c 3300; 3222 d 4200; 4181
e 5400; 5378 f 4300; 4212
g 38 400; 38 338 h 64 100; 64 150
i 47 800; 47 846

2 a 4422 b 379 c 2244

3 a 384 − 143 = 241
b 835 − 424 = 511
c 687 − 388 = 299

Answers for Workbook page 29

1   a 9   b 229   c 112
    d 152   e 36   f 2001

2 $(90 - 40) - 25 = 50 - 25 = 25$

3 9

4 $1356 - 773 = 583; 1356 - 583 = 773$

5 For example:

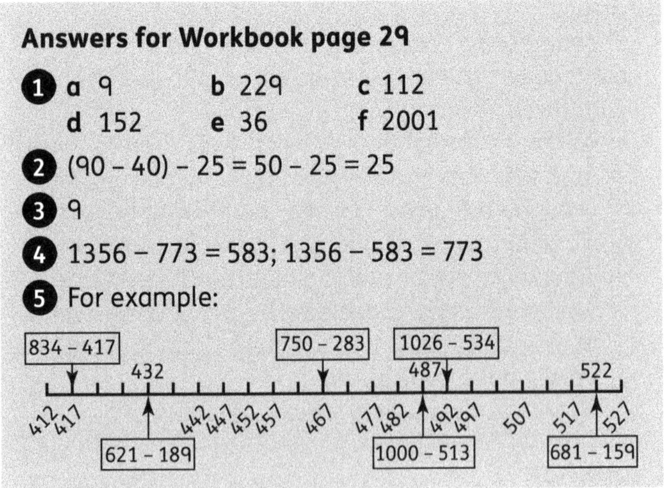

Use inverse operations

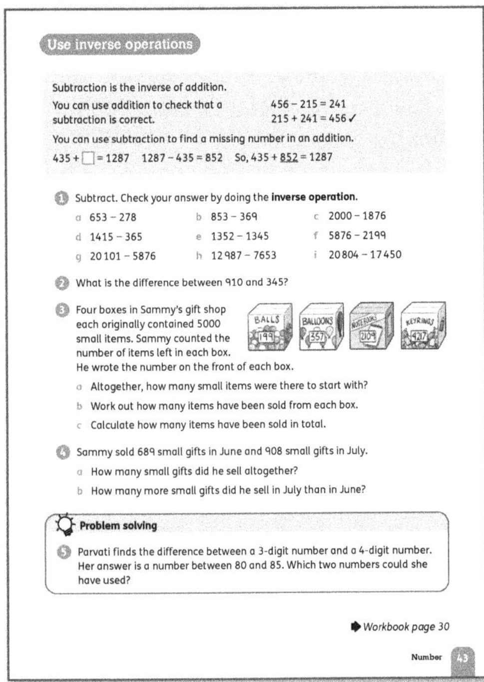

Warm-up

To give the children a break from number and calculation
work, choose any 'Geometry and measures' activity
(page 27) as a mental warm-up for this lesson.

Focus

- If you have time, and access to an outdoor area, play
  a quick movement game to revise the concept of
  inverse operations cancelling each other out.
  - Let the children stand outside, spaced out.
    Ask them to hop forward a number of times.
  - Then ask them how they could hop back to their
    starting point. Elicit that if you hop forwards three
    hops, you need to hop back three hops to get to
    the starting point.
  - Repeat a few times, including starting with
    hopping back and then hopping forward to get to
    the starting point.

- Work through the explanation on **Pupil book
  page 43** with the class. Make sure they understand
  that subtraction is the inverse of addition and how
  to use inverse operations to check solutions and find
  missing values. The children can work independently
  on questions 1–5, and you can go through the
  answers with the class.
- <u>Problem-solving:</u> As a hint for question 5, ask the
  children to think about the best strategy to use.

Follow-up

- Let the children complete **Workbook page 30**
  independently to informally assess their
  understanding of the concepts in this lesson.
- Ask different children to share their notes from
  question 2 with the class. Discuss any interesting
  ideas and resolve any disagreements.

Interesting mistakes

- Children may recognise some situations as
  subtraction. For example, they may realise that if
  there are 250 passengers and 45 get off the boat,
  they can subtract to find out how many are left.
- However, the children may be unable to solve
  problems that involve comparing two amounts or the
  missing part of a whole. Encourage them to draw bar
  models to help visualise the problem and see what
  operation is involved (page 21).

Answers for Pupil book page 43

1   a 375   b 484   c 124   d 1050
    e 7   f 3677   g 14 225   h 5334
    i 3354

2 565

3   a 20 000   b 4801; 4643; 2891; 783
    c 13 118

4   a 1597   b 219

5 For example, 1080 and 999

Answers for Workbook page 30

1   a 650   b 644   c 526   d 602
    e 222   f 199   g 413

2 For example: He decomposed a number as
a sum of numbers using place values and
he used rounding and compensating. These
two strategies are useful as we can do the
calculations faster.

Addition and subtraction problems

Addition and subtraction problems

Some problems involve both adding and subtracting. You may need to do two or more steps to find the answer. Drawing a diagram can help you to work out whether to add or subtract.

A ship can carry 20 000 containers. At Port A, 12 500 containers are loaded onto the ship. At Port B, 355 containers are offloaded and 7456 containers are loaded. At Port C, no containers are offloaded. What is the maximum number of containers that can be loaded at Port C?

Draw a diagram to show the problem. Work in steps to find the solution.

A maximum of 399 containers can be loaded at Port C.

1. Another ship can carry 19 462 containers.
   a. 4238 containers are loaded on Monday and another 3874 are loaded on Tuesday. How many more containers can the ship carry?
   b. The ship arrives at Port A fully loaded. 9463 containers are offloaded and 2081 are loaded. How many containers are on the ship when it leaves Port A?

2. A ship has two areas for containers. Each area holds 1049 containers. On the ship's first journey, 1980 containers are loaded at Port A. 479 are offloaded at Port B. At Port C, none are offloaded, but more are loaded. When the ship leaves Port C, there are 2057 containers on board. How many were loaded at Port C?

3. Write a multi-step word problem to match each bar model. Show your worked solution for each problem.

44 Number

Materials

Squared paper; rulers; drawing equipment.

Warm-up

Use any suitable 'Mental problem-solving' activity (pages 23–24) as a warm-up for this lesson.

Focus

- Let the children work in pairs. Turn to **Pupil book page 44**. Give them time to read the problem, study the bar model and check the calculations.
- Hold a class discussion about how the bar model shows the problem.
- For question 1, make sure the children understand that parts a and b relate to the ship in the stem of the question that can hold 19 462 containers. For both questions 1 and 2 the children should draw bar models and use these to help them solve the problems.
- For question 3, the children can work together to make up interesting problems.

Interesting mistakes

- Children sometimes use bar models incorrectly. One of the mistakes that children make is to represent equal quantities using unequal bars. For example, they might show a comparison problem like this:

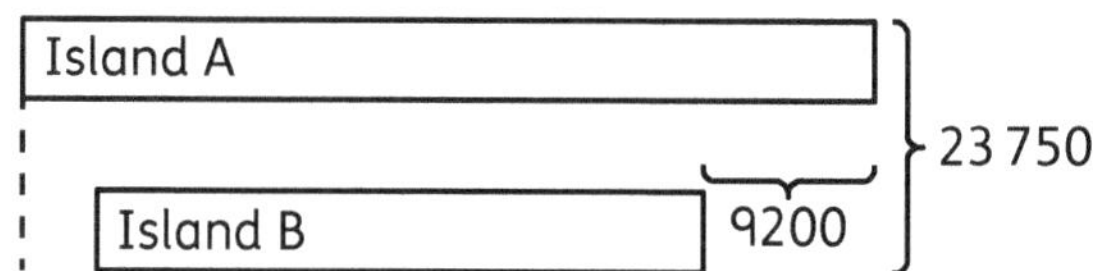

- To help, always ensure the children draw a vertical line on the left and start the bars in the problem against that line. In this way, they can compare equal quantities with equal length bars, and the difference is correctly represented as a gap at the other end.

- In our bar models, we place the bars so they touch to help children compare them.

> ### Answers for Pupil book page 44
>
> **1** a 11 350   b 12 080
>
> **2** 556
>
> **3** a, b Individual answers. e.g. a A ship carries 19 462 containers. At the first port 9863 containers are unloaded. At the second point 4789 containers are unloaded. How many containers are left on the ship?
>
> b A ship is loaded with 14 500 containers. 8097 of these are red and the rest are blue.
>
> At the port, some containers are unloaded and some new ones are put on the ship. Now there are 11 541 red containers. How many blue ones are there?

Multiply by 10, 100 and 1000

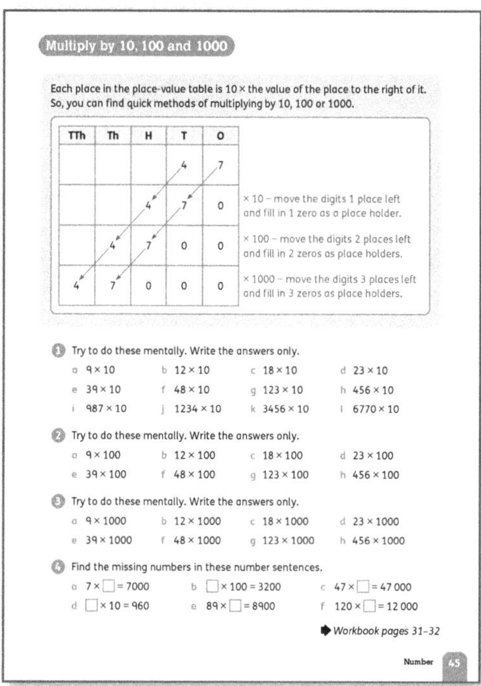

Multiply by 10, 100 and 1000

Each place in the place-value table is 10 × the value of the place to the right of it. So, you can find quick methods of multiplying by 10, 100 or 1000.

TTh	Th	H	T	O	
			4	7	
		4	7	0	× 10 – move the digits 1 place left and fill in 1 zero as a place holder.
	4	7	0	0	× 100 – move the digits 2 places left and fill in 2 zeros as place holders.
4	7	0	0	0	× 1000 – move the digits 3 places left and fill in 3 zeros as place holders.

1. Try to do these mentally. Write the answers only.
   a 9 × 10   b 12 × 10   c 18 × 10   d 23 × 10
   e 39 × 10   f 48 × 10   g 123 × 10   h 456 × 10
   i 987 × 10   j 1234 × 10   k 3456 × 10   l 6770 × 10

2. Try to do these mentally. Write the answers only.
   a 9 × 100   b 12 × 100   c 18 × 100   d 23 × 100
   e 39 × 100   f 48 × 100   g 123 × 100   h 456 × 100

3. Try to do these mentally. Write the answers only.
   a 9 × 1000   b 12 × 1000   c 18 × 1000   d 23 × 1000
   e 39 × 1000   f 48 × 1000   g 123 × 1000   h 456 × 1000

4. Find the missing numbers in these number sentences.
   a 7 × ☐ = 7000   b ☐ × 100 = 3200   c 47 × ☐ = 47 000
   d ☐ × 10 = 960   e 89 × ☐ = 8900   f 120 × ☐ = 12 000

➧ *Workbook pages 31–32*

Number 45

Materials

Large place-value table (page 21); place-value cards (page 21) or 0–9 digit cards; calculators.

Warm-up

- As a starter for this lesson, have a number talk (pages 17–18) about the relationship between powers of 10. For example, 10 is 10 times the size of 1, and 100 is 10 times the size of 10. Focus on language such as: 10 times more, 10 times greater, and so on.
- Encourage the children to relate this to different numbers. For example, ask:
  - *What number is 100 times greater than 3?* (300)
  - *What number is ten times the size of 15?* (150)

Focus

- Revise multiplication by 10 using place-value tables and calculators. Record the results and make sure the children understand that to multiply by 10 we move the digits one place to the left.
- Next, ask: *What happens when you multiply by 10 and then multiply by 10 again?*
- Discuss the results. Explain that this is the equivalent of multiplying by 100. Do some examples with the class to establish that to multiply by 100 you move the digits two places to the left.
- Write 13 × 1000 = 13 000. Ask the children to explain why the digits move three places left when you multiply by 1000. (1000 = 10 × 10 × 10)
- Have a class discussion about why it is mathematically incorrect to say that you add 0 to a number when you multiply it by 10 (use an example, such as 3 + 0 = 3 but 3 × 10 = 30, to make sure they understand this).
- Turn to **Pupil book page 45** and ask the children to read through the explanation and examples on their own before completing questions 1–4. Encourage them to work mentally using jottings if necessary.

Follow-up

- Use **Workbook page 31** to consolidate multiplication and division by powers of 10. Let the children check each other's answers when they are finished.
- **Workbook page 32** explores multiplication patterns arising from multiples of powers of 10 (for example, 20, 200, 2000). Let the children work in pairs to complete the questions and to talk about how they can use known facts together with place value to multiply when multiplying by a multiple of 10, 100 or 1000.
- After they have worked through the questions, have a class discussion about using known facts and place value to multiply by a multiple of 10, 100 or 1000.

Support

The main errors in this work are likely to be as a result of incorrect tables facts. Revise the multiplication and division facts related to the 2 to 12 times tables as necessary.

Answers for Pupil book page 45

1   a 90   b 120   c 1800   d 230
    e 390   f 480   g 1230   h 4560
    i 9870   j 12 340   k 34 560   l 67 700

2   a 900   b 1200   c 1800   d 2300
    e 3900   f 4800   g 12 300   h 45 600

3   a 9000   b 12 000   c 18 000   d 23 000
    e 39 000   f 48 000   g 123 000   h 456 000

4   a 7 × 1000 = 7000   b 32 × 100 = 3200
    c 47 × 1000 = 47 000   d 96 × 10 = 960
    e 89 × 100 = 8900   f 120 × 100 = 12 000

Answers for Workbook page 31

1

Number	× 10	× 100	× 1000
9	90	900	9000
17	170	1700	17 000
28	280	2800	28 000
39	390	3900	39 000
66	660	6600	66 000
89	890	8900	89 000
101	1010	10 100	101 000
145	1450	14 500	145 000
865	8650	86 500	865 000
435	4350	43 500	435 000
234	2340	23 400	234 000
1234	12 340	123 400	1 234 000
4076	40 760	407 600	4 076 000
5999	59 990	599 900	5 999 000
9800	98 000	980 000	9 800 000

2   a 4120; 100; 41.2   b 100 000; 1000; 10 000

Answers for Workbook page 32

1

8 × 2 = 16	6 × 3 = 18	9 × 4 = 36
8 × 20 = 160	6 × 30 = 180	9 × 40 = 360
8 × 200 = 1600	6 × 300 = 1800	9 × 400 = 3600
8 × 6 = 48	6 × 8 = 48	9 × 5 = 45
8 × 60 = 480	6 × 80 = 480	9 × 50 = 450
8 × 600 = 4800	6 × 800 = 4800	9 × 500 = 4500

Pattern: Add a zero to the product for each zero in the multiple of 10 or 100.

2

80 × 2 = 160	60 × 3 = 180	90 × 4 = 360
80 × 20 = 1600	60 × 30 = 1800	90 × 40 = 3600

80 × 200 = 16 000
60 × 300 = 18 000
90 × 400 = 36 000
80 × 2000 = 160 000
60 × 3000 = 180 000
90 × 4000 = 360 000

80 × 6 = 480	60 × 8 = 480	90 × 5 = 450
80 × 60 = 4800	60 × 80 = 4800	90 × 50 = 4500
80 × 600 = 48 000	60 × 800 = 48 000	

90 × 500 = 45 000
80 × 6000 = 480 000
60 × 8000 = 480 000
90 × 5000 = 450 000

Pattern: Add a zero to the product for each zero in the numbers in the calculation.

Combine multiplication and addition or subtraction

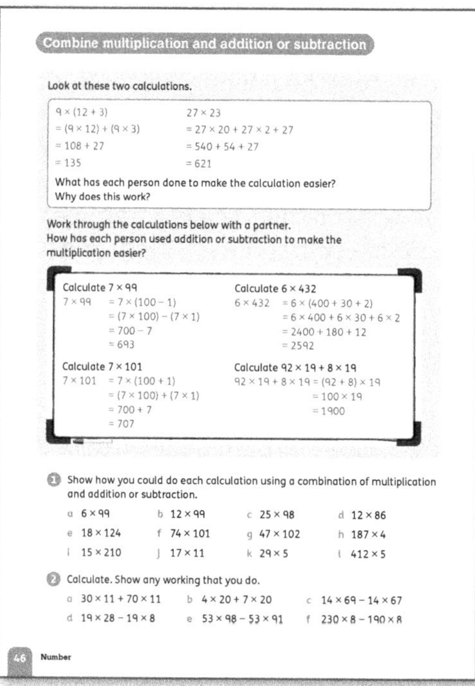

Materials
Calculators.

Warm-up
Use any suitable 'Mental problem-solving' activity involving multiplication (pages 23–24) as a starter for this lesson.

Focus
- Use the examples on **Pupil book page 46** to start a talk with the class about how we can use different strategies to make calculation easier.
- Begin by looking at the strategies shown. Make it clear that different people find different things easier, and that one strategy may be easier for them than another.
- The example box at the top of the page asks: How has each person used addition or subtraction to make the multiplication easier? (Where the number is close to 100, they have written it as 100 + a small number or 100 subtract a small number. They have multiplied 100 and the number and then added/subtracted to find the answer. To calculate 6×432 they have decomposed 432 into $400 + 30 + 2$, multiplied each of these by 6, then added them all together. To calculate $92 \times 19 + 8 \times 19$ they have seen that 92 of a number plus 8 of the same number makes $92 + 8 = 100$ of that number, so they have multiplied by 100.)
- The children can answer questions 1 and 2 independently. Choose children to explain their strategy for a given question, and discuss any other strategies used.

Interesting mistakes
- Children may need additional support to calculate 12×14 because they cannot see how to split either number into a product of its factors. When faced with a problem like this, ask the child: *How can we write 12 as a multiplication?* (1×12, 2×6 and 3×4).
- Then ask: *Could writing 12 as a multiplication help you to multiply 12×14?* Take their suggestions and model different options for them, for example: $3 \times 4 \times 14 = 3 \times 56$.
- Say, for example: *I know 2 times 14 is 28 and double that is 56. That gives me 3×56 which is the same as 3×50 plus 3×6. I know that 3×50 is 150 and 3×6 is 18, so the answer is 168.*

Answers for Pupil Book page 46

1

a $6 \times 99 = 6 \times (100 - 1)$
$= (6 \times 100) - (6 \times 1)$
$= 600 - 6$
$= 594$

b $12 \times 99 = 12 \times (100 - 1)$
$= (12 \times 100) - (12 \times 1)$
$= 1200 - 12$
$= 1188$

c $25 \times 98 = 25 \times (100 - 2)$
$= (25 \times 100) - (25 \times 2)$
$= 2500 - 50$
$= 2450$

d $12 \times 86 = 12 \times (100 - 14)$
$= (12 \times 100) - (12 \times 14)$
$= 1200 - 168$
$= 1032$

e $18 \times 124 = 18 \times (100 + 20 + 4)$
$= 18 \times 100 + 18 \times 20 + 18 \times 4$
$= 1800 + 360 + 72$
$= 2232$

f $74 \times 101 = 74 \times (100 + 1)$
$= 74 \times 100 + 74 \times 1$
$= 7400 + 74$
$= 7474$

g $47 \times 102 = 47 \times (100 + 2)$
$= 47 \times 100 + 47 \times 2$
$= 4700 + 94$
$= 4794$

h $187 \times 4 = (100 + 80 + 7) \times 4$
$= 100 \times 4 + 80 \times 4 + 7 \times 4$
$= 400 + 320 + 28$
$= 748$

i $15 \times 210 = 15 \times (200 + 10)$
$= 15 \times 200 + 15 \times 10$
$= 3000 + 150$
$= 3150$

j $17 \times 11 = 17 \times (10 + 1)$
$= 17 \times 10 + 17 \times 1$
$= 170 + 17$
$= 187$

k $29 \times 5 = (30 - 1) \times 5$
$= (30 \times 5) - (1 \times 5)$
$= 150 - 5$
$= 145$

l $412 \times 5 = (400 + 10 + 2) \times 5$
$= 400 \times 5 + 10 \times 5 + 2 \times 5$
$= 2000 + 50 + 10$
$= 2060$

2 a $30 \times 11 + 70 \times 11 = (30 + 70) \times 11$
$= 100 \times 11$
$= 1100$

b $4 \times 20 + 7 \times 20 = (4 + 7) \times 20$
$= 11 \times 20$
$= 220$

c $14 \times 69 - 14 \times 67 = 14 \times (69 - 67)$
$= 14 \times 2$
$= 28$

d $19 \times 28 - 19 \times 8 = 19 \times (28 - 8)$
$= 19 \times 20$
$= 380$

e $53 \times 98 - 53 \times 91 = 53 \times (98 - 91)$
$= 53 \times 7$
$= 371$

f $230 \times 8 - 190 \times 8 = (230 - 190) \times 8$
$= 40 \times 8$
$= 320$

Written multiplication

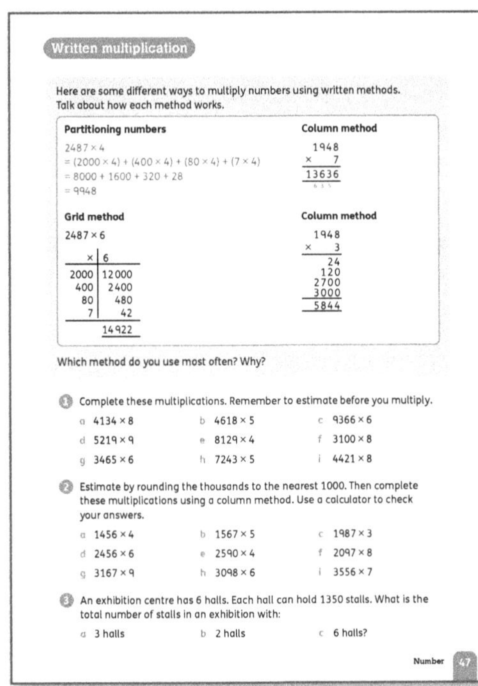

Materials
Incomplete multiplication tables (page 25); place-value tables (page 21); counters and/or squared paper; calculators.

Warm-up
Use 'Fill in the gaps' (page 25) as a mental warm-up for this lesson.

Focus
- Turn to **Pupil book page 47**. Let the children work in groups to discuss the different representations.
- After the children have discussed the methods, take feedback.
- Provide place-value tables and counters and/or squared paper to help the children keep their calculations in columns if necessary and let them work through questions 1–3 in their groups.
- Remind them to estimate before they calculate. Walk around the class to make sure they are doing so.

Support
- If necessary, revise written multiplication, as follows. Write some 2-digit by 1-digit multiplications on the board. For each, put a choice of three answers, including only one correct answer. Ask the children to look at the answers and, without doing the calculation, say which answer is correct.
- Discuss how they decided (for example, by estimating, using a known fact, or finding multiples of the multiplier).
- You should continue to allow for mental strategies and different approaches to multiplication, including informal methods and jottings, but at the same time, you should introduce and model more compact and efficient written methods where possible. Do not force children to use these methods until they are comfortable with them.

Challenge
- Write criteria for solving a word problem on the board, for example:
  - It must be solved by multiplying a 4-digit by a 1-digit number.
  - It must have a product less than 19 000.
- Let the children write word problems that match each and give them to a partner to solve.

Answers for Pupil book page 47

1 a 33 072   **b** 23 090   **c** 56 196
d 46 971   **e** 32 516   **f** 24 800
g 20 790   **h** 36 215   **i** 35 368
2 a 5824   **b** 7835   **c** 5961
d 14 736   **e** 10 360   **f** 16 776
g 28 503   **h** 18 588   **i** 25 892
3 a 4050   **b** 2700   **c** 8100

Multiply by 2-digit numbers

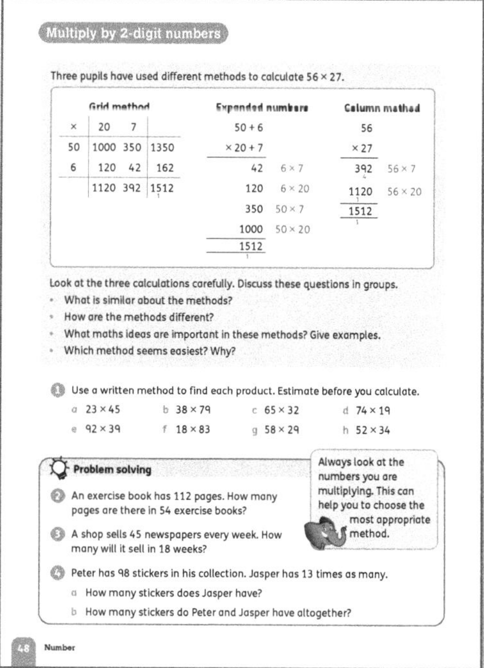

Warm-up

Use any 'Rounding and estimating' activity (page 23) as a mental warm-up for this lesson.

Focus

- Display a grid like this for the class to consider:

	300	40	2		
50	15 000	2000	100	⟶	17 100
4	1200	160	4	⟶	1364

- Ask: *What do you notice? What do you wonder?*
- Give the children a few minutes to write down things they notice and things they wonder about.
- Ask the children to share what they noticed. List the children's contributions for the class to see.
- Allow time for as many children as possible to contribute. Avoid praising, restating, clarifying or asking questions.
- Next, ask the children to share what they wondered, encouraging them to use a question format, for example: *Would this work for greater numbers?* List these as well.
- Choose some of the questions to investigate with the class or in smaller groups. You could add some starter questions yourself to get the children thinking and talking. For example, ask:
  - *What calculation is this? What is the answer?*
  - *What is the greatest number that could go into each cell? Why?*
  - *What words could we use to talk about this operation?*
  - *How else could I model this?*
  - *How would I do this as a written calculation?*

- Spend some time discussing the questions and the children's answers.
- Give them time to talk in groups before asking groups to share their answers to the questions in the example box.
- Turn to **Pupil book page 48**. Ask the children to read the examples and the methods used. The first question asks: What is similar about the methods? (They all decompose 27 and multiply by 20 and 7; decompose numbers to make easier calculations, then add the totals together.)
  The second question asks: How are the methods different? (The grid method and the expanded numbers method also decompose 56 as well as 27. Their layouts are different.)
  The third question asks: What maths ideas are important in these methods? Give examples. (Decomposition, for example $20 = 10 + 7$; $27 \times 6 = 20 \times 6 \times 7 + 6$, order of operations)
  The last question asks: Which method seems easiest? Why? (Individual answers.)
- Work through several examples with the class and discuss the strategies they use to find the answers before asking them to complete question 1.
- Problem solving: Spend some time discussing questions 2–4. Ask the children what strategies they think will be most efficient for solving them and why.

Interesting mistakes

- Some children may multiply each column as a separate single-digit multiplication. So, for example, they might work out 32×4 as 4×2 and 4×3.
- This incorrect thinking works in some cases (like this one) but fails when there is a number to carry and particularly when they have to multiply by 2-digit numbers.

3 2	4 4	3 2
× 4	× 4	× 1 3
128 This works.	1616 This fails.	3 6 This fails.

- When children estimate first, they can quickly see that the product is not reasonable. In the second example, estimating would give $40 \times 4 = 160$. In the third example, estimating 30×10 would give 300.
- Modelling the calculations using place-value counters will help the children to see that this method is not mathematically correct. Writing the multiplication in a grid will also help their understanding.

Answers for Pupil book page 48

1 a 1035	b 3002	c 2080
d 1406	e 3588	f 1494
g 1682	h 1768	
2 6048	**3** 810	
4 a 1274	b 1372	

Long multiplication

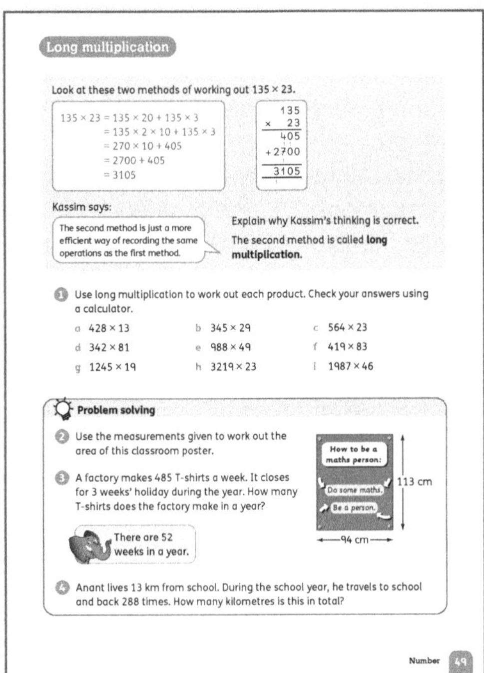

Materials

Calculators; squared paper; large sheets of paper; coloured pens.

Warm-up

Have a times table quiz as a mental starter for this lesson.

Focus

- Long multiplication is just a compact method of recording multiplication by a number with two or more digits. It is important that children don't see this as something new or different – both the grid method and the expanded notation method are contained in this way of recording.
- Turn to **Pupil book page 49**. Let the children read through the examples and work out how long multiplication is like the expanded notation method and the grid method. Discuss this as a class and encourage the children to share their ideas. The example box at the top of the page says: Explain why Kassim's thinking is correct. (The second method involves the same calculations as the first: 135×20, 135×3 and adding the results of these calculations, but involves less writing and takes up less room on the page.)
- Give the children squared paper to help them set out their working for question 1. They can check their answers using a calculator.
- <u>Problem-solving:</u> For question 2 you may need to revise the area of a rectangle formula. As a hint for questions 3 and 4, ask the children to estimate by rounding before they calculate.

Support

- Ask the children to make a poster showing different methods of multiplying with examples. This helps them think about what they have learnt and how to use this knowledge.
- You could ask them to include a set of hints for solving word problems involving multiplication.

Challenge

Questions involving missing digits can be used as problem-solving activities. For example, ask: *What are the missing numbers in these multiplications? How did you work this out?*

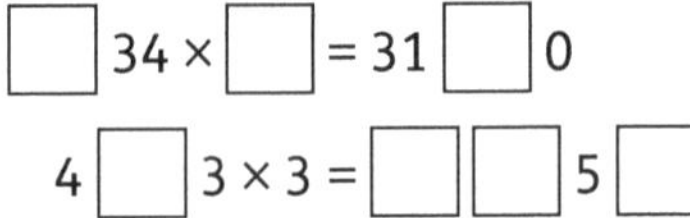
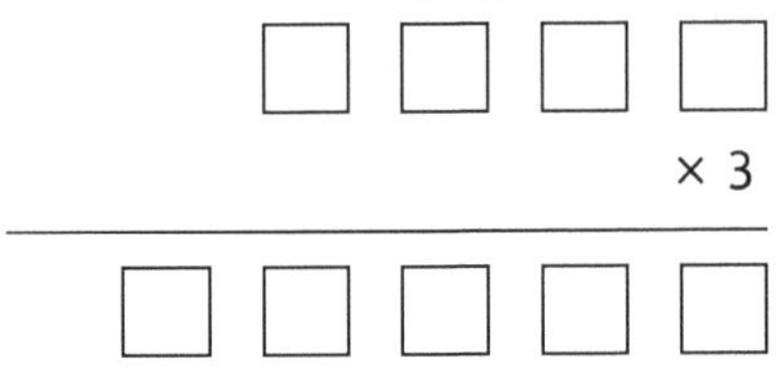

The digits 0 to 9 are used once each in this calculation. Can you work out which digit goes where?

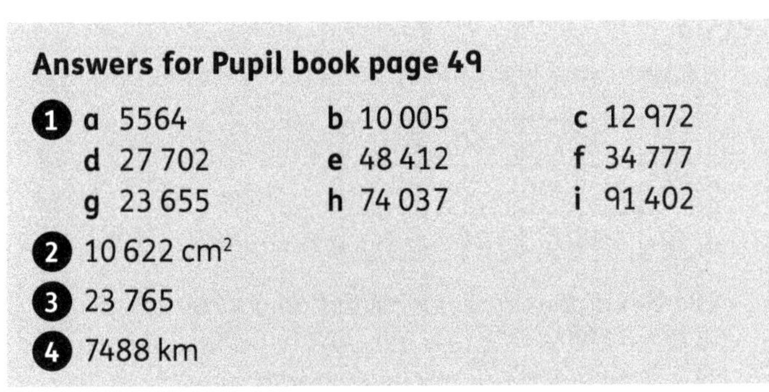

(A possible answer is $6819 \times 3 = 20\,457$.)

Answers for Pupil book page 49

1
- a 5564
- b 10 005
- c 12 972
- d 27 702
- e 48 412
- f 34 777
- g 23 655
- h 74 037
- i 91 402

2 10 622 cm²

3 23 765

4 7488 km

Revisit mental strategies for division

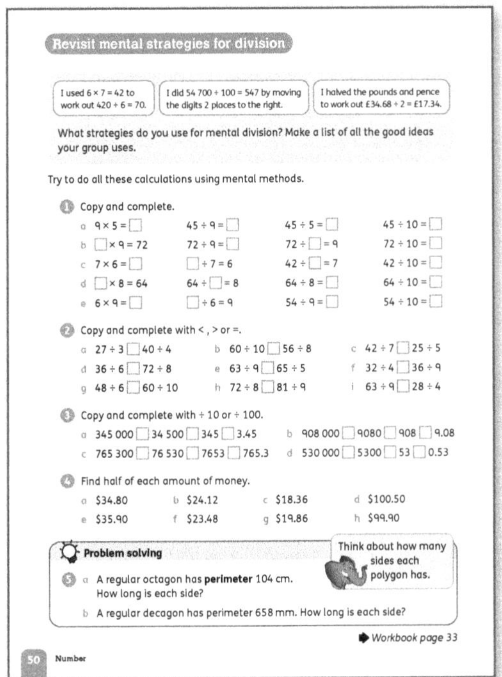

Warm-up

- Use the set A division facts questions on **Workbook page 33** as a starter for this lesson.
- Remember the focus is on accuracy rather than speed.
- Check the answers as a class.

Focus

- Turn to **Pupil book page 50**. Give the children time to read through the examples. The explanation box at the top of the page asks: What strategies do you use for mental division? (For example, using times table facts and related division facts, recognising multiples of 10, visualising a place-value table and moving digits columns to the right to divide by 10, 100, halving to divide by 2, halving twice to divide by 4, dividing by 10 then doubling to divide by 5.) Ask groups to choose two ideas to share with the rest of the class.
- Let the children work through questions 1–4 independently without using a calculator.
- <u>Problem solving</u>: For question 5, remind the children that they can refer back to Unit 4 if they cannot remember the names and properties of the polygons. Also remind them of the meaning of 'perimeter'.

Follow-up

- Use Sets B to E in question 1 on **Workbook page 33** over the next few lessons as starters, quick checks or homework.
- After completing all the sets of divisions, discuss question 2 as a class.

Challenge

- Provide a context in which the children would need to know how to do multiplication and division questions. Here are some examples:
  - an alien is visiting the school who wants to know how we do maths on Earth
  - a child in the class has been ill and now needs to catch up on missed calculation work.
- Ask the children to work together to write a clear explanation of how to approach calculations.

Interesting mistakes

- Some children may have fairly good recall of multiplication facts but need additional support with division facts. For example, a child may know that $6 \times 7 = 42$ but not use this to work out that $42 \div 7 = 6$.
- Work with arrays may help the children to visualise the relationship. Ask questions to guide the children. For example, you could show them an array of 6 rows of 8 dots and ask:
  - *How many dots are there altogether?* (48)
  - *What multiplication facts give us 48?* (6×8 and 8×6)
  - *How many rows is the array divided into?* (6 rows of 8)
  - *Can you give me a division fact for that?* ($48 \div 6 = 8$ dots per row)
  - *What if we divide it into columns?* (We get 8 columns with 6 dots in each.)
  - *How do we write that as a division fact?* ($48 \div 8 = 6$)
  - *How do the numbers 6, 8 and 48 give us a fact family?* (List the four facts.)

Answers for Pupil book page 50

1
a $9 \times 5 = 45$; $45 \div 9 = 5$; $45 \div 5 = 9$; $45 \div 10 = 4.5$
b $8 \times 9 = 72$; $72 \div 9 = 8$; $72 \div 8 = 9$; $72 \div 10 = 7.2$
c $7 \times 6 = 42$; $42 \div 7 = 6$; $42 \div 6 = 7$; $42 \div 10 = 4.2$
d $8 \times 8 = 64$; $64 \div 8 = 8$; $64 \div 8 = 8$; $64 \div 10 = 6.4$
e $6 \times 9 = 54$; $54 \div 6 = 9$; $54 \div 9 = 6$; $54 \div 10 = 5.4$

2
a $27 \div 3 < 40 \div 4$ b $60 \div 10 < 56 \div 8$
c $42 \div 7 > 25 \div 5$ d $36 \div 6 < 72 \div 8$
e $63 \div 9 < 65 \div 5$ f $32 \div 4 > 36 \div 9$
g $48 \div 6 > 60 \div 10$ h $72 \div 8 = 81 \div 9$
i $63 \div 9 = 28 \div 4$

3
a $\div 10$; $\div 100$; $\div 100$ b $\div 100$; $\div 10$; $\div 100$
c $\div 10$; $\div 10$; $\div 10$ d $\div 100$; $\div 100$; $\div 100$

4
a $17.40   b $12.06   c $9.18   d $50.25
e $17.95   f $11.74   g $9.93   h $49.95 m

5
a 13 cm   b 65.8 mm

Answers for Workbook page 33

1

Set A	Set B	Set C	Set D	Set E
17	10	9	4	0.4
34	100	40	40	4
4	1000	100	4.25	0.425
40	7	45	7	0.7
38	70	86	70	7
25	700	74	8	0.8
140	9	3.6	7	0.7
80	90	4.9	70	7
306	91	9.9	12	1.2
204	900	12.3	9	0.9
90	910	14.5	8	0.8
135	13	23.4	90	9
55	130	97.6	80	8
120	132	123	9.25	0.925
180	1300	104.5	90	0.9
500	1320	100.9	11	1.1
500	18	0.5	7	0.7
600	181	0.15	110	11
2500	182	0.29	12	1.2
3800	1820	0.97	13	1.3

2 Individual answers.

Divide by 10, 100 and 1000

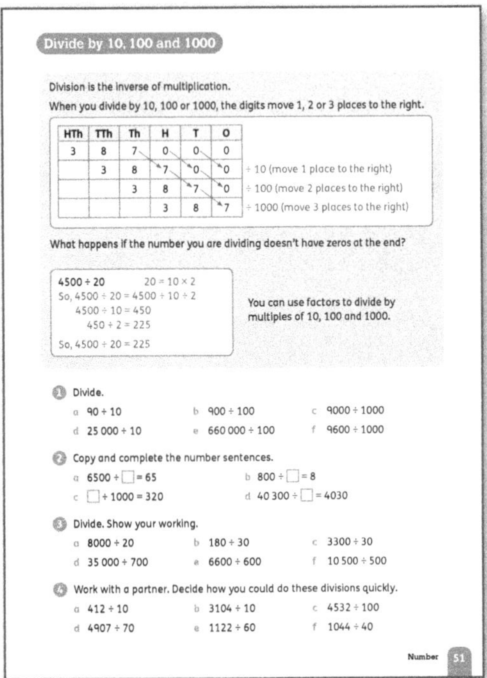

Divide by 10, 100 and 1000

Division is the inverse of multiplication.
When you divide by 10, 100 or 1000, the digits move 1, 2 or 3 places to the right.

HTh	TTh	Th	H	T	O	
3	8	7	0	0	0	
	3	8	7	0	0	÷ 10 (move 1 place to the right)
		3	8	7	0	÷ 100 (move 2 places to the right)
			3	8	7	÷ 1000 (move 3 places to the right)

What happens if the number you are dividing doesn't have zeros at the end?

$4500 \div 20 \qquad 20 = 10 \times 2$
So, $4500 \div 20 = 4500 \div 10 \div 2$
$\qquad 4500 \div 10 = 450$
$\qquad 450 \div 2 = 225$
So, $4500 \div 20 = 225$

You can use factors to divide by multiples of 10, 100 and 1000.

1 Divide.
a $90 \div 10$ — b $900 \div 100$ — c $9000 \div 1000$
d $25\,000 \div 10$ — e $660\,000 \div 100$ — f $9600 \div 1000$

2 Copy and complete the number sentences.
a $6500 \div \square = 65$ — b $800 \div \square = 8$
c $\square \div 1000 = 320$ — d $40\,300 \div \square = 4030$

3 Divide. Show your working.
a $8000 \div 20$ — b $180 \div 30$ — c $3300 \div 30$
d $35\,000 \div 700$ — e $6600 \div 600$ — f $10\,500 \div 500$

4 Work with a partner. Decide how you could do these divisions quickly.
a $412 \div 10$ — b $3104 \div 10$ — c $4532 \div 100$
d $4907 \div 70$ — e $1122 \div 60$ — f $1044 \div 40$

Number 51

Materials
Large place-value table (page 21); counters.

Warm-up
- Have a number talk (pages 17–18) about the relationship between scaling numbers down by powers of 10. For example, 1 is $\frac{1}{10}$ of 10, and 10 is $\frac{1}{10}$ of 100.
- Focus on language such as 10 times smaller, $\frac{1}{10}$ of the size and so on. Ask the children to relate this to different numbers. For example, ask: *What number is 100 times smaller than 3000?* (30) *What number is $\frac{1}{10}$ of 150?* (15)

Focus
- Remind the children that division is the inverse of multiplication, and use similar activities to 'Multiplying and dividing by 10'/'by 100 and 1000' (page 28) to establish the rules for dividing by 10, 100 and 1000 and multiples of these. Include some examples where the division will give a decimal.
- The example box at the top of the page asks: What happens if the number you are dividing doesn't have zeros at the end? (The numbers move into the column(s) to the right of the decimal point, and the answer is a decimal.)
- Let the children work independently on questions 1–3 on **Pupil book page 51**. They should try to do these orally, using a large place-value table, and jottings if necessary. For question 4 they should work in pairs.
- Check the children's answers as a class.

Challenge
- Ask the children to choose any two digits from 1 to 9 and then ask them to make a set of equivalent statements using × 10, × 100, × 1000, ÷ 10, ÷ 100, ÷ 1000, and zeros and decimal points as they need them, for example:
  - $23 \times 10 = 2300 \div 100$
  - $23 \times 100 = 2.3 \times 1000$
  - $230 \div 10 = 2300 \div 100$
- Encourage the children to try to make statements with a × and ÷ sign as well as statements with two × signs and two ÷ signs.
- Let the children use a calculator to check their work.

Interesting mistakes
- The rules for multiplying (and dividing) by 10 and 100 may confuse children who need additional support with their understanding of place value.
- Use manipulatives such as place-value cards (page 21), place-value tables (page 21) and base-ten blocks (page 20) to model calculations.
- You may also like to get the children to use a calculator to do a number of examples because many children seem to trust technology, and it will help them to internalise the patterns.

Answers for Pupil book page 51

1 a 9 — b 9 — c 9
d 2500 — e 6600 — f 9.6

2 a $6500 \div 100 = 65$
b $800 \div 100 = 8$
c $320\,000 \div 1000 = 320$
d $40\,300 \div 10 = 4030$

3 a $8000 \div 20 \qquad 20 = 10 \times 2$, so
$8000 \div 20 = 8000 \div 10 \div 2$
$8000 \div 10 = 800$
$800 \div 2 = 400$
So, $8000 \div 20 = 400$

b $180 \div 30 \qquad 30 = 10 \times 3$, so
$180 \div 30 = 180 \div 10 \div 3$
$180 \div 10 = 18$
$18 \div 3 = 6$
So, $180 \div 30 = 6$

c $330 \div 30 \qquad 30 = 10 \times 3$, so
$330 \div 30 = 330 \div 10 \div 3$
$330 \div 10 = 33$
$33 \div 3 = 11$
So, $330 \div 30 = 11$

d $35\,000 \div 700 \qquad 700 = 100 \times 7$, so
$35\,000 \div 700 = 35\,000 \div 100 \div 7$
$35\,000 \div 100 = 350$
$350 \div 7 = 50$
So, $35\,000 \div 700 = 50$

e $6600 \div 600 \qquad 600 = 100 \times 6$, so
$6600 \div 600 = 6600 \div 100 \div 6$
$6600 \div 100 = 66$
$66 \div 6 = 11$
So, $6600 \div 600 = 11$

f $10\,500 \div 500 \qquad 500 = 100 \times 5$, so
$10\,500 \div 500 = 10\,500 \div 100 \div 5$
$10\,500 \div 100 = 105$
$105 \div 5 = 21$
So, $10\,500 \div 500 = 21$

4 a 41.2 — b 310.4 — c 45.32
d 70.1 — e 18.7 — f 26.1

Short division

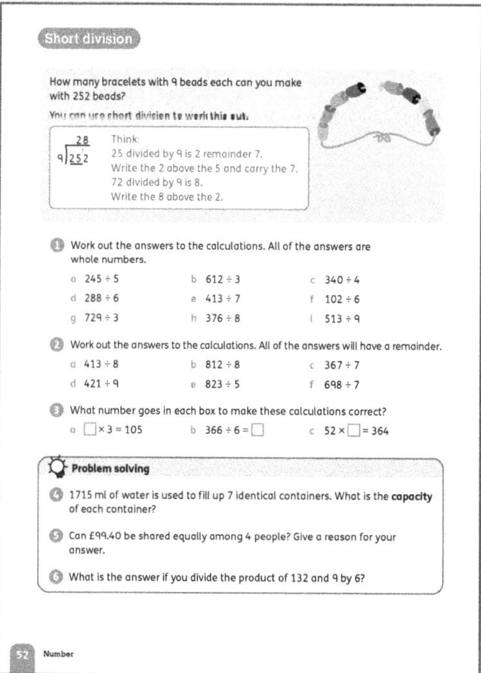

Materials
Calculators, if necessary.

Warm-up
As a starter for this lesson, prepare a set of mixed multiplication and division questions. Put the children into groups or teams and ask different groups in turn to give the answer. Allow individuals to check answers with other children in their group so that this does not become a pressurised activity.

Focus
- Remind the class that they used short division in Level 5. Ask the children to share what they remember.
- Turn to **Pupil book page 52** and look together at the problem and worked example to revise how to set out short division using a written method (division house).
- If necessary, do a few more examples with the class before asking them to complete questions 1–6.
- Before doing question 2, establish how to record the remainder (for example, r 4).
- For question 3, remind the children that division is the inverse of multiplication. Ask them to explain what this means and how it can help them answer the questions.
- <u>Problem solving</u>: Check that the children know the meaning of 'capacity'. Let them work together to discuss and then solve questions 4–6.
- Remind them to draw bar models if they need to (page 21). Stress that when a problem involves money or measures the answer will have units.

Interesting mistakes
- Some children may need additional support with written methods owing to misunderstanding how they work or misremembering the stages.
- To help the children, always encourage them to 'have a conversation in their head' about what they are doing at each stage of the calculation and model this with examples.
- Give the children examples of correct and incorrect worked examples and ask them to mark them and identify where mistakes have been made. Allow and encourage informal methods and jottings to support children's understandings as they work towards formal, written methods.

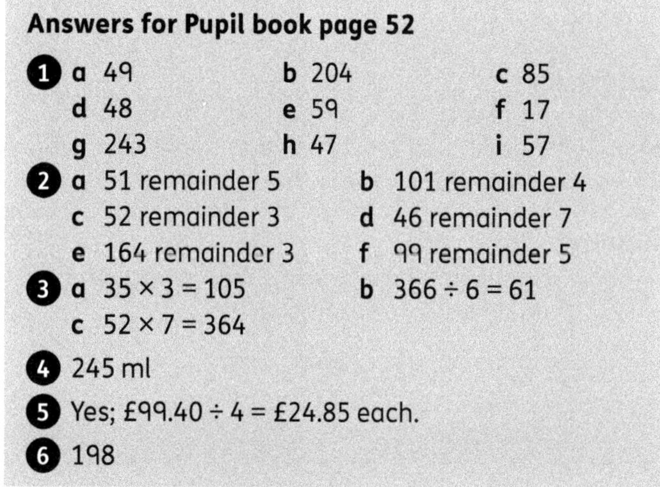

Answers for Pupil book page 52

1
a	49	b	204	c	85
d	48	e	59	f	17
g	243	h	47	i	57

2
a	51 remainder 5	b	101 remainder 4
c	52 remainder 3	d	46 remainder 7
e	164 remainder 3	f	99 remainder 5

3
| a | $35 \times 3 = 105$ | b | $366 \div 6 = 61$ |
| c | $52 \times 7 = 364$ | | |

4 245 ml

5 Yes; $£99.40 \div 4 = £24.85$ each.

6 198

Revisit division with remainders

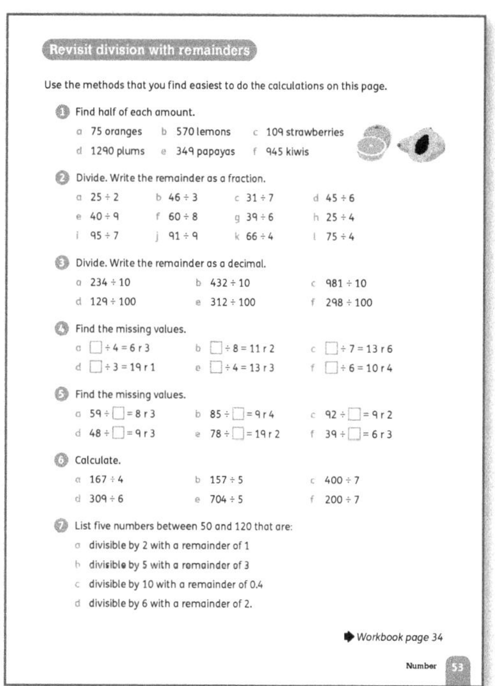

Materials
Calculators, if necessary.

Warm-up

Have a quick class quiz to revise multiplication and division facts as a warm-up for this lesson (you could use some of the sets of questions from **Workbook page 33**).

Focus

- Use questions 1–7 on **Pupil book page 53** as an informal assessment task. Let the children work independently to complete the questions.
- Observe the children as they work and note whether anyone has difficulties, particularly with expressing the remainder in different ways.
- Check their answers as a class and encourage the children to say what they found easy and what they found challenging.

Follow-up

Use the multiplication and division sets on **Workbook page 34** for additional practice and/or homework over the next few days. Discuss question 2 as a class.

Support

- Show the children how to divide by 2 in parts to find half of an amount. For example:

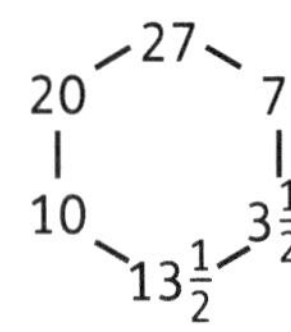

Answers for Pupil book page 53

1   a   37.5 oranges   b   285 lemons
    c   54.5 strawberries   d   645 plums
    e   174.5 papayas   f   472.5 kiwis

2   a $12\frac{1}{2}$ b $15\frac{1}{3}$ c $4\frac{3}{7}$ d $7\frac{1}{2}$
    e $4\frac{4}{9}$ f $7\frac{1}{2}$ g $6\frac{1}{2}$ h $6\frac{1}{4}$
    i $13\frac{4}{7}$ j $10\frac{1}{9}$ k $16\frac{1}{2}$ l $18\frac{3}{4}$

3   a   23.4   b   43.2   c   98.1   d   1.29
    e   3.12   f   2.98

4   a $27 \div 4 = 6\,r\,3$ b $90 \div 8 = 11\,r\,2$
    c $97 \div 7 = 13\,r\,6$ d $58 \div 3 = 19\,r\,1$
    e $55 \div 4 = 13\,r\,3$ f $64 \div 6 = 10\,r\,4$

5   a $59 \div 7 = 8\,r\,3$ b $85 \div 9 = 9\,r\,4$
    c $92 \div 10 = 9\,r\,2$ d $48 \div 5 = 9\,r\,3$
    e $78 \div 4 = 19\,r\,2$ f $39 \div 6 = 6\,r\,3$

6   a $41\frac{3}{4}$ b $31\frac{2}{5}$ c $57\frac{1}{7}$
    d $51\frac{1}{2}$ e $140\frac{4}{5}$ f $28\frac{4}{7}$

7   a   For example: 53, 75, 97, 101 and 115
    b   For example: 58, 73, 88, 103 and 118
    c   For example: 54, 84, 94, 104 and 114
    d   For example: 56, 62, 80, 92, 98 and 110

Answers for Workbook page 34

1

Day 1	Day 2	Day 3	Day 4	Day 5
30	50	6	1	42
9	4	16	4	8
18	54	9	40	42
8	9	35	8	5
16	8	40	110	90
5	7	10	144	6
36	16	56	9	16
6	3	10	54	27
72	27	18	7	32
6	42	9	24	30
63	8	30	48	96
9	24	100	8	48
81	6	54	24	8
12	64	54	9	6
36	5	7	63	9
45	72	32	10	1
54	6	5	4	99
9	28	49	36	4
72	5	8	35	3
100	56	36	12	36

2   a   The remainder cannot be 7 in a division by 6 because 7 can be divided by 6.
    b   The remainder of a division of a number by 6 could be 0, 1, 2, 3, 4 and 5.

Remainders with fractions

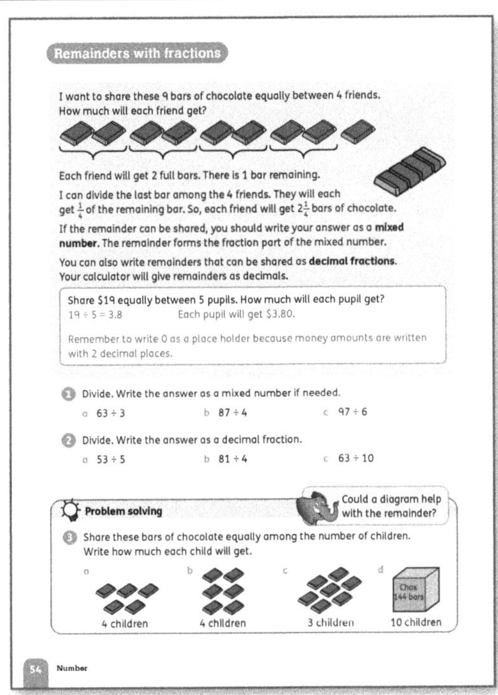

Materials

Rulers; calculators.

Warm-up

To give the children a break from number work and calculation, use any of the 'Geometry and measures' activities (page 27) as a starter for this lesson.

Focus

- Remind the children that for certain problems they have to think about the answer and about what to do with any remainder, for example:
  - *I have 3 cakes to share between 2 families. How much cake will each family get?*
- The answer of 1 remainder 1 does not make sense here, because surely you would cut the third cake in half and give each family $1\frac{1}{2}$ cakes.
- Use concrete examples to explain further, if possible. Include the term 'decimal fractions'.
- Get the children to investigate division with decimal remainders by asking them to draw five different straight lines in their books. They should then measure the lines in millimetres and work to show how they could divide each line into different numbers of parts. For example, they could divide:
  - the first line into two equal parts
  - the second line into three equal parts
  - the third into five equal parts
  - the fourth line into eight equal parts
  - the last line into ten equal parts.
  They can check their answers using a calculator.
- Spend some time discussing solutions and get the children to explain their strategies using lines you draw on the board and measure in centimetres.
- Let the children work independently to complete the divisions in questions 1 and 2 on **Pupil book page 54**. Check that they can express the remainders as fractions.
- <u>Problem solving</u>: Encourage the children to use the diagrams in the question or draw their own to help them answer question 3.

Challenge

- Let the children investigate how their calculator will give a remainder and how they can use this to work out the remainder as a whole number, fraction or decimal.
- For example, $68 \div 8$ will give 8.5. Discuss what this means $\left(8 \text{ and } 0.5 \text{ of } 8, \text{ which is } \frac{4}{8} \text{ or } \frac{1}{2}\right)$.
- Other examples, such as $63 \div 11$ give results such as 5.7272727… . Let the children work out how to use the calculator answer to find the remainder:
  - In this case, they know 11 goes into 63 five whole times. They can work out $5 \times 11 = 55$, which means there is 8 remainder.
  - Rounding the calculator remainder to 0.73×11 will give them 8.03 (which rounds to 8).
  - If they round to 0.7 they will get 7.7, which also rounds to 8.

Answers for Pupil book page 54

1 a 21   b $21\frac{3}{4}$ c $16\frac{1}{6}$

2 a 10.6   b 20.25   c 6.3

3 a $1\frac{1}{4}$ b $1\frac{1}{2}$

    c $2\frac{1}{3}$ d $14\frac{2}{5}$

Round the remainder

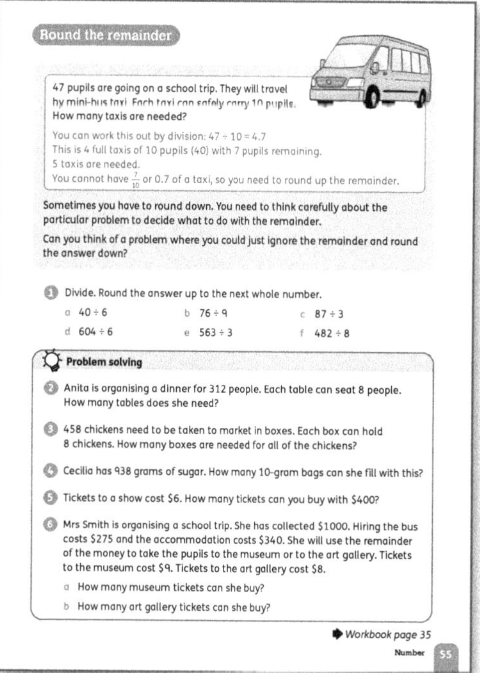

Materials

Calculators, if necessary.

Warm-up

- Go around the class asking each child to give a times table multiplication or division question for the next child to answer.
- If you haven't completed the multiplication and division sets on **Workbook page 34**, you could use one of those.

Focus

- Read through the problem on **Pupil book page 55** with the class. The example box at the top of the page asks: Can you think of a problem where you could just ignore the remainder and round the answer down? (*Any problem where you are working out how many boxes/packs you could fill with a given number of items. For example: How many boxes of 6 eggs could you fill with 20 eggs?* (3 ignoring the 2 eggs left over)
- Discuss what you do when you could just ignore the remainder, for example:
  - *If you buy a pack with 27 batteries and you need to put 2 batteries into each calculator, how many calculators can you put batteries into?*
  - In this case, you get an answer of 13 remainder 1.
  - To answer the question, you can ignore the remainder and say 13 calculators.
- For question 1, the children should work independently.
- <u>Problem solving</u>: Let the children work in pairs on questions 2–6. They should think about what they will do with the remainder (if there is one).

Follow-up

- Use **Workbook page 35** to check that the children can divide and work with remainders in different ways.
- Let the children complete the work independently.
- Check their answers as a class, asking the children to rate what they did well and what they still need to pay some more attention to.

Challenge

- Prepare a set of division cards (or let the children work in pairs to make these) in which the remainder has to be interpreted in some way. This could mean:
  - rounding up to give the answer (for example needing 5 taxis as on **Pupil book page 55**)
  - ignoring the remainder (as in the example above)
  - using the remainder as the answer (a question like: How many children would not be in a group?)
- Let the children choose five different cards and solve the problems, saying what they did with the remainder.

Interesting mistakes

Children may get confused about which number gets divided (they may incorrectly see $40 \div 10$ as equivalent to $10 \div 40$). You can use practical activities such as dividing a length of string to show that these are not the same thing.

Answers for Pupil book page 55

1 a 7   b 9   c 29
    d 101   e 188   f 61
2 39
3 58
4 93
5 66
6 a 42   b 48

Answers for Workbook page 35

1 a $114\frac{1}{6}$ [Provided as an example]
   b $97\frac{6}{7}$ c $85\frac{5}{8}$ d $166\frac{2}{5}$
   e $138\frac{2}{3}$ f $92\frac{4}{9}$
2 a 49.4   b 123.5   c 61.75
   d 163.8   e 409.5   f 204.75
3 a 231   b 146   c 45
   d 80   e 42   f 81
   g 100   h 99   i 91
4 $263 \div 6 = 43$ remainder 5, so he can make 43 sets.

Round to estimate answers

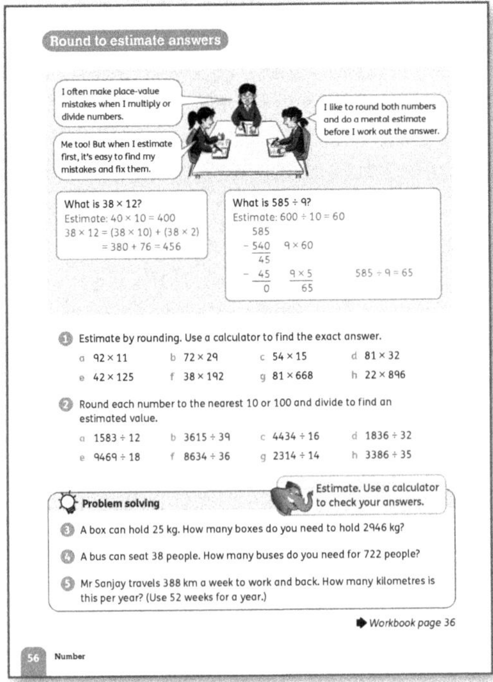

Materials

Calculators.

Warm-up

Use any of the 'Rounding and estimating' activities (page 24) as a mental warm-up for this lesson.

Focus

- Turn to **Pupil book page 56**. Read the speech bubbles with the class. Ask: *Does this ever happen to you? How is estimating important in maths?* Take feedback.
- Work through the examples on **Pupil book page 56** to revise using rounding to estimate when you multiply and divide.
- For question 1, the children should write down the rounded calculation and the estimate. Only after this should they use a calculator to do each calculation and write the actual answers.
- For question 2, only the estimated values are required.
- <u>Problem solving</u>: The children can work through questions 3–5 in pairs. Make sure they write an estimate before they do any calculation work. They use a calculator to check their answers.

Follow-up

Use **Workbook page 36** for additional practice and support and to ensure the children can divide by single digit numbers, can round and estimate answers, and can check using inverse operations.

Answers for Pupil book page 56

1
  a estimate: 900; exact answer: 1012
  b estimate: 2100; exact answer: 2088
  c estimate: 1000; exact answer: 810
  d estimate: 2400; exact answer: 2592
  e estimate: 4000; exact answer: 5250
  f estimate: 8000; exact answer: 7296
  g estimate: 56 000; exact answer: 54 108
  h estimate: 18 000; exact answer: 19 712

2
  a 160   **b** 90   **c** 220   **d** 60
  e 475   **f** 215   **g** 230   **h** 85

3 118

4 19

5 20 176 km

Answers for Workbook page 36

1 764, 382 [Provided as an example]; 191, 95.5; 945, 315; 105; 35

2

Calculation	Estimate by rounding	Actual answer	Check using inverse operation
343 ÷ 7	50	49	49 × 7 = 343
288 ÷ 9	30	32	32 × 9 = 288
256 ÷ 8	30	32	32 × 8 = 256
785 ÷ 5	80	157	157 × 5 = 785
976 ÷ 8	100	122	122 × 8 = 976
245 ÷ 7	20	35	35 × 7 = 245
810 ÷ 9	80	90	90 × 9 = 810
102 ÷ 3	33	34	34 × 3 = 102
888 ÷ 2	450	444	444 × 2 = 888
954 ÷ 3	320	318	318 × 3 = 954

Long division

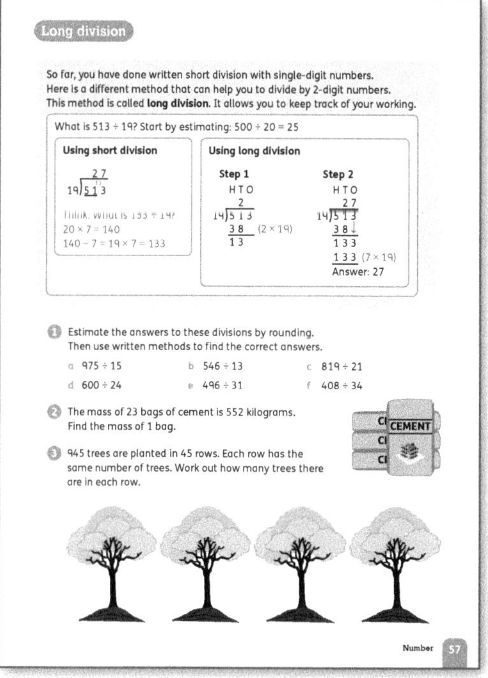

Materials

Squared paper.

Warm-up

Use any of the 'Rounding and estimating' activities (page 23) as a mental warm-up for this lesson.

Focus

- In this lesson, the children will extend their strategies for division to the division of 3-digit numbers by 2-digit numbers using long division.
- Ensure the children have a good understanding of division as both repeated subtraction and grouping before moving on to formal written methods and algorithms for long division.
- Children should not be forced to use formal written methods before they are ready.
- Use **Pupil book page 57** to teach the class how to carry out and record long division. Stress that as with long multiplication this is just a different way of keeping track and recording your working.
- Repeat using different numbers as necessary.
- Make sure you estimate before you calculate to encourage the children to do the same.
- Explain to the class that some people like to list the multiples of the number they are dividing by alongside the calculation. For example:
  - If you are dividing by 19 (as in the example on **Pupil book page 57**) you might list 19, 38, 57, 76, 95, 114, 133 and 152.
  - Ask the children how this is helpful and try to get them to see that in the first step, for example, they are going to divide 19 into 51.
  - By listing the multiples in this way, they can see that it only goes into 51 two times, 3 times is too much. (57)
- Let the children work in pairs on questions 1–3. Using squared paper will help them line up the numbers in their division calculations.

Support

Provide examples of incomplete calculations and let children suggest how they could complete these. Here are some examples you could use:

a
```
        1
  11 ) 2 3 6
        2 2 ↓
        1 6
```

b
```
        1
  23 ) 4 5 0
        2 3 ↓
        2 2 0
```

c
```
        1
  15 ) 2 5 7
        1 5 ↓
        1 0 7
```

d
```
        . .
  29 ) 1 3 4
            29 × 4 = 116
```

e
```
  48 ) 7 8 4
        4 8
```

f
```
  19 ) 9 8 6
        9 5
            19 × 5 = 95
```

Interesting mistakes

- Children may require additional support because they feel the need to use formal written methods too early. You can avoid this by continually modelling and encouraging the use of different strategies and methods.
- Often, children need additional support with the process of long division because they cannot see how it relates to earlier work on repeated subtraction. Remind them that what they are doing here is dividing in parts (chunks) and working out what is left to divide.

Answers for Pupil book page 57

1 a 65　　　　b 42　　　　c 39
　　d 25　　　　e 16　　　　f 12
2 24 kg
3 21

More division

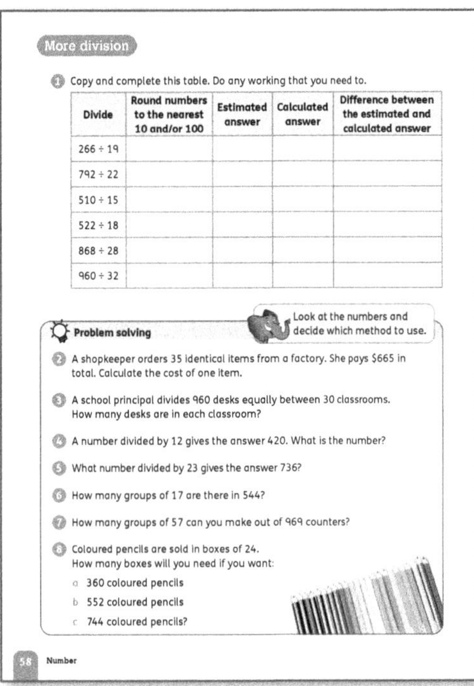

Materials

Large sheets of paper; coloured pens; calculators, if necessary.

Warm-up

- As a mental warm-up for this lesson, ask the children to work out where the number cards go on this grid.

480	÷		=	
÷		÷		÷
	÷		=	
=		=		
	÷		=	

80	80	60	60	40	30	20	20

- Tell the children to use the numbers on the cards and that each division must work across and down.
- Let the children think about the problem and indicate when they have some answers. Let different children give their solutions, explaining their strategies and thinking.

Focus

- Turn to **Pupil book page 58**. Work through question 1 to consolidate estimating and using long division. Check the children's answers before moving on.
- <u>Problem solving:</u> For questions 2–8, put the children into groups of three or four. Give each group a large sheet of paper and coloured pens. Tell the groups that each child must choose a different problem from the page.
- Next, they should work out the solution, asking for help if they get stuck. Then the group should make a poster to show how they got to each solution.
- Give the groups time to work and then let different groups show their solution posters. The children in the group should say why they chose their problem and explain how they solved it.
- If there are any problems that are left unsolved, you could allocate them for homework or use them at the end of this unit to check understanding.

Answers for Pupil book page 58

1

Divide	Round numbers to nearest 10 and/or 100	Estimated answer	Calculated answer	Difference between estimated and calculated answer
266 ÷ 19	300 ÷ 20	15	14	1
792 ÷ 22	800 ÷ 20	40	36	4
510 ÷ 15	500 ÷ 20	25	34	9
522 ÷ 18	500 ÷ 20	25	29	4
868 ÷ 28	900 ÷ 30	30	31	1
960 ÷ 32	1000 ÷ 30	$33\frac{1}{3}$	30	$3\frac{1}{3}$

2 $19　　**3** 32
4 5040　　**5** 16 928
6 32　　**7** 17
8 a 15　　b 23　　c 31

Mixed calculations

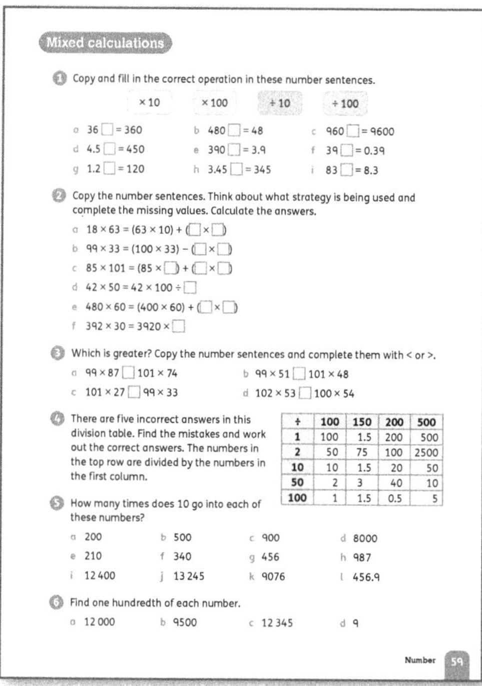

Warm-up

Use question 1 on **Pupil book page 59** as a starter for this lesson.

Focus

- Use questions 2–6 on **Pupil book page 59** to assess whether children can confidently multiply and divide and that they can move between these operations freely.
- Discuss the children's answers as a class and allow time for the children to explain how they decided what to do in each case, and what they did to find the answers.

Challenge

- Let the children collect data involving sets of numbers, for example: attendance at sporting events over a season, number of tourists visiting a place each month of the year, area or population figures in different provinces or states in your country.
- Let the children work out average figures using the data (in other words, find the mean) by calculating the total number and dividing by the number of pieces of data collected.
- They can do a short report on what they did and what the averages suggest.

Interesting mistakes

- Some children are reluctant to show their full working in calculations such as divisions involving regrouping; rather misguidedly, they think that putting down the minimum of working makes them appear more clever.
- Point out that by jumping steps and not showing working, they risk making unnecessary mistakes, and make it harder to see what has gone wrong. Learning from their mistakes then also becomes more difficult.

Answers for Pupil book page 59

1 a $36 \times 10 = 360$
 b $480 \div 10 = 48$
 c $960 \times 10 = 9600$
 d $4.5 \times 100 = 450$
 e $390 \div 100 = 3.9$
 f $39 \div 100 = 0.39$
 g $1.2 \times 100 = 1200$
 h $3.45 \times 100 = 345$
 i $83 \div 10 = 8.3$

2 a $18 \times 63 = (63 \times 10) + (63 \times 8) = 1134$
 b $99 \times 33 = (100 \times 33) - (1 \times 33) = 3267$
 c $85 \times 101 = (85 \times 100) + (85 \times 1) = 8585$
 d $42 \times 50 = 42 \times 100 \div 2 = 2100$
 e $480 \times 60 = (400 \times 60) + (80 \times 60) = 28\,800$
 f $392 \times 30 = 3920 \times 3 = 11\,760$

3 a $99 \times 87 > 101 \times 74$
 b $99 \times 51 > 101 \times 48$
 c $101 \times 27 < 99 \times 33$
 d $102 \times 53 > 100 \times 54$

4 $150 \div 1 = 150$; $500 \div 2 = 250$; $150 \div 10 = 15$; $200 \div 50 = 4$; $200 \div 100 = 2$

5 a 20 b 50 c 90
 d 800 e 21 f 34
 g 45.6 h 98.7 i 1240
 j 1324.5 k 907.6 l 45.69

6 a 120 b 95 c 123.45
 d 0.09

Multiplication and division problems

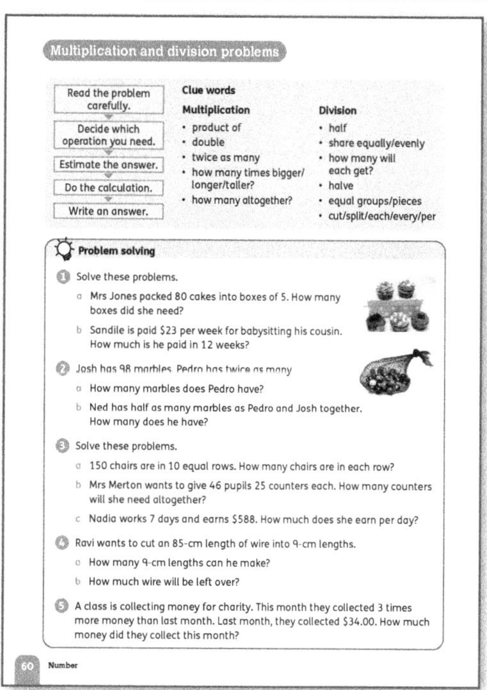

Warm-up

Use any of the 'Mental problem-solving' activities (pages 23–24) as a mental warm-up for this lesson.

Focus

- Problem-solving: Many of the difficulties children have with worded problems stem from them not understanding the problem rather than being unable to do the mathematics involved in finding a solution.
- Use **Pupil book page 60** to develop and discuss the steps involved in solving problems and also to explore the language in worded problems and what different terms mean.
- Spend some time as a class, reading and discussing questions 1–5. Encourage the children to explain how they will solve them before asking them to do so.

Answers for Pupil book page 60

1. a 16   b 276
2. a 196   b 147
3. a 15   b 1150   c $84
4. a 9   b 4 cm
5. $102

Multi-step problems

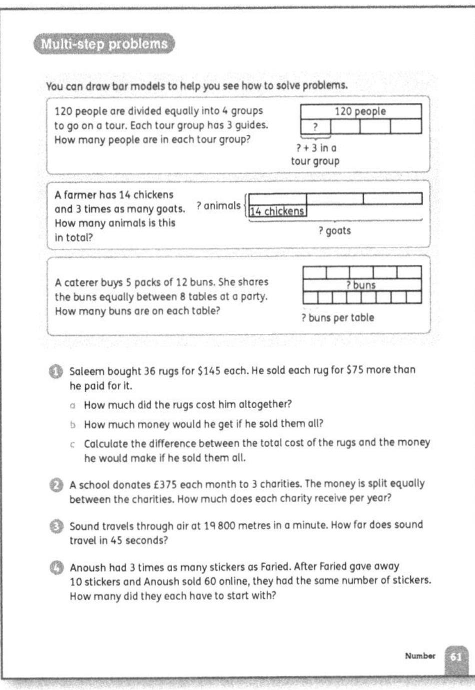

Materials

Drawing equipment; rulers.

Warm-up

Use any suitable 'Calculation skills' activity (pages 24–26) as a mental warm-up for this lesson.

Focus

- This lesson provides the children with the opportunity to use bar models (page 21) to show problems that involve two or more steps and different operations.
- Work through the examples on **Pupil book page 61** with the class. Focus on how we use bars to show different operations and parts of the problem.
- Ask questions to make sure the children can see what each part of the bar model represents.
- Ask the children to work in pairs to answer questions 1–4 on **Pupil book page 61**. They should use any strategy that makes sense to them. Encourage the children to draw bar models to show the problems.
- Observe the children as they work to see what strategies they use. Help anyone who is stuck.
- After enough time for the children to solve the problems, choose a few pairs who used different strategies to share their ideas with the class. Once they have presented their strategy, ask questions to help them think critically about what they did. For example, ask:
  - *Did your strategy make it easy to find a solution? What worked well?*
  - *What didn't work so well?*
  - *What would you change if you had to solve this problem again? Why?*
- Remind the class that one strategy is not necessarily better than another, but that in maths we aim to find a compact and efficient way of recording our work.

Interesting mistakes

- Some children may try to use formal written methods before they fully understand the concepts involved in multiplication and/or division because they think they have to.
- This makes them focus too hard on remembering rules and as a result they ignore the other strategies that they can use when they get stuck.
- Encourage the children to use earlier, more secure, mental strategies to increase confidence and continue to discuss why different strategies are useful in different situations.
- Stress the importance of estimating and jottings to support thinking and improve calculation skills.

Answers for Pupil book page 61

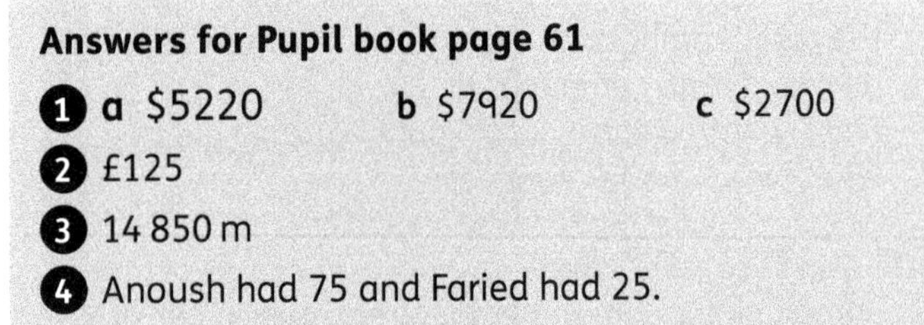

1. a $5220   b $7920   c $2700
2. £125
3. 14 850 m
4. Anoush had 75 and Faried had 25.

Use all or some of these questions and problems to check how well the children have understood the concepts in this unit.

- *How can you quickly multiply or divide by a number ending in zero?* (Multiply by the digit before the zero and then multiply by 10 – move all the digits one place to the left.)
- *What is 35 multiplied by 20/30/50?* (700/1050/1750)
- *Give a calculation such as 40 × 70. How can you use a known fact to find this product?* (for example: 4 × 7 = 28, so 40 × 70 = 28 × 10 × 10 = 2800)
- *How can you use factors to make multiplication easier?* (Multiply the number by the first factor and then multiply the answer by the second factor.)
- *Is multiplying by 6 and then multiplying by 10 the same as multiplying by 60? Why or why not?* (It is the same: 6 × 10 = 60.)
- *Three children did the division 21 ÷ 4. Their answers were 5 r 1, $5\frac{1}{4}$ and 5.25. Which answers are correct?* (They are all correct.)
- *What is 360 divided by 20?* (18)
- *What is 213 × 3?* (639)
- *What is 423 × 8?* (3384)
- *What is 3926 × 4?* (15 704)
- *What are the multiples of 7 up to 12 × 7?* (7, 14, 21, 28, 35, 42, 49, 56, 63, 70, 77, 84)
- *What is (give a number) multiplied by 10/100/1000?*
- *What is 72 ÷ 6?* (12)
- *What is the remainder from the division sum 85 ÷ 7?* (1)
- *What is 456 ÷ 4?* (114)
- *In this money problem using pounds and pence, my calculator gives an answer of 6.2. What does this mean?* (£6.20)
- Show a worked calculation. *Look at this calculation. What did this child do? Is this a good strategy? Why or why not? What other methods could you use to do this calculation?*
- Give a problem. *Explain the method you used to find your answer.*
- Give a division problem. *In this division problem, there is a remainder of 4. What should you do with it?*
- Give children some correct and some incorrect calculations. *Which of these calculations are correct/ incorrect? What has this person done wrong? How could you help them to correct it?*

Answers to Pupil book pages 62–63

1 a 916 500; 1 098 600; 1 138 900; 1 285 200; 2 766 900; 8 514 900

   b Venezuela

   c Argentina

   d 9 000 000 km²

   e All of the digits of the actual area are greater than 5, so when you round the actual number, it always gives a number greater than the actual number.

2 a 583 912   b 3 583 912

   c 370 471   d 5 839 120

3 a 7   b 9

4 a 21   b 31, 37, 41, 43, 47, 53 or 59

   c 25, 35 or 49

5 a 1600   b 90

6 a B

   b 18 : (3 × 2) = 12. The multiplication was done first.

7 a x = radius; y = diameter; z = centre

   b 2.5 cm; the radius is half of the diameter

8 a A = −20; B = −12; C = −6; D = −2; E = 5; F = 13

   b A and E

9 $a = 60°$; $b = 30°$; $c = 108°$; $d = 48°$; $e = 42°$; $f = 132°$; $g = 18°$

10 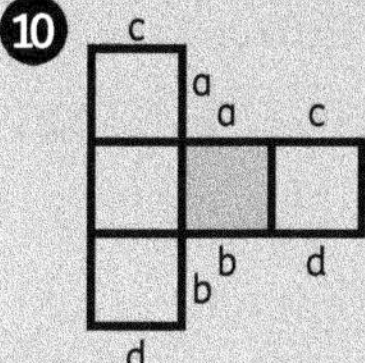

11 4236 + 6675 = 10 911

12 a 437 784   b 654   c 8913

13 a an equilateral triangle

   b a hexagon

   c a rhombus

Learning objectives

- Recognise when fractions can be simplified and use common factors to simplify fractions, use common multiples to express fractions with the same denominator.

- Compare and order fractions, including fractions greater than 1, using reasoning and choose between reasoning and writing fractions with a common denominator as a comparison strategy.

- Understand that a fraction can be represented as a division of the numerator by the denominator (proper and improper fraction) and calculate decimal fraction equivalents for a simple fraction.

- Add and subtract fractions with different denominators and mixed numbers, using the concept of equivalent fractions.

- Understand that proper and improper fractions can act as operators.

- Multiply and divide fractions by whole numbers.

- Multiply simple pairs of proper fractions, writing the answer in its simplest form.

Key words

equivalent fraction   numerator   denominator
mixed number   improper fraction   simplify
cancelling   lowest terms   simplest form
common denominator

Unit introduction

Materials

Resource sheets (as shown).

Teaching guidance

- You will need to prepare and print copies of two resource sheets for each child (or pair of children) before the lesson.

- On sheet A place 9 to 12 boxes like this:

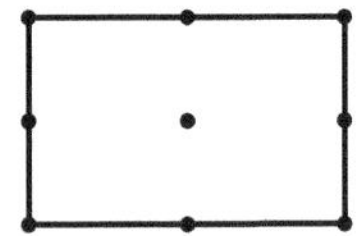

- On sheet B place 9 to 12 boxes like this:

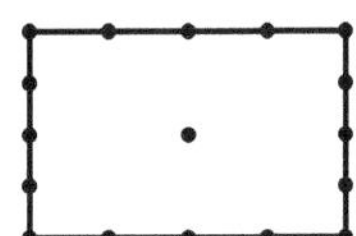

- Tell the class to draw two straight lines on each rectangle and then work out the size of each fraction they make.
- Each line must start and end on a point (dot).
  For example:

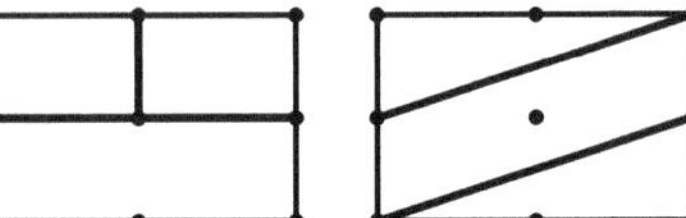

- Everyone will be able to experience success almost immediately but there is enough of a challenge for those who work quickly or who think differently. The second and third example above are much more challenging than the first.
- Give the children time to try out a few options. Then ask questions such as these:
  - *What is the smallest fraction you can make?* $\left(\frac{1}{8}\right)$
  - *What is the greatest fraction you can make?* $\left(\frac{3}{4}\right)$
  - *How many ways can you find to make halves?* (4)
  - *How many ways can you find to make quarters?* (6)
  - *Is it possible to divide the shape into thirds? How?* (no)
- Repeat with sheet B. It has more divisions (more dots). Let the children investigate using the rectangle.
- Ask them to pose some interesting questions of their own. Ask: *What do you wonder?* Share these with the class and choose a few to investigate or answer.

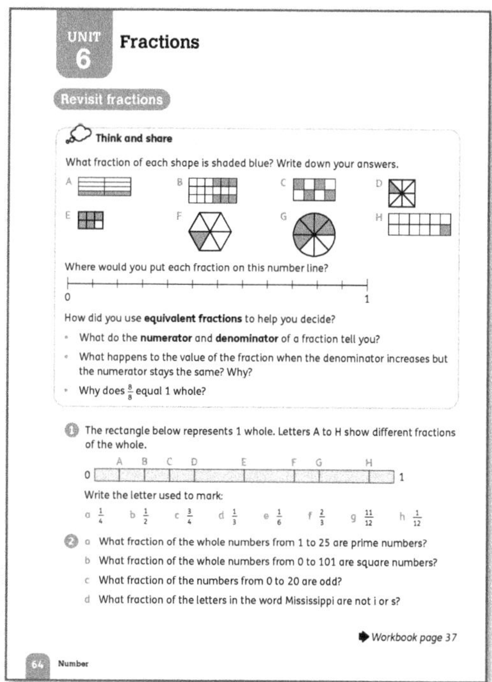

Materials

Colouring equipment.

Warm-up

Ask the children to complete **Workbook page 37** as a warm up and diagnostic assessment for this lesson. Check their answers and address any errors or misconceptions.

Focus

- Write a fraction on the board, for example $\frac{2}{3}$ or $\frac{4}{7}$. Ask the children if they can remember the correct name for the top number of the fraction and the bottom number of the fraction.
- If necessary, remind them of the words 'numerator' and 'denominator'.
- Ask the children to explain what the numbers mean and use their responses to establish that the denominator is the number of parts the whole has been divided into and the numerator tells us how many of these parts we have.
- Put the children into groups and turn to **Pupil book page 64**.
- <u>Think and share:</u> Let the groups answer the questions. Afterwards, hold a class discussion about the diagrams and the children's answers. The first question asks the children to explain how they used 'equivalent fractions' to put each fraction on a number line from 0 to 1 divided into 12 sections. (Write each fraction as an equivalent fraction with a denominator of 12.) The second question asks what the numerator and denominator of a fraction tell you. (The denominator tells you the number of equal parts the whole is divided into. The numerator tells you the number of these parts in the fraction.) The third question ask what happens to the value of the fraction when the denominator increases but the numerator stays the same. (The value of the fraction decreases because the whole is divided into a greater number of parts.) The last question asks why $\frac{8}{8}$ equals 1 whole. (The fraction consists of all 8 parts that the whole is divided into. Any fraction with numerator equal to the denominator is equal to 1 whole.)

- For question 1, discuss the number line with the class.
- Let the children work orally to locate the fractions before they write their own answers in their books.
- Before the class tackles question 2, ask the children to define prime numbers, square numbers and odd numbers. They can work on their own or in pairs.

Support

- Work with children who have difficulty with the concepts in small groups to revise the basic concepts and teach the vocabulary.
- Draw sets of shapes on the board divided into equal parts. For example, draw a shape divided into quarters and a copy of the same shape divided into eighths.
- Circle a fraction of one set, for example $\frac{1}{4}$, and ask the children how much of the other set must be shaded to make the two diagrams look the same.
- Ask the children to work out where the fraction $\frac{1}{4}$ would go on a number line showing 0, $\frac{1}{2}$ and 1. Let them explain why.

Interesting mistakes

- Many of the mistakes that children make with fractions arise from an over-reliance on the part of a whole model and this always being presented through shaded diagrams. Check that the children see fractions as numbers that can be ordered on a number line and compared.
- When children work with shaded diagrams they sometimes express the shaded part using a part : part model rather than a part of the whole model. For example, in a diagram like the one below they may say that the fraction is $\frac{2}{3}$ rather than $\frac{2}{5}$.

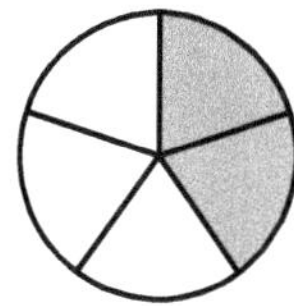

- If children do this, ask questions such as:
  - *How many parts is the shape divided into?* (5)
  - *How many of the five parts are shaded?* (2 of the 5).
  - *How do we write 2 out of 5 as a fraction?* $\left(\frac{2}{5}\right)$.
- They should also be aware that the 'whole' can be a quantity and so, for example, 12 out of 24 can be expressed as a fraction.

Answers for Pupil book page 64

1 a C   b E   c G
     d D   e B   f F
     g H   h A

2 a $\frac{9}{25}$ b $\frac{10}{102}$
     c $\frac{10}{21}$ d $\frac{3}{11}$

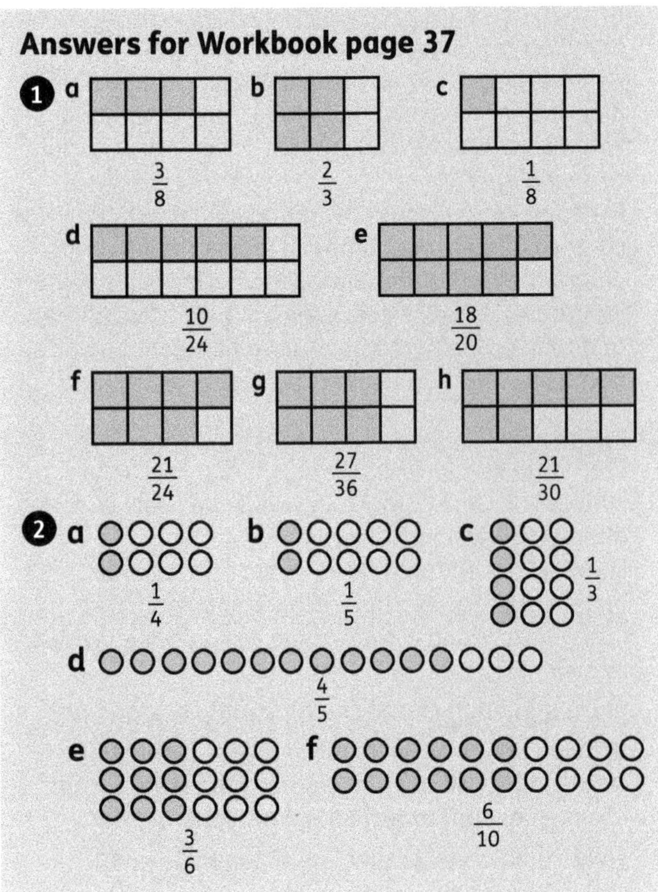

Answers for Workbook page 37

1
a — $\dfrac{3}{8}$
b — $\dfrac{2}{3}$
c — $\dfrac{1}{8}$
d — $\dfrac{10}{24}$
e — $\dfrac{18}{20}$
f — $\dfrac{21}{24}$
g — $\dfrac{27}{36}$
h — $\dfrac{21}{30}$

2
a — $\dfrac{1}{4}$
b — $\dfrac{1}{5}$
c — $\dfrac{1}{3}$
d — $\dfrac{4}{5}$
e — $\dfrac{3}{6}$
f — $\dfrac{6}{10}$

Mixed numbers and improper fractions

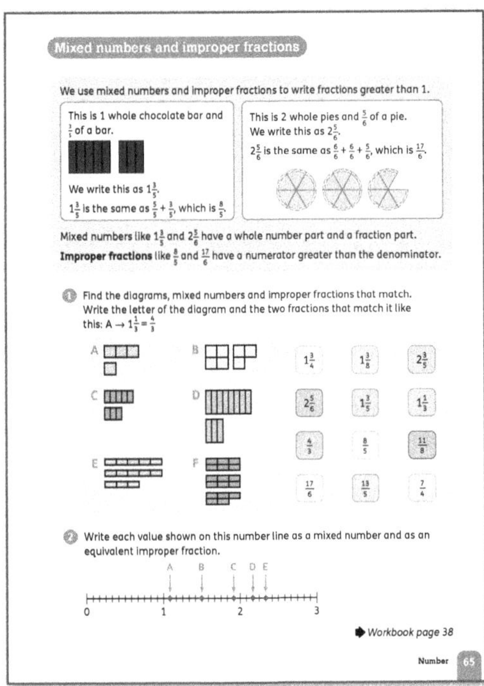

Materials

Real-life examples (or pictures) to show fractions greater than 1 (for example, a whole chocolate bar and part of another bar; a sandwich and half a sandwich; a full box of crayons and a box that is $\frac{3}{4}$ full).

Warm-up

- Much of the work on converting and 'simplifying' fractions relies on multiplication and division facts that the children already know, so you may want to do some warm-up activities to reinforce and practise these facts, such as the following:
  - Revise mixed numbers and improper fractions. Display whole objects and parts of objects, such as whole pizzas and slices, whole cakes and parts of cakes, pairs of shoes and an odd shoe, boxes full of crayons and a box with 3 of the 8 crayons (or another fraction).
  - Ask the children how they could express each amount. If necessary, model an example such as: *I have two and a half cakes.* Show the children how to write mixed numbers.
  - Use the same pictures to show how a mixed number can be expressed as an improper or 'top-heavy' fraction.

Focus

- Turn to **Pupil book page 65** and give the children time to read and discuss the two examples in pairs. Use question 1 to check that the children can express a number greater than 1 as both a mixed number and as an improper fraction.
- The children can work in pairs to talk this through if you think it will be helpful, but they need to write the answers in their own books afterwards.
- For question 2, discuss the number line with the class to make sure they see that all of the fractions shown are greater than 1. They can count the number of twelfths to represent each number as an improper fraction.

Follow-up

Use **Workbook page 38** to assess whether the children can recognise proper fractions, mixed numbers and improper fractions and that they can represent numbers on a number line.

Challenge

- Let the children work in pairs to design a game involving equivalent mixed numbers and improper fractions. (Dominoes and card games work well, but they might have their own ideas.)
- Let each pair show their game to the class and explain how it works. If you have time, let the children play the different games and rate them.

Support

- At this level, the children may find it easier to understand improper fractions and mixed numbers as concrete amounts. Use examples such as:
- *I cut some pizzas in half, I ate one half and now I have seven halves left. How many whole pizzas did I cut into halves?* (4)
- *I cut pieces of card into quarters. I gave each group 9 quarters. How many whole cards can they make?* (2)
- This helps them understand that a numerator can be greater than a denominator and that improper fractions can be rearranged to make equivalent mixed numbers.

Answers for Pupil book page 65

1. $A = 1\frac{1}{3}, \frac{4}{3}$ (or equivalent fraction); $B = 1\frac{3}{4}, \frac{7}{4}$ (or equivalent fraction);
$C = 1\frac{3}{5}, \frac{8}{5}$ (or equivalent fraction); $D = 1\frac{3}{8}, \frac{11}{8}$ (or equivalent fraction);
$E = 2\frac{3}{5}, \frac{13}{5}$ (or equivalent fraction); $F = 2\frac{5}{6}, \frac{17}{6}$ (or equivalent fraction)

2. $A = 1\frac{1}{12}$ or $\frac{13}{12}$; $B = 1\frac{1}{2}$ or $\frac{3}{2}$; $C = 1\frac{11}{12}$ or $\frac{23}{12}$; $D = 2\frac{1}{6}$ or $\frac{13}{6}$; $E = 2\frac{1}{3}$ or $\frac{7}{3}$

Answers for Workbook page 38

1. proper fractions: $\frac{1}{3}, \frac{11}{12}, \frac{7}{8}, \frac{1}{25}, \frac{2}{5}, \frac{5}{6}, \frac{9}{12}$

   mixed numbers: $1\frac{1}{2}, 8\frac{1}{2}, 8\frac{3}{10}, 4\frac{5}{9}, 7\frac{7}{8}, 2\frac{3}{4}, 3\frac{2}{5}$

   improper fractions: $\frac{15}{3}, \frac{18}{10}, \frac{8}{7}, \frac{17}{5}, \frac{9}{4}, \frac{11}{5}, \frac{3}{2}$

2. 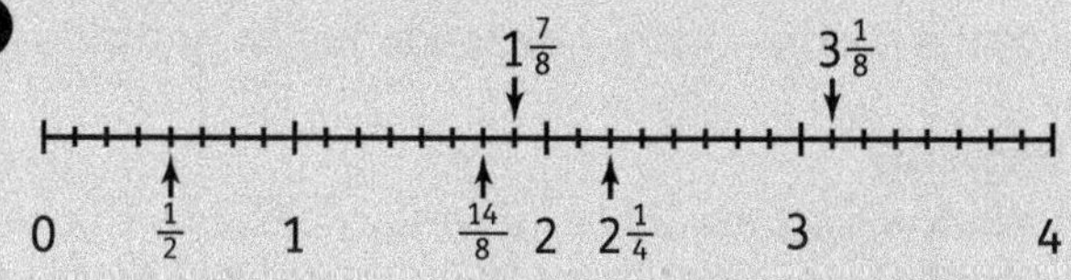

3. a $A = 1\frac{1}{4}$; $B = 1\frac{7}{8}$; $C = 2\frac{1}{8}$; $D = 2\frac{1}{2}$; $E = 3\frac{1}{2}$; $F = 3\frac{3}{4}$

   b Individual answers.

   c $A = \frac{5}{4}$; $B = \frac{15}{8}$; $C = \frac{17}{8}$; $D = \frac{5}{2}$; $E = \frac{7}{2}$; $F = \frac{15}{4}$

Convert between mixed numbers and improper fractions

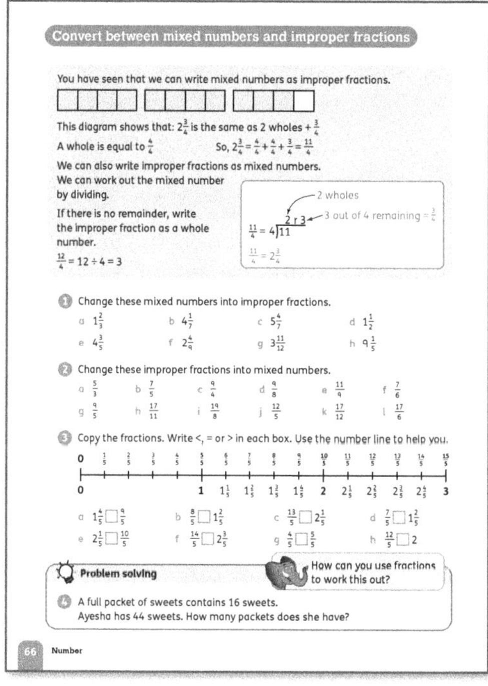

Materials
Cards with a selection of equivalent mixed numbers and improper fractions; counters.

Warm-up
Use activities that revise known multiplication and division facts (including remainders) as a mental starter for this lesson.

Focus
- Ask a group of children to stand up. Give each child a card showing an improper fraction or mixed number. Get the rest of the class to instruct the group to sort them into equivalent pairs.
- Hold up a card showing a mixed number. Invite a child to convert it to an improper fraction. Then hold up an improper fraction and ask the children to convert it to a mixed number.
- Encourage the children to explain how they did their conversions.
- Next, turn to **Pupil book page 66**. Work through the examples with the class to demonstrate how to convert between mixed numbers and improper fractions.
- Work through some similar examples with the children, before asking them to answer questions 1 and 2 independently.
- For question 3, spend some time talking about how you can use the number line to compare the two numbers. Make sure the children can tell you that it shows equivalent mixed numbers and improper fractions.
- Problem solving: For question 4, discuss the problem orally with the class and let the children suggest how they would work to solve it.
- If necessary, let the children model the solution using counters.

Interesting mistakes
Some children will grasp the division principle here and use it to convert from improper fractions to mixed numbers; others may find this too abstract and need to use diagrams and jottings. Continue to ask the children to share their thinking and explain their methods to expose them to different ways of working and thinking about converting fractions.

Answers for Pupil book page 66

1. a $\frac{5}{3}$ b $\frac{29}{7}$ c $\frac{39}{7}$ d $\frac{3}{2}$
   e $\frac{23}{5}$ f $\frac{22}{9}$ g $\frac{47}{12}$ h $\frac{46}{5}$

2. a $1\frac{2}{3}$ b $1\frac{2}{5}$ c $2\frac{1}{4}$ d $1\frac{1}{8}$
   e $1\frac{2}{9}$ f $1\frac{1}{6}$ g $1\frac{4}{5}$ h $1\frac{6}{11}$
   i $2\frac{3}{8}$ j $2\frac{2}{5}$ k $1\frac{5}{12}$ l $2\frac{5}{6}$

3. a $=$ b $>$ c $>$
   d $=$ e $>$ f $>$
   g $<$ h $>$

4. $2\frac{12}{16}$, i.e. 2 full packets and 1 packet with 12 sweets.

Simplify and find equivalent fractions

Materials
A large fraction wall showing equivalent fractions for classroom display (you can search for one online); coloured pencils.

Warm-up
Ask the children to find the highest common factor (HCF) of pairs of numbers as a mental starter for this lesson.

Focus
- Look at the fraction wall showing equivalent fractions and show the children how to use it to compare two fractions.
- Ask the children how they think they could use the fraction wall to order three fractions. Give them time to think about this and then ask them to share their ideas.
- Check that the children can use the fraction wall and that they understand the concept of an equivalent fraction by writing a fraction in the middle of the board. Go round the class asking the children to say an equivalent fraction and record answers. How many different fractions can they say? Are there any more? How do they know?
- Write statements on the board, such as '$\frac{2}{3} > \frac{3}{5}$'. Ask the children to say whether the statements are true or false and how they know.
- Work through the examples at the top of **Pupil book page 67** with the class, and confirm that to find an equivalent fraction we can multiply or divide the numerator and denominator by the same number. Explain that to get a fraction into its 'simplest form' we must divide. Use the terms 'cancelling', and 'lowest terms'.

- Write a fraction on the board, for example $\frac{10}{25}$, and ask the children to tell you it in its simplest form. Encourage them to explain how they found their answer; that is, 'I divided the numerator and denominator by 5.'
- Next, give the children a fraction such as $\frac{16}{24}$ and ask them to simplify it. Ask the children to explain how they found their answer. Many children are likely to do more than one division. For example, they will divide by 4 then 2. Establish whether anyone managed to simplify the fraction in one step.
- Either using their ideas, or by leading the children from the multi-step approach they took, highlight the advantages of doing just one division.
- Ask the children to list the factors of 16 and 24 and tell you what the HCF is. Point out that this is the same as the number used to simplify the fraction in one step. Repeat for other examples.
- Let the children work independently on questions 1–3.

Follow-up
Use **Workbook page 39** to informally assess whether the children can identify equivalent fractions. Compare their answers to question 2.

Challenge
- Let the children investigate how to use cross multiplication to see whether fractions are equivalent.
- To do this, they multiply the numerator of one fraction with the denominator of the other. If the products are equal, the fractions are equivalent. For example:

$$\frac{3}{5} \times \frac{24}{40} \qquad \begin{array}{l} 5 \times 24 = 120 \\ 3 \times 40 = 120 \end{array} \quad \checkmark \quad \text{Fractions are equivalent.}$$

$$\frac{4}{7} \times \frac{7}{10} \qquad \begin{array}{l} 7 \times 7 = 49 \\ 4 \times 10 = 40 \end{array} \quad \times \quad \text{Fractions are not equivalent.}$$

- The children can use the results to compare fractions as well. To do this, write the products next to the numerator you are multiplying by (but you can let children work that out for themselves). For example:

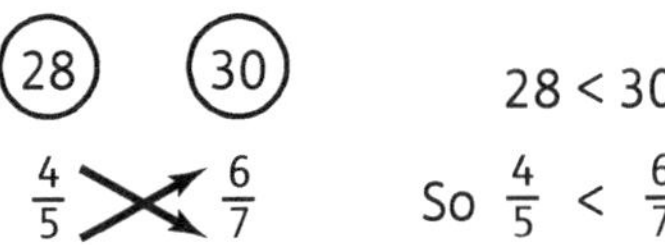

$$\widehat{28} \quad \widehat{30} \qquad 28 < 30$$
$$\frac{4}{5} \times \frac{6}{7} \qquad \text{So } \frac{4}{5} < \frac{6}{7}$$

- Let the children discuss why they think this method works (essentially it is a shortcut that uses 'common denominators'). They can find more information about this method online if they search for 'order-fractions-from-least-to-greatest'.

Interesting mistakes

- Children often decide that a fraction is greater than another by comparing just the denominators or just the numerators; often, they do not have a benchmark to help them understand the relative size of fractions.
- When children make mistakes with the size of fractions it is useful to ask them to compare the fractions to a fixed point such as 0, $\frac{1}{2}$ or 1. Then ask them to explain and justify their thinking, showing you their working so that you can address mistakes using real objects or number lines.

Answers for Pupil book page 67

1
a $\frac{1}{2} = \frac{3}{6}$ b $\frac{3}{4} = \frac{9}{12}$ c $\frac{3}{5} = \frac{6}{10}$ d $\frac{7}{8} = \frac{14}{16}$
e $\frac{7}{12} = \frac{28}{48}$ f $\frac{3}{5} = \frac{60}{100}$ g $\frac{4}{5} = \frac{8}{10}$ h $\frac{4}{7} = \frac{20}{35}$
i $\frac{3}{5} = \frac{6}{10}$ j $\frac{5}{8} = \frac{15}{24}$ k $\frac{2}{3} = \frac{4}{6}$ l $\frac{5}{12} = \frac{10}{24}$

2
a $\frac{6}{12}$ b $\frac{3}{12}$ c $\frac{4}{12}$ d $\frac{2}{12}$
e $\frac{9}{12}$ f $\frac{8}{12}$ g $\frac{10}{12}$ h $\frac{2}{12}$
i $\frac{4}{12}$ j $\frac{8}{12}$

3
a $\frac{1}{2}$ b $\frac{1}{5}$ c $\frac{1}{6}$ d $\frac{1}{4}$
e $\frac{3}{4}$ f $\frac{1}{2}$ g $\frac{5}{6}$ h $\frac{1}{2}$
i $\frac{3}{4}$ j $\frac{1}{3}$ k $\frac{3}{4}$ l $\frac{2}{3}$
m $\frac{2}{3}$ n $\frac{1}{3}$ o 5   p $\frac{4}{5}$
q $\frac{3}{10}$ r $\frac{1}{3}$ s $\frac{1}{2}$ t $\frac{1}{4}$
u $\frac{4}{7}$ v $\frac{7}{20}$ w $\frac{19}{20}$ x 8
y $\frac{8}{7}$

Answers for Workbook page 39

1 Individual colours.

2 equivalent to $\frac{1}{2}$: $\frac{2}{4}, \frac{1}{2}, \frac{5}{10}, \frac{50}{100}, \frac{1}{2}, \frac{11}{22}, \frac{500}{1000}, \frac{23}{46}, \frac{3}{6}, \frac{9}{18}, \frac{12}{24}, \frac{15}{30}, \frac{1}{2}$

equivalent to $\frac{1}{3}$: $\frac{3}{9}, \frac{10}{30}, \frac{15}{45}, \frac{5}{15}, \frac{9}{27}, \frac{200}{600}, \frac{2}{6}, \frac{100}{300}$

equivalent to $\frac{1}{4}$: $\frac{12}{48}, \frac{4}{16}, \frac{6}{24}, \frac{100}{400}, \frac{12}{48}, \frac{2}{8}, \frac{4}{16}, \frac{5}{20}$

equivalent to $\frac{3}{4}$: $\frac{24}{32}, \frac{75}{100}, \frac{27}{36}, \frac{9}{12}, \frac{18}{24}, \frac{750}{1000}$

equivalent to $\frac{1}{5}$: $\frac{20}{100}, \frac{2}{10}, \frac{4}{20}, \frac{6}{30}, \frac{3}{15}, \frac{200}{1000}$

equivalent to $\frac{4}{5}$: $\frac{800}{1000}, \frac{8}{10}, \frac{36}{45}, \frac{16}{20}$

3 For example:

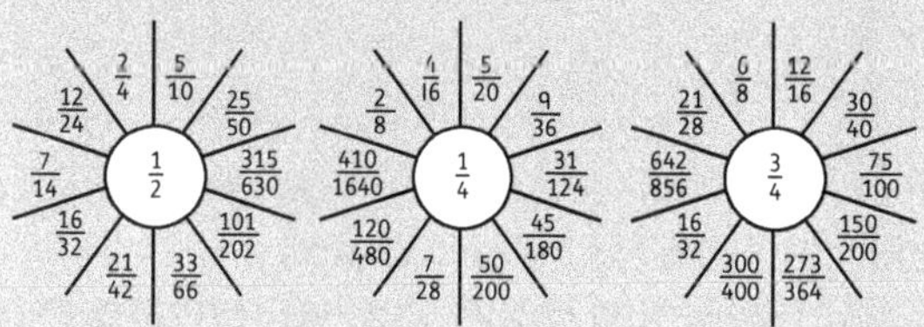

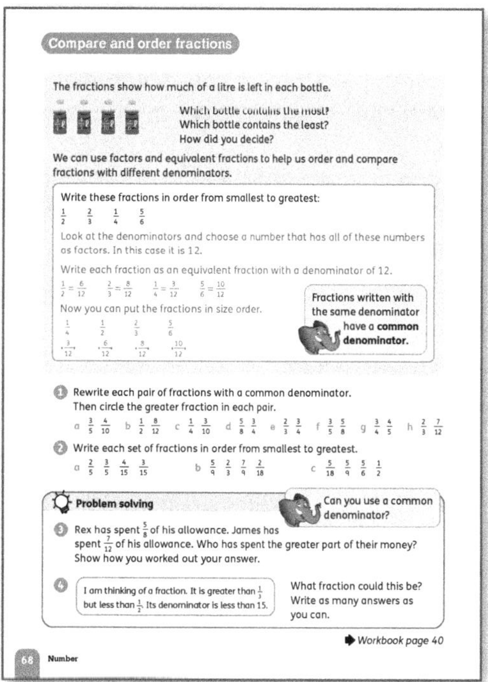

Materials

Fraction walls (you can search for these online); number lines; calculators.

Warm-up

Use a different activity in which the children find the HCF of a pair or set of numbers as a mental starter for this lesson.

Focus

- Remind the class that we can write fractions as their equivalents with different denominators without changing their value.
- Explain that we can use this principle to compare and order fractions. Let the children suggest how they would rewrite $\frac{2}{3}$ and $\frac{7}{10}$ to make it easier to compare them.
- With these numbers it is likely that the children will use the LCM of 3 and 10 and write the fractions as 30ths. However, they can use any common multiple to find a 'common denominater'.
- Turn to **Pupil book page 68**. Let the children think about the problem and then ask for their answers, encouraging them to say how they decided. The example box at the top of the page asks: Which bottle contains the most? $\left(\frac{9}{10}\,\ell\right)$ Which bottle contains the least? $\left(\frac{3}{4}\,\ell\right)$ How did you decide? (For example, draw diagrams or a number line to compare the fractions; compare 'how far' each fraction is from 1: $\frac{9}{10}$ is $\frac{1}{10}$ from 1, $\frac{5}{6}$ is $\frac{1}{6}$ from 1, $\frac{4}{5}$ is $\frac{1}{5}$ from 1, $\frac{3}{4}$ is $\frac{1}{4}$ from 1. Comparing these unit fractions shows that $\frac{1}{4}$ is the largest (as it has the smallest denominator), and so $\frac{3}{4}$ is furthest from 1.)
- Show the children how to convert a fraction to a decimal using a calculator, by dividing numerator by denominator, as an alternative method for comparing and ordering fractions.

- Work through the example on **Pupil book page 68** to show the class how to use equivalent fractions and common denominators to help them order a set of fractions. Make sure they understand that when you answer a question like this, you have to give the answer using the mixed set of fractions in the question.
- Children can work on questions 1 and 2 independently.
- <u>Problem solving:</u> Let the children work on questions 3 and 4 in pairs. They can use number lines and/or a fraction wall to help them.

Follow-up

- Use **Workbook page 40** as an informal assessment activity. Let the children complete the work on their own and have them mark their own work.

Challenge

- Display two fractions, for example $\frac{2}{3}$ and $\frac{3}{4}$. Ask the children to say a fraction that could go in between these two fractions and explain how they chose it.

Support

- When children have a good understanding of place value and decimal comparison but need additional support with fractions, they can use a calculator to express fractions as decimals and to use the decimals to compare and order the fractions.
- Understanding how to use a calculator in this way is empowering, allowing them to make connections and deepen their understanding of how the number system works.

Answers for Pupil book page 68

1 a $\left(\frac{3}{5}\right)\frac{4}{10}$ b $\frac{6}{12}\left(\frac{8}{12}\right)$ c $\frac{5}{20}\left(\frac{6}{20}\right)$

   d $\frac{5}{8}\left(\frac{6}{8}\right)$ e $\frac{8}{12}\left(\frac{9}{12}\right)$ f $\frac{24}{40}\left(\frac{25}{40}\right)$

   g $\frac{15}{20}\left(\frac{16}{20}\right)$ h $\left(\frac{8}{12}\right)\frac{7}{12}$

2 a $\frac{3}{15}$ $\frac{4}{15}$ $\frac{2}{5}$ $\frac{3}{5}$ b $\frac{2}{18}$ $\frac{5}{9}$ $\frac{2}{3}$ $\frac{7}{9}$

   c $\frac{5}{18}$ $\frac{1}{2}$ $\frac{5}{9}$ $\frac{5}{6}$

3 $\frac{5}{8} = \frac{15}{24}; \frac{7}{12} = \frac{14}{24}$

So, Rex has spent the larger part of the money.

4 For example, $\frac{5}{12}$

Answers for Workbook page 40

1 a $\frac{1}{4}$ $\frac{4}{15}$ $\frac{7}{12}$ $\frac{15}{30}$ b $\frac{1}{3}$

2 a $\frac{3}{4}$ b $\frac{2}{3}$ $\frac{9}{12}$ $\frac{19}{24}$

3 a $\frac{1}{2}$ b $\frac{1}{6}, \frac{1}{2}, \frac{5}{6}$ c $\frac{2}{7}, \frac{5}{7}$

   d $\frac{1}{4}, \frac{1}{2}, \frac{2}{3}, \frac{11}{12}$ e $\frac{1}{4}, \frac{5}{8}, \frac{8}{8}$ f $\frac{1}{4}, \frac{3}{4}, \frac{4}{4}, \frac{3}{2}$

4

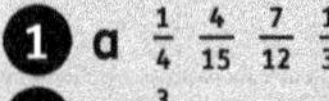
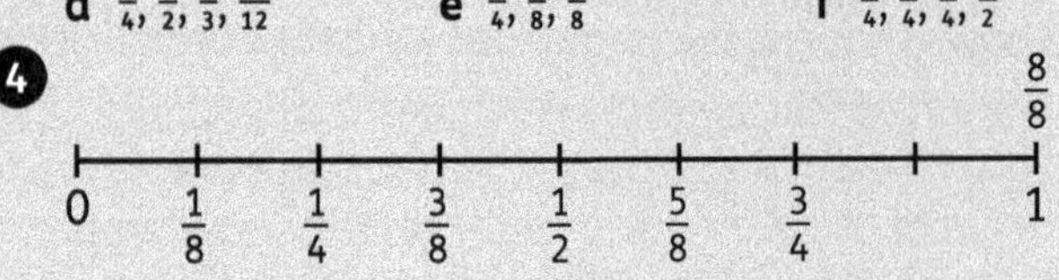

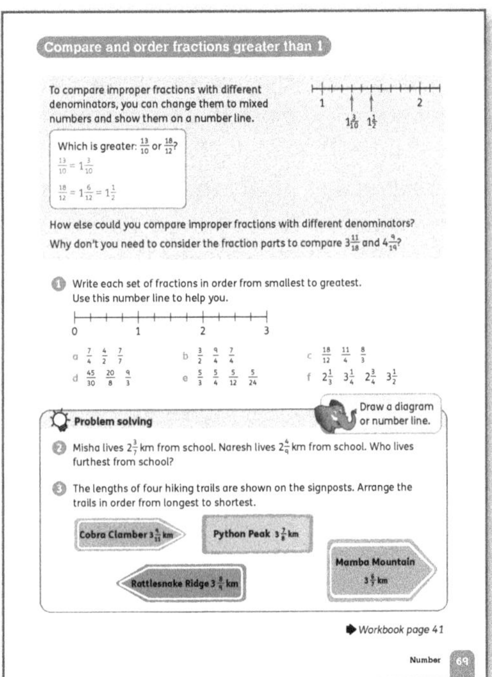

Materials

Number lines.

Warm-up

- Display a mixed number such as $3\frac{1}{5}$ and an improper fraction such as $\frac{18}{7}$. Ask which number is greater.
- Give them time to think about this and let them share their answers as well as how they decided.
- Draw a number line like the one below. Explain that we can locate $3\frac{1}{5}$ quite easily on this number line and mark it on the line.

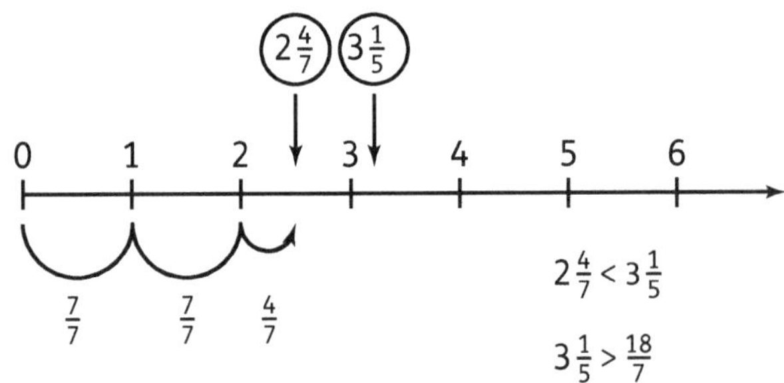

- Then say: *What about eighteen sevenths? I know that $\frac{7}{7}$ is 1, so I can count up in ones.* Show the jumps beneath the line (as in the example).
- Say: *Seven sevenths is one, another seven sevenths makes 2 and I still have 4 sevenths left. That takes me to $2\frac{4}{7}$. Now I can compare the numbers and see that $3\frac{1}{5}$ is greater than $2\frac{4}{7}$.*

Focus

- Turn to **Pupil book page 69**. Let the children work in groups to read through the example and to answer the questions. Have a class number talk (pages 17–18) using their ideas and suggestions. The example box at the top of the page asks the following questions: How else could you compare improper fractions with different denominators? (You could write them with a common denominator.) Why don't you need to consider the fraction parts to compare $3\frac{11}{18}$ and $4\frac{9}{19}$? (You can just compare the whole number parts to see that $3\frac{11}{18} < 4\frac{9}{19}$.)

- Turn to **Workbook page 41** to consolidate the children's ideas. The children complete a number line of equivalent fractions which they can then use to help them answer the other questions on the page.
- Check the children's answers as a class and resolve any problems that arise.
- Turn back to **Pupil book page 69**. Let the children work on their own to complete the questions.
- For question 1, remind the children that they can use the completed number line on **Workbook page 41** as well.
- <u>Problem solving:</u> Encourage the children to draw their own number lines or diagrams for questions 2 and 3.

Answers for Pupil book page 69

1 a $\frac{7}{7}, \frac{7}{4}, \frac{4}{2}$ b $\frac{3}{2}, \frac{7}{4}, \frac{9}{4}$ c $\frac{18}{12}, \frac{8}{3}, \frac{11}{4}$

d $\frac{45}{30}, \frac{20}{8}, \frac{9}{3}$ e $\frac{5}{24}, \frac{5}{12}, \frac{5}{4}, \frac{5}{3}$ f $2\frac{1}{3}, 2\frac{3}{4}, 3\frac{1}{4}, 3\frac{1}{2}$

2 a Naresh

3 Rattlesnake Ridge; Python Peak; Mamba Mountain; Cobra Clamber

Answers for Workbook page 41

1

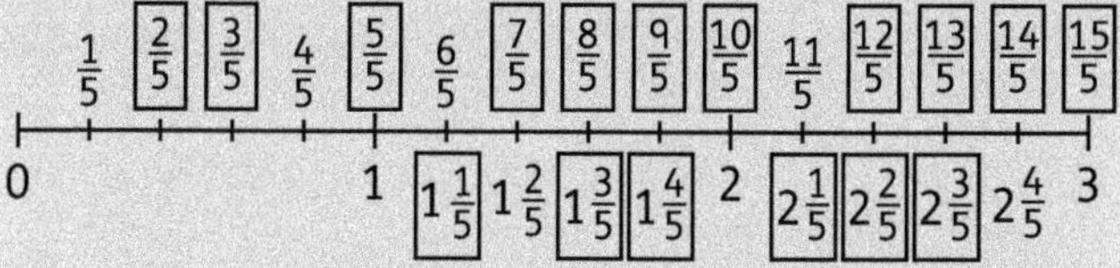

2 a >   b <   c >
d >   e <   f <
g =   h >

3 a $\frac{3}{4}$ $\frac{12}{30}$ $\frac{3}{7}$ $\frac{19}{20}$

b $\frac{9}{3} = 3$ $\frac{90}{40} = 2\frac{1}{4}$ $\frac{9}{8} = 1\frac{1}{8}$ $\frac{12}{8} = 1\frac{1}{2}$ $\frac{20}{19} = 1\frac{1}{19}$

c $1\frac{1}{19}$ $1\frac{1}{8}$ $1\frac{4}{8}$ 2   3

4 a $\frac{14}{6} = 2\frac{1}{3}, \frac{30}{5} = 6, \frac{28}{6} = 4\frac{2}{3}, 6, \frac{22}{12} = 1\frac{5}{6}, \frac{39}{12} = 3\frac{1}{4}$

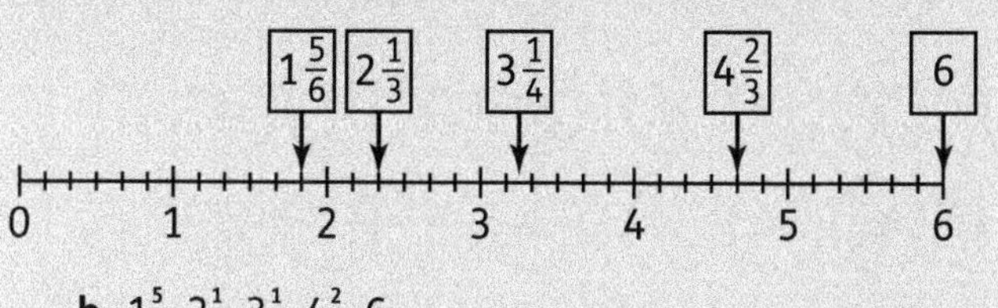

b $1\frac{5}{6}, 2\frac{1}{3}, 3\frac{1}{4}, 4\frac{2}{3}, 6$

Add and subtract fractions

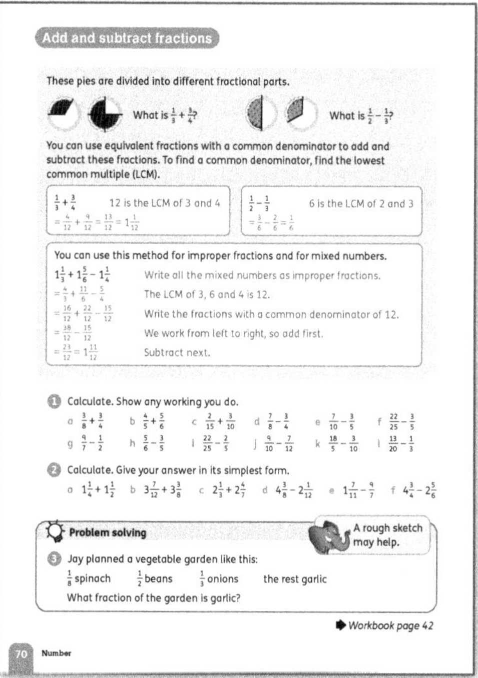

Materials
Fraction wall (you can search for this online); identical shapes cut into fractions that are multiples of each other; squared paper; number lines showing equivalent fractions.

Warm-up
Ask the children to find the LCM of different pairs of numbers. For example: 2 and 3; 3 and 4; 3 and 5; 3 and 6; 2 and 8; 3 and 9; 5 and 10; 4 and 12.

Focus
- Use fraction pieces of identical shapes to show $\frac{1}{3} + \frac{1}{3} = \frac{2}{3}$. Then show $\frac{1}{3} + \frac{1}{6}$. Ask the children how they could add these two fractions. Let them share their ideas.
- The children should be able to relate thirds to sixths and say that $\frac{1}{3}$ is equivalent to $\frac{2}{6}$ and $\frac{2}{6} + \frac{1}{6} = \frac{3}{6}$ (which can be simplified to $\frac{1}{2}$).
- Stress that we can only add fractions when they have the same denominators, so we use equivalent fractions to make the denominators the same. (Generally, we use the lowest common denominator (LCD), but it is not incorrect to use any common denominator. $\frac{1}{3} + \frac{1}{6}$ could also be written as $\frac{4}{12} + \frac{2}{12} = \frac{6}{12}$, which can be simplified to $\frac{1}{2}$).
- Repeat the procedure for a different pair of fractions, for example $\frac{1}{3} + \frac{5}{12}$.
- After completing a few examples with the class, turn to **Pupil book page 70**. Let the children work through the illustrated examples in pairs.

- Next, go through the formal worked example with the class.
- For question 1, tell the children to leave their answers as improper fractions where these occur.
- For question 2, you may need to point out that a mixed number in simplest form will have the fractional part simplified. For example, $1\frac{9}{12}$ can be simplified to $1\frac{3}{4}$ without changing its value in any way.
- <u>Problem solving:</u> For question 3, give the children squared paper to help them show the parts of the garden if they need it.

Follow-up

Use **Workbook page 42** to check that the children are able to apply what they have learnt to add and subtract fractions and mixed numbers in context.

Support

- Let children who need additional support with equivalent fractions continue to use the fraction wall or other representations to model the fractions and find the equivalents.
- You could also make sure they have a set of equivalent fraction number lines for reference.

Interesting mistakes

- The most common mistake that children make at this level is to assume that if they are adding the numerators, they must also add the denominators, so they may say: $\frac{2}{5} + \frac{2}{5} = \frac{4}{10}$.
- This indicates that the children do not fully understand the concept of a denominator. By using fraction pieces and number lines, they are able to see that the denominator stays the same; in other words, when you add one type of fraction your answer will be the same type of fractions. Using the names as you model the operations helps make this clear as well.
- Children may need additional support with calculations such as $1\frac{3}{5} - \frac{4}{5}$. Remind them that they can use a number line and count on or back, and that they can rewrite mixed numbers to make improper fractions. In this example, they actually have $\frac{5}{5} + \frac{3}{5} - \frac{4}{5} = \frac{8}{5} - \frac{4}{5}$.
- By getting the children to show you this on a number line, they will make sense of the operation.

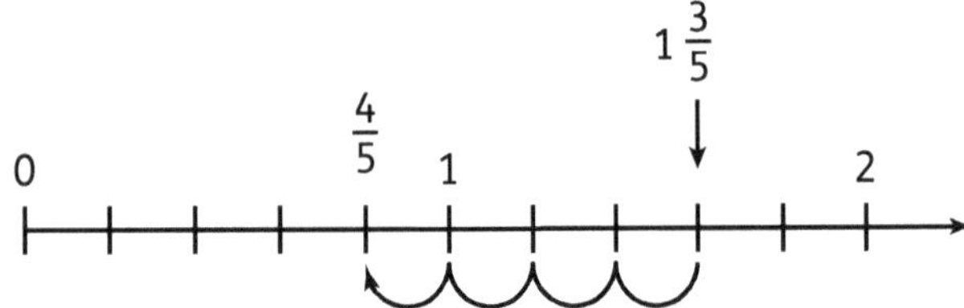

- Children may not realise that they can simplify the fractional part of a mixed number without affecting its value. Remind them that a number such as $3\frac{4}{10}$ means 3 wholes $+ \frac{4}{10}$, so the fraction can be treated separately and reduced to $\frac{2}{5}$.

Multiply fractions

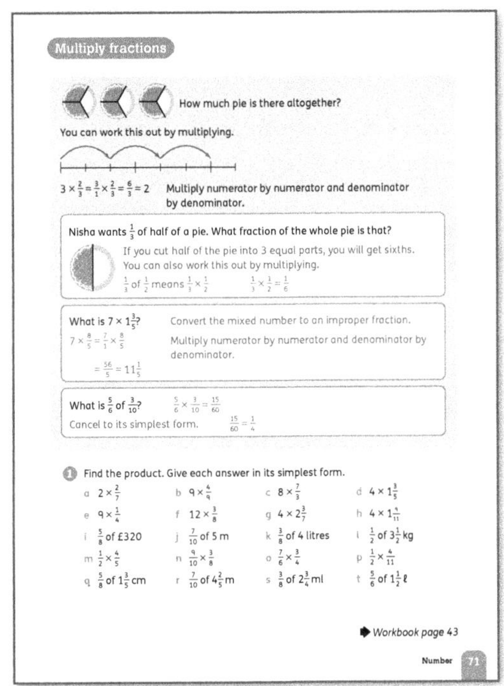

Materials

A set of circles divided into fractional parts; rulers; a resource sheet of divided circles for each child. This example shows ninths, twelfths, tenths and eighths.

Warm-up

Revise multiplication table facts as a mental starter for this lesson.

Focus

- The children multiplied fractions by whole numbers in Level 5. Use the printed sheets of fractional parts to revise this.

- Put the children into groups. Hand out a resource sheet to each group and ask them to shade the circles to show four different examples of a fraction multiplied by a whole number.
- For example, a group could shade a set of circles like this:

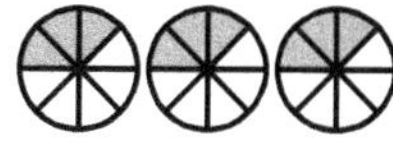

- The children should be able to recognise that this shows two different multiplications:
  - $\frac{3}{8} \times 3 = \frac{9}{8}$ (the shaded parts)
  - $\frac{5}{8} \times 3 = \frac{15}{8}$ (the unshaded parts).
- Remind them that we can write any whole number as a fraction with a denominator of 1 (whole). So, $3 \times \frac{3}{8} = \frac{3}{1} \times \frac{3}{8}$.
- Look at the examples and guide the children into seeing that we can think of this as multiplying the numerators by each other (3 × 3) and the denominators by each other (1 × 8).
- After completing a few examples, turn to **Pupil book page 71**. Read through the information with the class to consolidate what you have just done.
- Look at the second example problem $\left(\frac{1}{3} \text{ of } \frac{1}{2}\right)$. Encourage the children to use their rulers to show how much $\frac{1}{3}$ of the half pie would be. Then ask them to say what fraction this is of the whole pie. They should be able to say that $\frac{1}{3}$ of $\frac{1}{2}$ is $\frac{1}{6}$ of the whole pie.
- Remind the class that 'of' means multiply and then work through the examples slowly and carefully with the class. The children can then work independently on question 1. Remind them to simplify their answers where they can. Generally, for number calculations you would leave any answers as improper fractions (in simplest form) but where measures are involved, it is more common to give the answer as a mixed number.

Follow-up
- The children can complete **Workbook page 43** independently as a test to check how well they can multiply fractions and calculate fractions of amounts.
- Display the answers and let the children work in pairs to mark each other's answers.
- Discuss any interesting mistakes and/or questions that the children might have as a class.

Challenge
- You may like to explore how to cancel as you work so that you do not have to simplify in the end. However, if the children do not have a good conceptual framework for this, it is likely that they will think they can cross multiply instead of working in rows.
- Show the class two different methods of working out the same product. For example:

$$\frac{5}{8} \times \frac{7}{5} = \frac{35}{40} \div \frac{5}{5} = \frac{7}{8}$$

$$\frac{\cancel{5}}{8} \times \frac{7}{\cancel{5}} = \frac{7}{8}$$

- Ask the children to try this out and to explain how it works. It is important that they realise that cancelling is the same as dividing by $\frac{5}{5} = 1$.

Answers for Pupil book page 71

1   a $\frac{4}{7}$ b 4   c $18\frac{2}{3}$

d $6\frac{2}{5}$ e $2\frac{1}{4}$ f $4\frac{1}{2}$

g $9\frac{5}{7}$ h $7\frac{3}{11}$ i £200

j 3.5 m   k 1.5 litres   l 1.75 kg

m $\frac{2}{5}$ n $\frac{27}{80}$ o $\frac{7}{8}$

p $\frac{2}{11}$ q 1 cm   r $3\frac{2}{25}$ ml

s $1\frac{1}{32}$ ml   t $1\frac{1}{4}$ litres

Answers for Workbook page 43

1

×	3	5	$\frac{1}{2}$	$\frac{3}{5}$	$\frac{7}{9}$
$\frac{2}{5}$	$\frac{6}{5}$	2	$\frac{1}{5}$	$\frac{6}{25}$	$\frac{14}{45}$
$\frac{3}{8}$	$\frac{9}{8}$	$\frac{15}{8}$	$\frac{3}{16}$	$\frac{9}{40}$	$\frac{7}{24}$
$\frac{4}{9}$	$\frac{4}{3}$	$\frac{20}{9}$	$\frac{2}{9}$	$\frac{4}{15}$	$\frac{28}{81}$
$\frac{7}{8}$	$\frac{21}{8}$	$\frac{35}{8}$	$\frac{7}{16}$	$\frac{21}{40}$	$\frac{49}{72}$
$\frac{9}{10}$	$\frac{27}{10}$	$\frac{9}{2}$	$\frac{9}{20}$	$\frac{27}{50}$	$\frac{7}{10}$
$\frac{3}{2}$	$\frac{9}{2}$	$\frac{15}{5}$	$\frac{3}{4}$	$\frac{9}{10}$	$\frac{7}{6}$
$\frac{7}{5}$	$\frac{21}{5}$	7	$\frac{7}{10}$	$\frac{21}{25}$	$\frac{49}{45}$
$\frac{11}{9}$	$\frac{11}{3}$	$\frac{55}{9}$	$\frac{11}{18}$	$\frac{11}{15}$	$\frac{77}{81}$

2   a $\frac{1}{6}$ b $\frac{3}{4}$ litres

c $\frac{7}{12}$ d $\frac{4}{5}$

Divide fractions by whole numbers

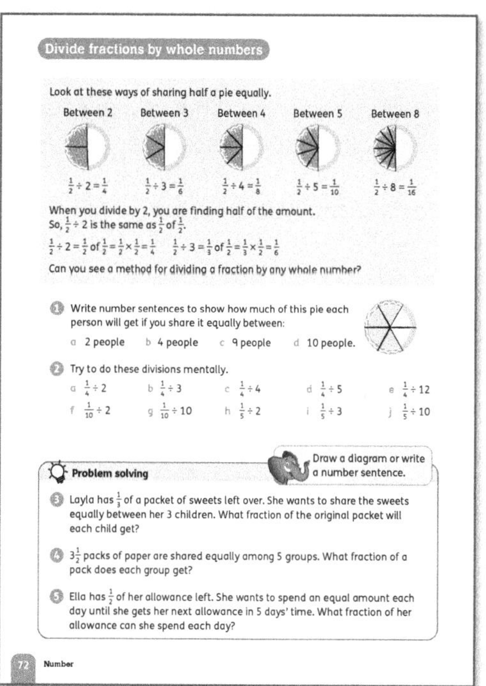

Materials

Objects showing fraction parts, such as $\frac{1}{2}$ metre of string, $\frac{3}{4}$ of a cake or a diagram or cards showing fractions of shapes; sheet of fractions of different shapes (see Support for examples).

Warm-up

Do some mental division fact calculations as a starter for this lesson.

Focus

- Turn to **Pupil book page 72**. Use the pictures of the pies to talk about how you share half a pie among different numbers of people. The explanation box at the top of the page asks: Can you see a method for dividing a fraction by any whole number? (Multiply the denominator by the whole number.)
- Remind the children that equal sharing is division in mathematics.
- Discuss the diagrams and the calculations with the class. Let the children share their ideas about dividing a fraction by a whole number.
- Give at least one fractional model to each group and ask them to use it to prove that the ideas they have discussed work.
- Let the groups show what they do to prove this.
- Next, show the class a bar model showing $\frac{1}{4}$. Say: $\frac{1}{4}$ of the whole is shaded. How could I divide this quarter into 3 equal parts?

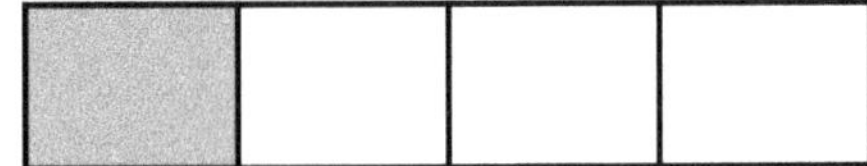

- Show the children how to draw lines on the model to represent the division, like this:

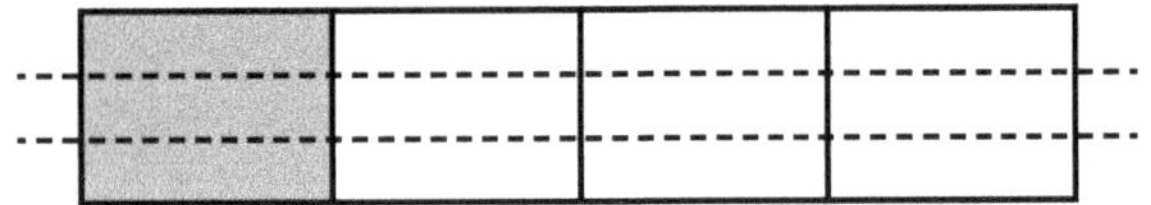

- Check they agree that you have divided the shaded part into 3 smaller equal parts. Then ask: How can we work out, mathematically, what fraction one part of the quarter is of the whole bar?
- Show them how to think about the bar as an area model like this:

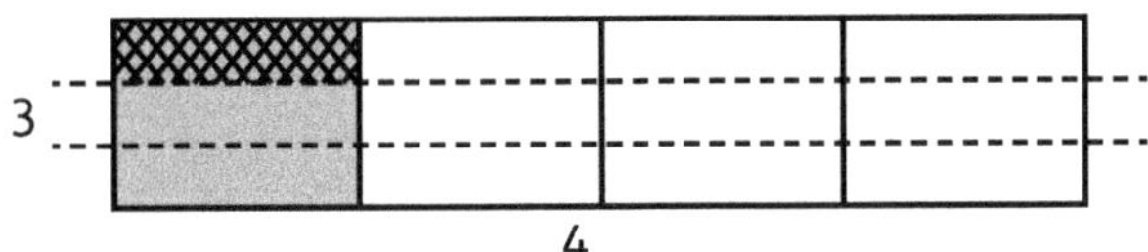

- Point out that by dividing each quarter into 3 equal parts you have ended up with 12 parts in the whole.
- This means that the one part you are interested in $\left(\frac{1}{3}\text{ of }\frac{1}{4}\right)$ is 1 of the 12 pieces.
- This type of modelling is very useful as it shows the children visually why you can write $\frac{1}{4} \div 3 = \frac{1}{4} \times \frac{1}{3} = \frac{1}{12}$.
- Repeat this with different fractions, for example $\frac{1}{7}$ divided by 8, to show how you can use the bar model

to represent the division without actually drawing all the lines. For example:

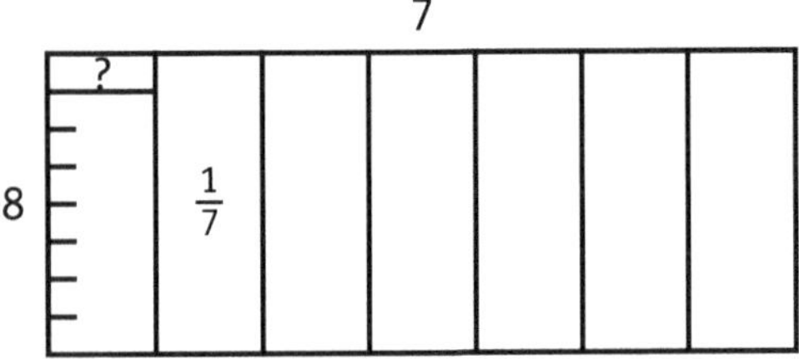

$$7 \times 8 = 56$$

$$\frac{1}{7} \div 8 = \frac{1}{56}$$

- When you are sure the children understand this concept, let them complete questions 1 and 2 on **Pupil book page 72**.
- Problem solving: Encourage them to use bar models to help them solve the problems in questions 3–5.

Support

- Give the children a sheet of fractions of different shapes. For example:

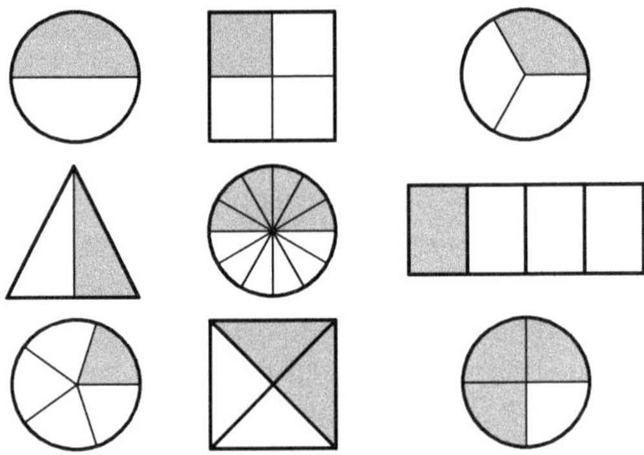

- Explain that they are going to share the shaded part equally between two people.
- Ask: What are you sharing or splitting up? Let them write this below each shape, for example: 'I am sharing $\frac{1}{2}$'; 'I am sharing $\frac{1}{4}$'.
- Next, ask them to draw lines on each shaded part to show how they would share it equally between two people.
- Now ask them to work out what fraction of the shape each person will get.
- This visual reinforcement will help the children understand what they are doing.

Answers for Pupil book page 72

1   a $\frac{1}{3}$ b $\frac{1}{6}$
   c $\frac{2}{27}$ d $\frac{1}{15}$

2   a $\frac{1}{8}$ b $\frac{1}{12}$ c $\frac{1}{16}$
   d $\frac{1}{20}$ e $\frac{1}{48}$ f $\frac{1}{20}$
   g $\frac{1}{100}$ h $\frac{1}{10}$ i $\frac{1}{15}$
   j $\frac{1}{50}$

3 $\frac{1}{9}$

4 $\frac{7}{10}$

5 $\frac{1}{10}$

Work with fractions

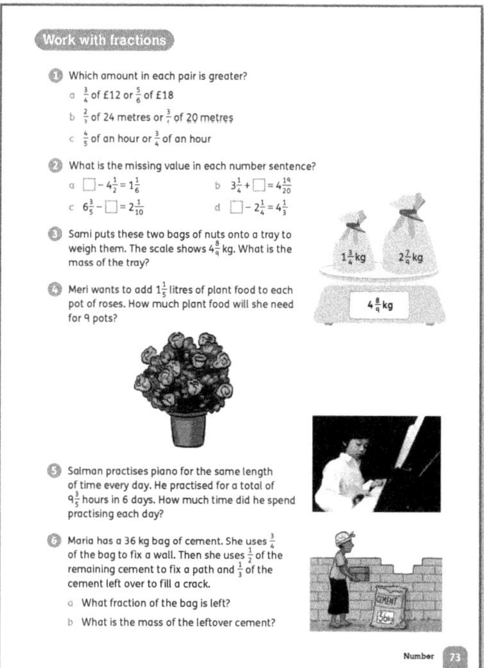

Warm-up

Revise calculating a fraction of a quantity from Level 5, using 'Odd one out' (page 25).

Focus

- In this lesson, the children are expected to apply what they have learnt about fractions in this unit to solve different problems.
- For question 1, the children have to be able to find a fraction of a quantity before they can do the comparisons.
- For question 2, remind the children that addition and subtraction are inverse operations.
- For question 3, the children discuss the problem. Encourage them to draw a bar model (page 20) to show the problem situation. For example:

$$4 \tfrac{8}{9} \text{ kg}$$

$1 \tfrac{3}{4}$	$2 \tfrac{7}{9}$	?

- For question 4, remind the children that they can convert mixed numbers to equivalent improper fractions if it will make them easier to work with, but they do not have to. They can multiply whole number parts and fractional parts separately.
- For question 5, a bar model like this might help the children see what to do:

$$9 \tfrac{3}{5} \text{ hrs}$$

?					

- Remind the children that they are dividing a fraction $\left(9\tfrac{3}{5} \text{ hours} = \tfrac{48}{5} \right)$ by a whole number.

- Question 6 is a multistep problem, so the children should realise that they need more than one bar in the model. For example:

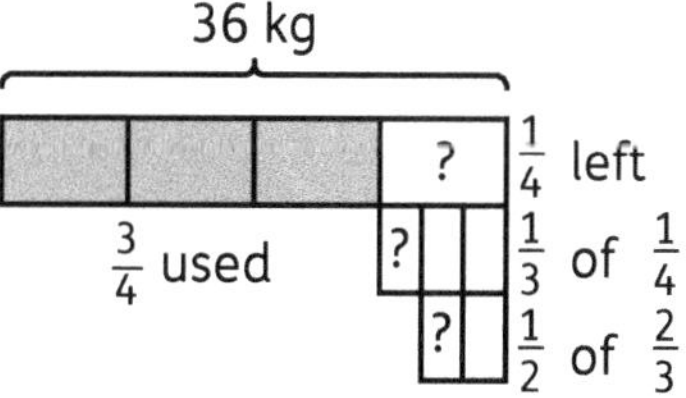

Challenge

- Let the children make a set of dominoes using fraction calculations and their solutions.
- Provide a frame like the one below for them to make the set and then ask them to cut them up and shuffle them to play.
- They can work on their own set and combine their dominoes with a partner to play.

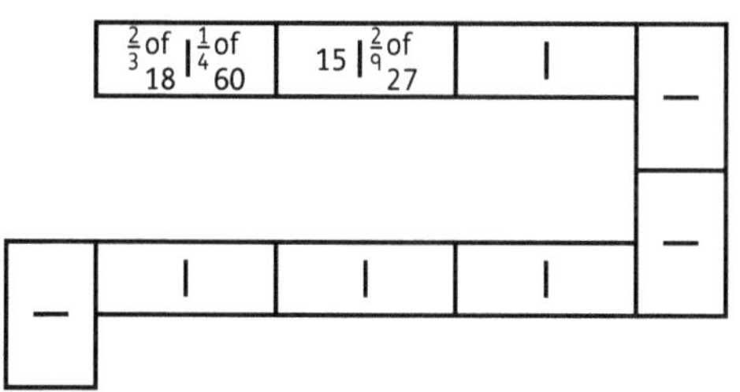

End-of-unit check

Use all or some of these questions and problems to check how well the children have understood the concepts in this unit.

- *What did you find easiest in this unit? What made it easy?*
- *What did you find most challenging? Why was it challenging?*
- *What are three things that are important to remember when you work with fractions? Why are these things important?* (To find equivalent fractions, multiply or divide the numerator and denominator by the same number. To add or subtract fractions, the denominators must be the same. In an improper fraction, the numerator is larger than the denominator, and the fraction can be converted to a mixed number.)
- *Give a fraction that is equivalent to $\tfrac{3}{5}$ but has a denominator of 10. How did you work it out?* $\left(\tfrac{6}{10} \right.$ Multiply both the numerator and denominator by 2.$\left. \right)$

- *Give a fraction that is equivalent to $\frac{3}{5}$ but has a denominator of 25. How did you do it?* $\left(\frac{15}{25}.\right.$ Multiply both the numerator and denominator by 5.$)$
- *Give a fraction that is equivalent to $\frac{2}{3}$.* (For example: $\left.\frac{4}{6}, \frac{6}{9}, \frac{20}{30}\right)$
- *How would you order these fractions from smallest to greatest?* $\frac{3}{5}, \frac{3}{4}, \frac{5}{8}, \frac{7}{10}$ $\left(\right.$ Write as equivalent fractions with the same denominator. $\left.\frac{3}{5}, \frac{5}{8}, \frac{7}{10}, \frac{3}{4}\right)$
- *How did you cancel this fraction to its simplest form?* (Divide the numerator and denominator by the same numbers until there are no common factors.)
- *How do you know when you have the simplest form of a fraction?* (The numerator and denominator have no common factors.)
- *How would you explain to someone how to order a set of fractions and mixed numbers? What tips would you give them?* (Convert the fractions to equivalent fractions with the same denominator. Compare the whole number parts first. If the whole number parts are the same, compare the fraction parts.)
- Show two fractions. *What fraction lies between these two fractions? How did you make your choice?*
- *What fraction of the children in our class wear glasses/don't wear glasses/have long hair/don't have long hair?*
- *How would you change $1\frac{1}{5}$ to an improper fraction?* $\left(1 = \frac{5}{5}, \frac{5}{5} + \frac{1}{5} = \frac{6}{5}\right)$
- Display a mixed number. *How could you explain how to change this mixed number into an improper fraction?* (Multiply the whole number part of the mixed number by the denominator of the fraction part. Add the result to the numerator of the fraction part and put that over the denominator of the fraction part.)
- Display the number $2\frac{7}{5}$. *What is wrong with this number? Why?* $\left(\right.$The numerator in the fraction part is greater than the denominator and there is also a whole number part. The number should be either $3\frac{2}{5}$ or $\frac{17}{10}.\right)$
- *What is $\frac{2}{7} + \frac{3}{7}$?* $\left(\frac{5}{7}\right)$
- *If you subtract $\frac{3}{8}$ from $\frac{7}{8}$, what are you left with?* $\left(\frac{4}{8} = \frac{1}{2}\right)$
- *What is $\frac{4}{5} + \frac{1}{3}$?* $\left(1\frac{2}{15}\right.$ or $\left.\frac{17}{15}.\right)$
- *How would you subtract $\frac{3}{8}$ from $\frac{3}{4}$?* $\left(\right.$Write both fractions with denominator 8 and subtract the numerators: $\left.\frac{6}{8} - \frac{3}{8} = \frac{3}{8}.\right)$
- *What is $2\frac{3}{4} - 1\frac{7}{8}$?* $\left(\frac{7}{8}\right)$
- *What is $3 \times \frac{2}{9}$?* $\left(\frac{6}{9} = \frac{2}{3}\right)$
- *How can you draw a diagram to show four lots of $\frac{2}{7}$?* (For example: four 7-part bars, each with 2 parts shaded.)
- *What is $3 \times 1\frac{1}{2}$?* $\left(4\frac{1}{2}\right)$
- *What is $\frac{6}{7} \times \frac{2}{3}$?* $\left(\frac{12}{21} = \frac{4}{7}\right)$
- *What is (give a fraction) of (give an amount)?*
- *Write a word problem that has the answer $\frac{3}{2}$.*
- *Make up a two- or three-step problem using fractions. Show how you would solve it.*

UNIT 7 Position, direction and movemen

Learning objectives
- Read and plot coordinates on the full coordinate grid (all four quadrants).

- Draw and translate simple shapes on the coordinate grid and reflect them in the axes.

- Reflect 2D shapes in a given mirror line (vertical, horizontal and diagonal on square grids).

- Rotate shapes clockwise and anti-clockwise around a vertex.

Key words
coordinates   *x*-axis   *y*-axis   quadrants
reflection   translation   mirror line   horizontal
vertical   (full) rotation

Materials
Large world map showing lines of latitude and longitude; 8 by 8 grid.

Teaching guidance
- Display a large world map with lines of latitude and longitude shown.
- Talk about how the lines form a grid on the map (depending on the map projection, some will be shown as curves) and how we can use grid 'coordinates' to locate places.
- In mapping terms, the coordinates are given in degrees (and sometimes smaller units called minutes and seconds) with latitude first, followed by longitude.
- Point out that the equator is the 0° line of latitude. The lines are labelled from 0 to 90 above and below the equator, so the coordinate always has N or S following it to show whether it is above or below the equator.

- The 0° line of longitude runs through Greenwich in the UK. Lines of longitude are labelled from 0 to 180 east and west of the 0° line, so these coordinates always have E or W to show where they are located. (Understanding this will help the children with negative coordinates below x and to the left of y on the grid).
- Choose some simple coordinates and ask the children which country these points are in.
  For example, use:
  - 0; 100 E (Indonesia)
  - 20 N; 100 E (Thailand)
  - 23 S; 140 E (Australia)
  - 40 N; 80 E (China)
  - 20 N, 100 W (Mexico)
- Then let the children make up a set of coordinates for six different countries. The children can locate countries using a partner's set of coordinates.
- You could get the children to give coordinates for some cities in different countries. For example, ask them to give the position of a city in India.

Revisit the coordinate grid

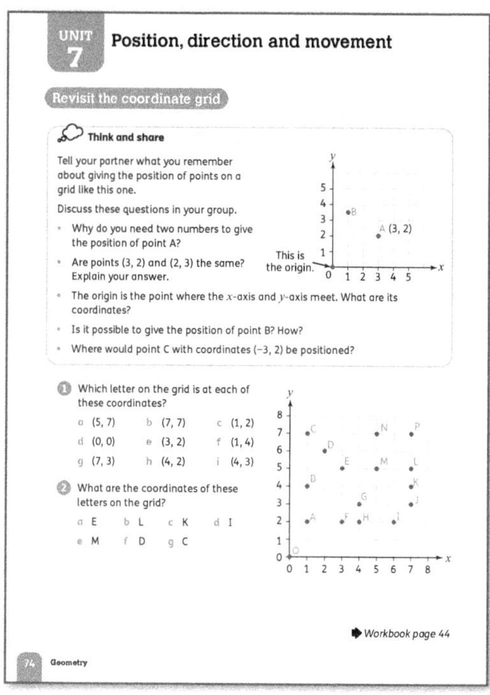

- give the coordinates of point B which is not at a point of intersection of the grid lines. (1, 3.5) The last question asks the children to say where (−3, 2) would be found. (Starting at the origin move 3 squares to the left and 2 squares up.)
- The children should manage to locate points when they are given coordinates and be able to give the coordinates of points on a grid.

Focus
- Let the children work in pairs on questions 1 and 2.
- Use **Workbook page 44** to consolidate the work in this lesson and to check that the children can work confidently in the first quadrant.
- Check their answers in class, with the children checking each other's work.

Challenge
- Play a game of 'Gridlock' to revise working with coordinates in the first quadrant.
- Divide the children into pairs (these can change after the first game). Each pair will need two spinners numbered 1–6 and two different-coloured pens plus a copy of a board like:

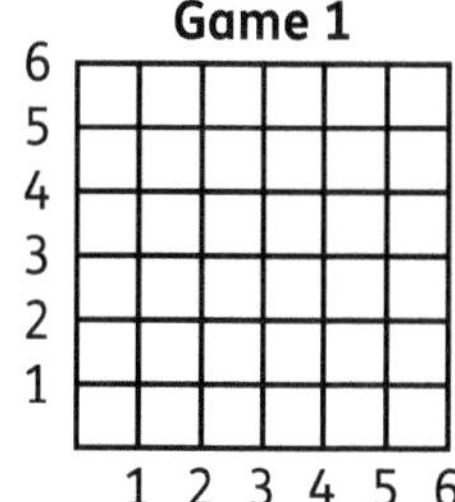

  - The children take turn to spin both spinners. They use the numbers (in any order) to form a coordinate pair. So, if they get 2 and 3, they make (2, 3) or (3, 2).
  - The child draws a dot at that position on the grid.
  - The aim of the game is to be the first person to get three dots in a row ('horizontal', 'vertical' or diagonal).
  - If both positions are occupied, the player cannot use their turn.
  - If the game ends with no one getting three in a row, it is a tie.
- Give the children time to play one or two games.
- Hold a discussion about the strategies they used to try and get three in a row while also preventing their partner from getting a row of three.

Support
- Prepare an 8 by 8 grid and work in small groups with the children to help them read and plot points.
- Place a counter on a point and ask: *Where is my counter?*
- Let the children give the coordinates and make sure they know that they have to give the x-coordinate first. It may help to remind them that x comes before y in the alphabet.
- Move the counter and repeat this for a number of positions.

Materials
8 by 8 grids; counters.

Warm-up
- <u>Think and share:</u> This is a revision lesson, so use the Think and share discussion on **Pupil book page 74** as a warm-up for this lesson. Remind the children of the terms 'x-axis' and 'y-axis'. The first question asks why two numbers are needed to give the position of point A. (You need to know the distance from both the x-axis and the y-axis to locate a unique point.) The second question asks whether the points (2, 3) and (3, 2) are the same. (No, because (2, 3) is the point 2 squares to the right and 3 squares up from the origin while (3, 2) is 3 squares to the right and 2 squares up.) The third question asks for the coordinates of the origin. ((0, 0)) The children are then asked to

- Give the children a position and ask them to place the counter on that point. Let them decide whether it has been placed correctly.
- Use the correct vocabulary as you work with the children to make sure they understand these terms: axis, coordinates, pair of coordinates, position, point and grid.

Interesting mistakes

- Children may be haphazard about how they write coordinate pairs. For example, they may write 2, 5 and leave out the brackets, or they may write (2.5) using a point instead of a comma.
- If they do this, bring it to their attention by asking the class or each group: *What is the problem when we write coordinates like this?* (It looks like a number sequence; It looks like a decimal; and so on.)

Answers for Pupil book page 74

1
  a N   b P   c A
  d O   e F   f B
  g J   h H   i G

2
  a (3, 5)   b (7, 5)   c (7, 4)
  d (6, 2)   e (5, 5)   f (2, 6)
  g (1, 7)

Answers for Workbook page 44

1 and **2**

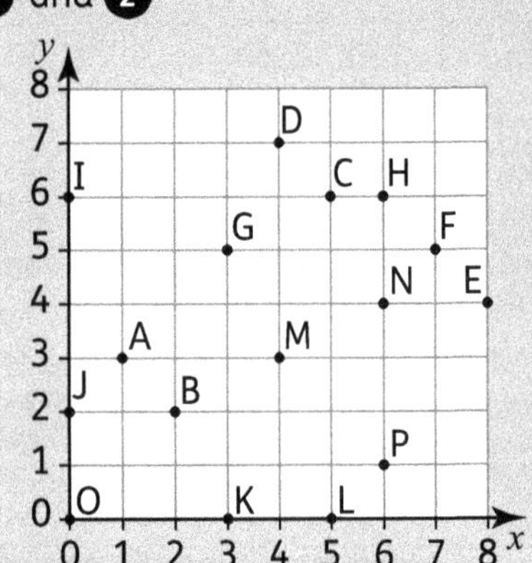

3 a and b

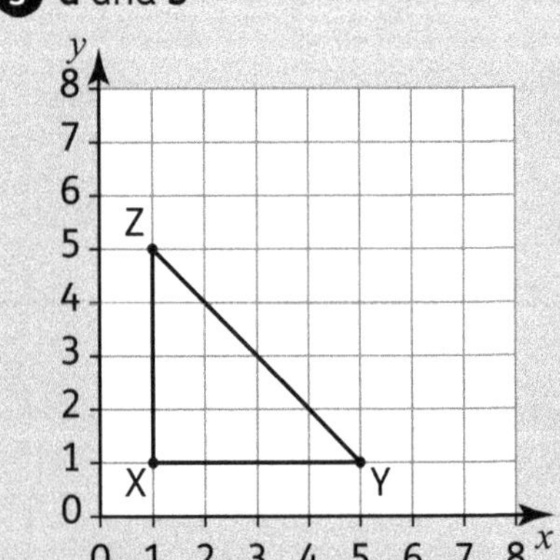

  c It is a right-angled triangle. The sides XY and XZ are perpendicular, i.e., form a right angle because they are parallel to the x- and y-axis, respectively.

4 Individual answers.

Extend the grid

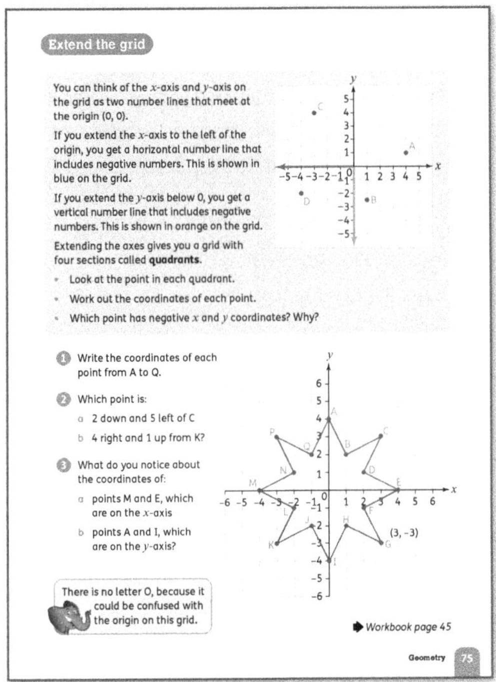

Materials

Number line that extends below zero; copies of an extended grid; counters.

Warm-up

Use a number line that extends below zero and do some skip-counting activities (forwards and backwards) as a starter for this lesson.

Focus

- Introduce the term 'quadrants'. In this lesson the children will extend their ideas to include points in all four quadrants.
- Display a 5 × 5 grid for the class. Point to the x-axis and ask the class how it is similar to a number line. Let them share their ideas.
- Extend the x-axis to the left of the origin (0, 0) and mark the first five points without labelling them.
- Ask the class what numbers we could use to label these points.
- They should be able to say that numbers to the left of 0 on a number line are negative, so the points are −1, −2 and so on.
- Place a few counters on the grid at different positions in the second (top left) quadrant.
- Ask the children to find the coordinates of each counter. Remind them that they still give (x, y) but now that x-coordinate is negative.
- Point to the y-axis. Tell the class you are going to extend it below the origin and draw the first five points, again without labelling them. Ask the class to say what each point is.
- Explain that you now have a grid with four different quadrants. Put one counter in each quadrant, placing them at (2, 3), (−2, 3), (−2, −3) and (2, −3).

- Let the class work out the coordinates of each counter. Discuss the x- and y-coordinates in each quadrant. Ask questions, such as:
  - *What can you say about the coordinates in this quadrant?*
  - *Where will the coordinates always both be positive?*
  - *Where will the coordinates always both be negative?*
  - *If a point has a positive x-coordinate, in which quadrants could it be?*
- Do some activities on the extended grid, asking the children to give the coordinates of a counter in different positions and letting them work in groups to place counters correctly when you give them coordinate pairs.
- Play a game of 'Where is it now?' with the class. Give them the extended grid and counters to model the movements.
  - Give a starting point such as (−2, 4) and let the children place a counter in that position on their grids.
  - Then give a series of movements (some single, some double) and let the children work out where the counter ends up. For example, say: *Move two places left. Move three places up. Move one left and two down.*
  - When you have given a few sets of movements ask: *Where is the counter now?*
  - The children give the coordinates. Let them discuss any mistakes and what they might have done wrong if they ended up in the wrong position.
- Turn to **Pupil book page 75**. Let the children work through the information text and questions 1–3 to consolidate what they have learnt.
- The explanation box at the top of the page asks: Which point has negative x and y coordinates? Why? (*Point D, because its coordinates are (−4, −2), it is in the quadrant where both x and y values are negative.*)
- Make sure the children know that they should record the letters and the coordinates of each point, for example: A (0, 4).
- Check that the children remember what a 'translation' is. Revise translation if necessary.
- Discuss their answers to question 3.

Follow-up

- Use **Workbook page 45** to assess whether the children can plot and write coordinates of in all four quadrants.
- Let the children compare their answers with a partner and discuss any mistakes or disagreements. Compare answers to question 3 as a class.

Interesting mistakes

Initially, children may need additional support to locate the position of points when one or both of the coordinates are negative. Generally, they will get better at this with practice and you should continue to give them opportunities to find and plot coordinates on the extended grid.

Answers for Pupil book page 75

1 A = (0, 4); B = (1, 2); C = (3, 3); D = (2, 1); E = (4, 0); F = (2, −1); G = (3, −3); H = (1, −2); I = (0, −4); J = (−1, −2); K = (−3, −3); L = (−2, −1); M − (−4, 0); N − (−2, 1); P = (−3, 3); Q = (−1, 2)

2 a N (−2, 1)   **b** H (1, −2)

3 a Their x-coordinates are the same number but one is negative and one is positive. Their y-coordinates are zero.
b Their y-coordinates are the same number but one is negative and one is positive. Their x-coordinates are zero.

Answers for Workbook page 45

1
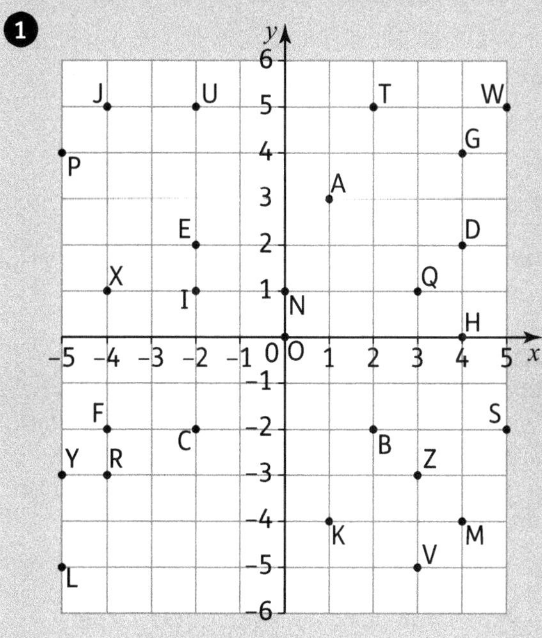

2 Q = (3, 1); R = (−4, −3); S = (5, −2); T = (2, 5); U = (−2, 5); V = (3, −5); W = (5, 5); X = (−4, 1); Y = (−5, −3); Z = (3, −3)

3 a
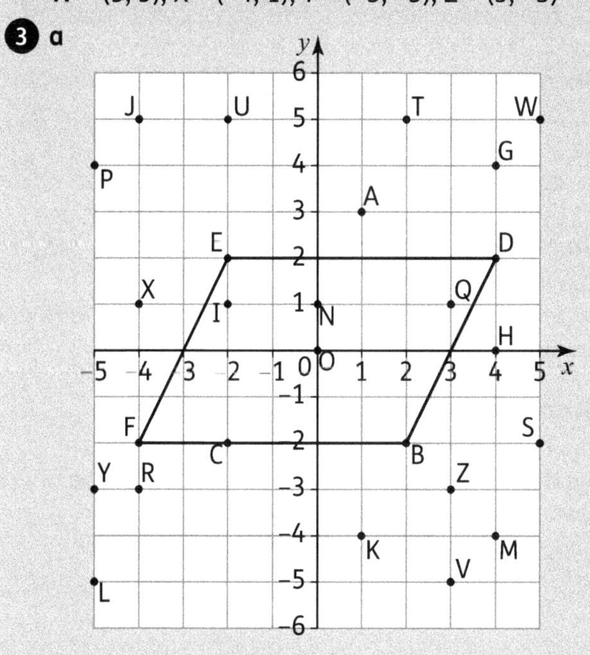

b (−2, 2), (4, 2), (2, −2) and (−4, −2)

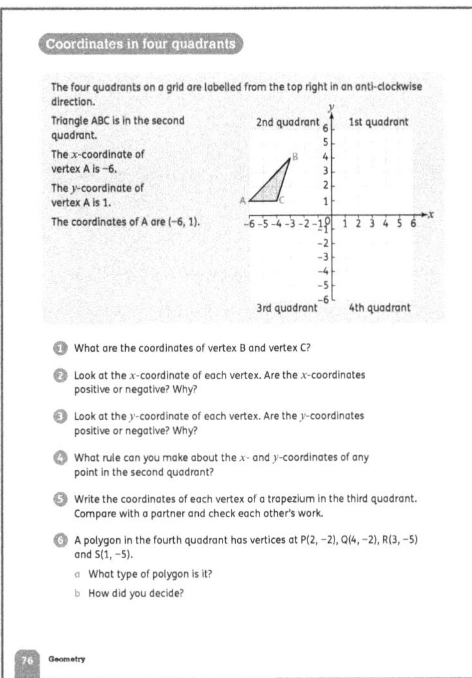

Materials

Copies of an extended grid; counters.

Warm-up

Revise properties of triangles and quadrilaterals. For example, you could name a shape and ask the children to give the properties of its sides and/or angles. You could also give a set of properties and ask the children to say what shape it could be.

Focus

- Turn to **Pupil book page 76**. Read through the information with the class.
- Explain that the quadrants are labelled anti-clockwise. Point out that the x-coordinate comes first, so when the x-axis is extended it forms the second quadrant (remind them that they did this in the previous lesson).
- When the y-axis is extended, it makes two more quadrants. As you have labelled the top left one as the second, it makes sense to continue numbering in the anti-clockwise order.
- Let the children work in pairs to discuss questions 1–4. However, encourage them to write their own answers.
- For questions 5 and 6, the children can use a grid and counters to model the shape.

Challenge

- Give children a grid from −5 to 5 on both axes.
- Ask them to draw a square ABCD in one of the quadrants (they can choose which one).
- Next, ask them to write the coordinates of each vertex in the first row of a copy of this table.

ABCD	A (__, __)	B (__, __)	C (__, __)	D (__, __)
Change the sign of the x-coordinate				
Change the sign of the y-coordinate				
Change the sign of both coordinates				

- Let the children complete the table by changing the signs of the coordinates as indicated. They should draw the new square ABCD after each change.
- Let them discuss what they find out, then share their ideas with the class.

Answers for Pupil book page 76

1. B = (−3, 4); C = (−4, 1)
2. They are all negative because each vertex is in the left side of the y-axis.
3. They are all positive because each vertex is above the x-axis.
4. A point in the second quadrant has its x-coordinate negative and its y-coordinate positive.
5. Individual answers.
6. a It is a parallelogram.
   b It has two sets of parallel sides of different lengths and no right angles.

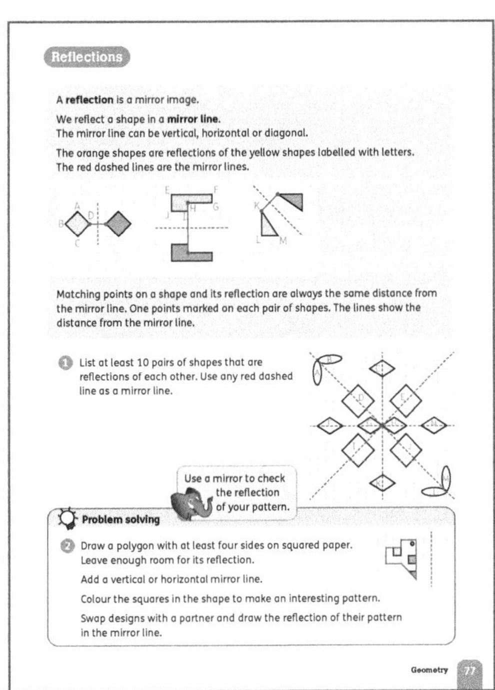

Materials

Cut-out shapes; small mirrors; squared paper or coordinate grids to work on; coloured pens.

Warm-up

Use any suitable 'Place value and number sense' activity (pages 22–23) for a mental warm-up.

Focus

- Demonstrate how to use a mirror to reflect a shape. Place the mirror on the axis of a grid and show the children how squared paper can help them to draw the 'reflection' of shapes.
- Point to a vertex on one of the original shapes and the corresponding corner on the reflection. Ask children to look at how far away each point is from the 'mirror line' and establish they are the same distance away, and if you join them with a line, this line is perpendicular to the mirror line.
- Repeat with other vertices.
- Turn to **Pupil book page 77** and work through the explanation and examples with the class.
- You may need to model the reflections using cut-out shapes and a mirror.
- Many children need additional support to visualise reflections and will often trace and cut out shapes or fold along the mirror line to find the position of the transformed shape. Allow and encourage the children to explore practical methods to help them.
- For question 1, make sure the children can see that the axes and the two diagonal lines are all mirror lines before they start.
- Problem solving: For question 2, if you use small $\left(\frac{1}{2}\,\text{cm}\right)$ squared grids or graph paper, the children can make more intricate patterns.

Interesting mistakes

When reflecting shapes, children sometimes do not appreciate that each point of the image must be the same distance away from the mirror line in a perpendicular direction; they draw a copy of the shape on the other side of the line without thinking about the orientation of the image. Using a mirror to check their work can help prevent this.

Answers for Pupil book page 77

1 A and B, E and I, D and J, G and H, C and K, L and M, F and N, A and L, B and M, C and F

2 Individual answers.

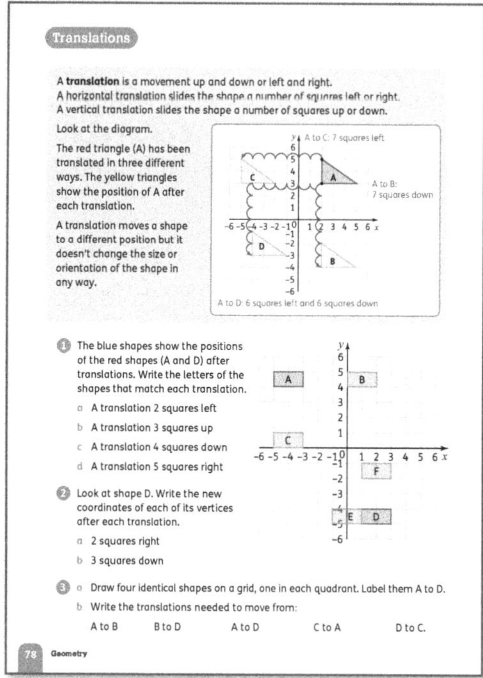

Materials

Large coordinate grid for display; coordinate grids for the children to work on; squared paper; rulers; coloured pens.

Warm-up

Use any 'Place value and number sense' activity (pages 22–23) as a starter for this lesson.

Focus

- Display a large grid. Draw a shape on the grid and then perform a translation on it and draw the new shape.
- Ask the children to describe the transformation.
- Draw another shape on the grid and translate the same shape at a different position.
- Ask the children to take it in turns to describe different ways of moving the original shape until it lies on top of the second shape. What is the fewest number of moves they can make?
- Draw a shape on the grid and let one child give instructions on where to move it to, while another child follows those instructions.
- Revise the term 'translation' as a movement up or down and left or right. Explain that translations are also called slides because the shape moves without changing its orientation in any way.
- Turn to **Pupil book page 78**. Read through the information with the class while they follow on the grid. Use letters to refer to the shapes so that it is clear which ones you are talking about. Ask the children to describe the translation from A to B and A to C as well as the reverse translations.

- For question 1, explain that both A and D are the original (red) shapes. The others show translations of either of these shapes.
- For question 2, allow the children to make a small shape and move it on the grid if necessary.
- For question 3, hand out a grid with four quadrants for the children to work on. They can work on their own to draw the shapes, and then compare their answers with a partner's.

Challenge
- Let the children investigate how coordinates change with different translations and try to find patterns that will allow them to work out the coordinates without moving or drawing the shapes. (For example, when a shape moves one square to the right all the x-coordinates increase by one. The y-coordinates stay the same.)

Support
- Ask pairs of children to sit opposite each other with a screen, such as an open file, between them. Give each child squared paper.
- Each child must draw a shape on their paper and translate it to a new position.
- The children take turns to describe to their partner where to draw the shape and how to move it to its new position.
- Then they each reveal their work to check that their shape matches their partner's.

Interesting mistakes
- Children might try to describe translations as diagonal moves.
- Use the concept of a right-angled triangle to show them that a move down and left (or up and right) along the sides adjacent to the right angle will take them to the same spot as a move along the hypotenuse.
- Next, explain that we have to describe the moves using only the 'horizontal' and 'vertical' shift. (This is important for later work with matrices and vectors.)

Answers for Pupil book page 78
1   **a** D to E   **b** D to F
    c A to C   **d** A to B
2   **a** (3, −4), (5, −4), (5, −5), (3, − 5)
    b (1, −7), (3, −7), (3, −8), (1, − 8)
3   **a, b** Individual answers.

Rotations

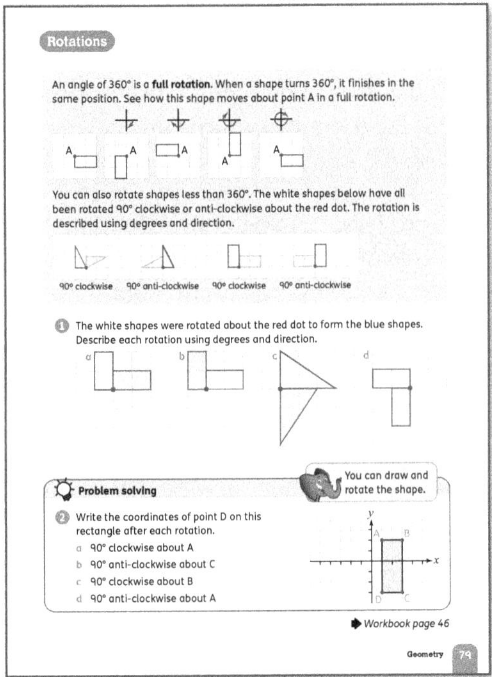

Materials
Tacks or push pins; small cards; tracing paper.

Warm-up
Use any of the 'Mental problem-solving' activities (pages 23–24) as a starter for this lesson.

Focus
- The children have not worked with 'rotation' previously. They need to understand that a rotation is a turn.
- Use cut-out shapes and push pins to demonstrate how a shape can be rotated (multiples of 90° only) around a vertex.
- Draw a four-point compass on the board and use it to remind the children that a full turn ('full rotation') is 360°, a half turn or straight line is 180° and a quarter turn is 90°. They should realise that a turn can be clockwise or anti-clockwise.
- Let the children each draw a shape on the small cards. Show them how to stick the push pin through one of the vertices of the shape.
- Next, do some practical rotation activities. For example, say:
  - *Turn your shape a quarter of a turn clockwise.*
  - *Turn the shape 180° clockwise.* (Use this to show them that a 180° turn clockwise or anti-clockwise ends up in the same position.)
- Explain that we call a turn a rotation and that we need to describe it correctly using the direction of the turn and the amount of turn (in degrees).

- Turn to **Pupil book page 79**. Let the children read through the examples in pairs and make the rotations with their own shapes if necessary.
- If the children need additional support to visualise rotations, encourage them to trace and rotate the shapes to work out the answers in question 1.
- <u>Problem solving:</u> For question 2, the children can make a rectangle and rotate it on their own grids to work out the answers, or they could draw the rotations on a copy of the grid.

Follow-up
- Use **Workbook page 46** to consolidate the work on rotations.
- Hold a class discussion to share ideas on how to make rotation work easier.

Answers for Pupil book page 79

1 a 90° clockwise or 270° anticlockwise
   b 270° clockwise or 90° anticlockwise
   c 90° clockwise or 270° anticlockwise
   d 270° clockwise or 90° anticlockwise

2 a (−4, 2)   b (3, −5)
   c (−2, 4)   d (6, 2)

Answers for Workbook page 46

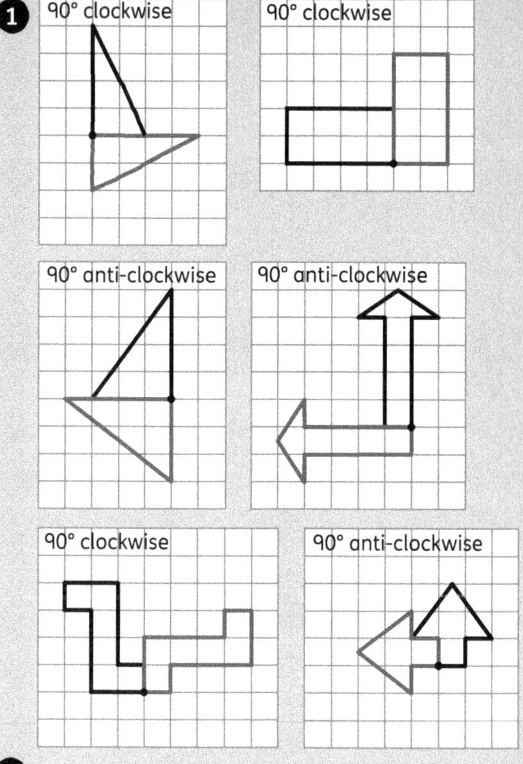

1 90° clockwise / 90° clockwise / 90° anti-clockwise / 90° anti-clockwise / 90° clockwise / 90° anti-clockwise

2 e.g. First, rotate only one side of the shape, then work out where to draw the remaining sides.

Find matching shapes

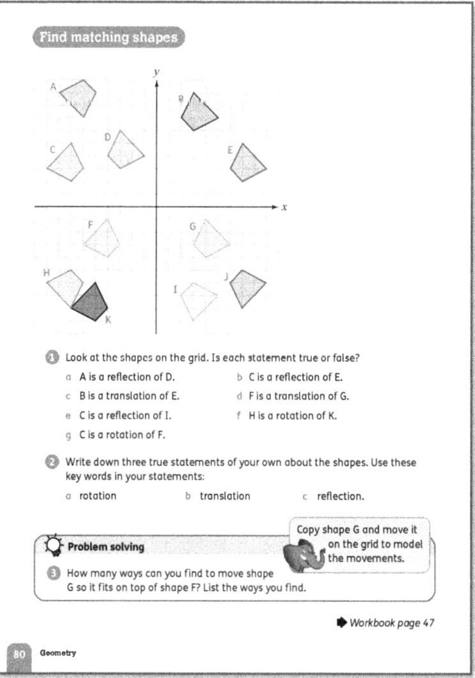

Materials
Coordinate grid with a shape on it for display; card; scissors; rulers.

Warm-up
Use any 'Place value and number sense' activity (pages 22–23) as a starter for this lesson.

Focus
- Begin with a number talk (pages 17–18). Display a shape on a grid for the class and ask them what the shape will look like if it is reflected.
- Guide the class to recognise that if the shape is reflected it will be a mirror image of the original shape.
- Repeat this for translations and make sure the children recognise that in a translation the shape is in the same orientation but it's just moved to a different position. Similarly, for a rotation, guide them to recognise that the shape might be lying in a different orientation, but that it may not be a mirror image.
- Discuss how this information can help them work out how a shape was transformed on a grid. Let the children share their ideas, demonstrating with shapes on the grid if necessary.
- Use **Pupil book page 80** to check that the children are able to work with combinations of reflections, translations and rotations of shapes. They can work independently or in pairs. Observe the children as they work and assist them as necessary.

- For questions 1 and 2, remind the children that they can make a small copy of the shape and move it on the grid if they are not sure of the answers.
- <u>Problem solving</u>: For question 3, discuss with the children how they will describe each method of movement. Let them share their ideas and agree a format for each kind of transformation, for example:
  - If we are using a reflection, we will say what line is the mirror line.
  - If we are using a translation, we will give the number of squares and the direction of each movement.
  - If we use a rotation, we will give the direction and number of degrees.

Follow-up

Use **Workbook page 47** for additional practice and to consolidate transforming shapes and drawing them on coordinate grids.

Challenge

- Give the children a grid with the shapes of a tangram (A to G) drawn in the first two quadrants (see below).
- Ask the children to transform the given shapes in different ways to make one of the pictures below in the space in the centre of the grid.
- Next, they should list the shapes, and for each write the transformation or combination of transformations they used to move it into position.

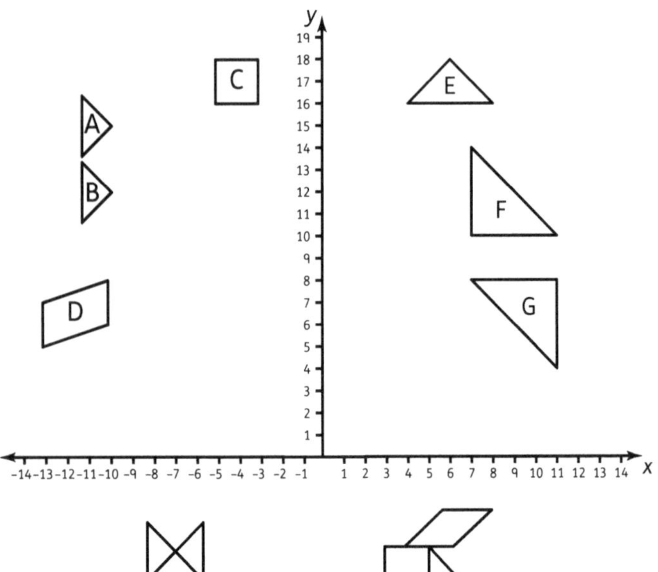

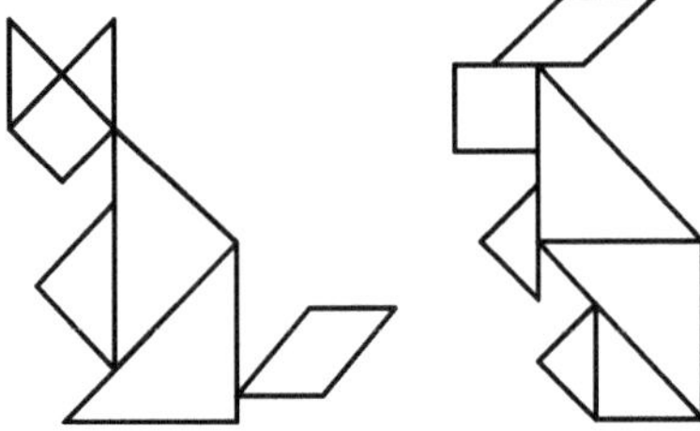

Support

Ask the children to write down a true statement about each shape on the grid. They should use one of the terms (rotation, translation or reflection) in each statement. Let them compare and check each other's work.

Interesting mistakes

The children need to be familiar with a lot of terminology in this area of maths and will sometimes confuse the names of transformations. Give the children plenty of practice in describing movements using the correct words.

Answers for Pupil book page 80

1   **a** false   **b** true   **c** true
    d false   **e** true   **f** true
    g false

2   **a, b, c** Individual answers.

3   For example: by reflection in the y-axis; by rotation about the origin 90° anticlockwise then translation 7 squares to the left and 7 squares down; by reflection in the x-axis then rotation about the origin 180°; by reflection in the x-axis, then reflection in the y-axis, then reflection in the x-axis; by rotation about the origin 180° then reflection in the x-axis.

Answers for Workbook page 47

1   **a**

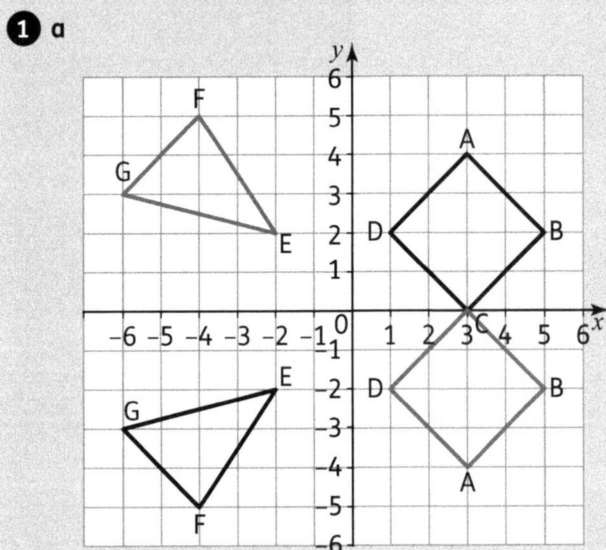

b

	Before reflection	After reflection
A	(3, 4)	(3, −4)
B	(5, 2)	(5, −2)
C	(3, 0)	(3, 0)
D	(1, 2)	(1, −2)
E	(−2, −2)	(−2, 2)
F	(−4, −5)	(−4, 5)
G	(−6, −3)	(−6, 3)

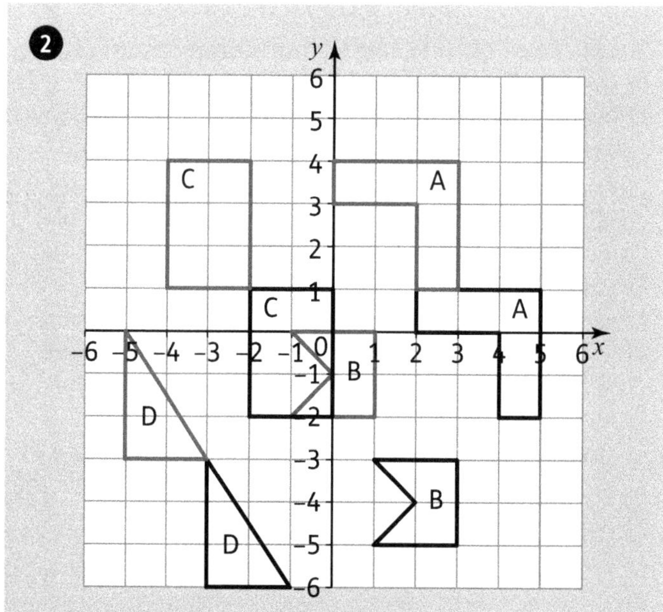

Work with coordinates

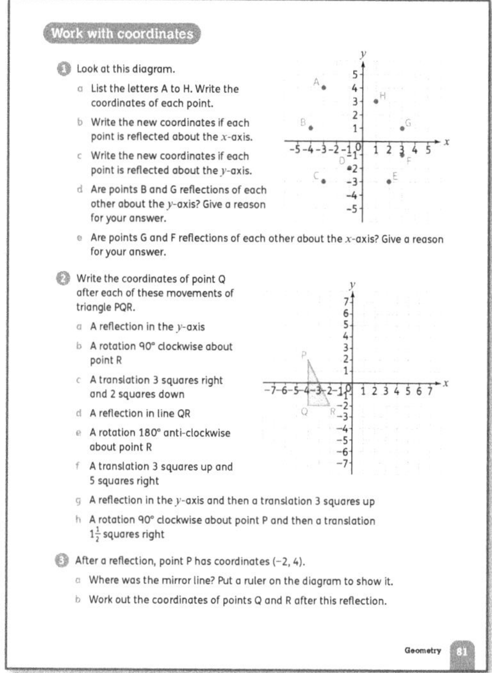

Materials
Card; scissors; grids.

Warm-up
Use any 'Mental problem-solving' activity (pages 23–24) as a starter for this lesson.

Focus
- This lesson combines what the children have learnt about coordinates with what they have learnt about transformations.
- The children can work independently to complete the questions on **Pupil book page 81**. Observe them as they work, assisting as necessary. For example, use card to model the shapes on the grids.
- Check the children's work as a class. Discuss any mistakes and address questions from the children.

Challenge
- Ask the children to look at the table of coordinates on **Workbook page 47** question 1b and try to write some rules for reflecting points (including vertices of shapes) about the axes.
- Ask the children to share their ideas, explaining why they work.

Answers for Pupil book page 81

1 **a** A = (−3, 4); B = (−4, 1); C = (−3, −3);
D = (−1, −2); E = (2, −3); F = (3, −1);
G = (3, 1); H = (1, 3)

b A′ = (−3, −4); B′ = (−4, −1); C′ = (−3, 3);
D′ = (−1, 2); E′ = (2, 3); F′ = (3, 1);
G′ = (3, −1); H′ = (1, −3)

c A″ = (3, 4); B″ = (4, 1); C″ = (3, −3); D″ = (1, −2);
E″ = (−2, −3); F″ = (−3, −1);
G″ = (−3, 1); H″ = (−1, 3)

d No. Point B is 4 squares from the y-axis and the point G is 3 squares from the y-axis.

e Yes. They are at the same distance from the x-axis.

2 **a** (4, −2)   **b** (−2, 0)   **c** (−1, −4)
d (−4, −2)   **e** (0, −2)   **f** (1, 1)
g (4, 1)   **h** (−6$\frac{1}{2}$, 2)

3 **a**

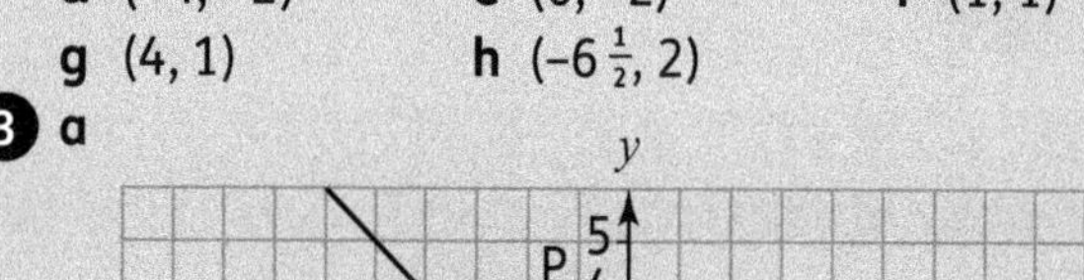

b Q(2, 4) and R(2, 2)

End-of-unit check

Use all or some of these questions and problems to check how well the children have understood the concepts in this unit.
- *What do we call the horizontal line in a pair of axes?* (x-axis)
- *What do we call the vertical line in a pair of axes?* (y-axis)
- *At which number do the axes intersect?* (Clarify that 'intersect' means 'cross', if necessary.) (0) *What do we call this point?* (the origin)
- *Find (0, 0) on this grid.*
- *Show where point (4, 5) is on your grid.*
- *Show a grid with points in all four quadrants. What is at (give coordinates of a point)?*
- Show a shape and a reflection/translation/rotation of the shape on a grid. *Look at the two positions of this shape. Describe the reflection/translation/rotation of the shape from position 1 to position 2.*

- Show a shape on a grid. *Move the shape following these instructions.* Give instructions such as: *Move 2 squares right and 3 squares up. Rotate the shape 90° anticlockwise. Reflect the shape in the x-axis.*
- Draw a shape and a mirror line on a squared grid. *Where will this shape be if you reflect it about this line? Why?* (Each point on the shape is the other side of the mirror line and the same distance from it.)
- Show a grid with a number of shapes on it (like the one on **Pupil book page 80**). *Which shape is a reflection/translation/rotation of shape A?*
- Show a shape in the third quadrant. *What are the coordinates of each vertex after reflection in the x-axis/y-axis?*
- Show two shapes on a grid where one is larger than the other. *How do you know these shapes are not reflections/translations/rotations of each other?* (When a shape is reflected, translated or rotated, the shape does not change.)
- Display the statement: 'When you reflect a shape, all the coordinates of the vertices will be different from those of the original.' *Draw an example to show that this statement is not always true.* (for example: any shape and mirror line where one or more of the vertices of the shape lie on the mirror line)
- *A square has its vertices at (–2, 1), (2, 1), (2, –3) and (–2, –3). Without drawing it or looking at a grid, give the coordinates of a point inside the square or on one of the sides.* (x-coordinate between –2 and 2, y-coordinate between –3 and 1) *Will the origin be inside the square?* (Yes.) *Will point (–3, 1) be inside the square?* (No.)

UNIT 8 Decimals

Learning objectives

- Identify the value of each digit in numbers given to 3 decimal places.

- Multiply and divide numbers by 10, 100 and 1000, giving the answers to 3 decimal places.

- Compose, decompose and regroup decimals using standard and non-standard partitioning.

- Multiply 1-digit numbers with up to 2 decimal places by whole numbers.

- Use written division methods in cases where the answer has up to 2 decimal places.

- Round decimals to the nearest tenth or hundredth.

- Solve problems which require answers to be rounded to specified degrees of accuracy.

- Recall and use equivalences between simple fractions, and decimals, including in different contexts.

- Understand that a fraction can be represented as a division of the numerator by the denominator and calculate decimal fraction equivalents for a simple fraction.

Key words

decimal point   thousandths

Unit introduction

Materials
Food labels (examples shown below); calculators.

Teaching guidance
Show the class some food labels like these:

Vitamin Boost
Per 5 g sachet

Vitamin C	500.0 mg
Glutathione	0.5 mg
Vitamin B	2.0 mg
ATP	0.5 mg
Calcium	100.0 mg

Also contains:

| Sucrose | 2.82 g |
| Lactose | 0.60 g |

Beans in tomato sauce
(per 100 g)

Carbohydrates		12 g
	Sugar	1.5 g
Total fat		0.6 g
	Saturated	0.1 g
	Transfats	<0.1 g
	Monounsaturated	<0.1 g
	Polyunsaturated	0.3 g
	Cholesterol	<1 mg
Fibre		4.49 g
Sodium		126 mg

- Explain that most countries have strict laws about food safety, including products such as vitamins and other supplements. Manufacturers have to state exactly what is in their products and this information is included on the label. Food technologists work out how much of each ingredient there is in different kinds of foods and they use decimals to give measurements in fraction of grams, milligrams and even micrograms.
- Ask the class why they think measurements are given in decimals rather than fractional amounts such as $5\frac{1}{2}$ grams or $3\frac{4}{5}$ grams.
- Let the children share their ideas, guiding them to the realisation that most scientific or technical measurements use decimals, and that, because most scientists use computers and calculators, it makes more sense to use the decimal forms. Revise the term 'decimal point'.
- Look together at the vitamin sachet label. Ask the children to discuss the ingredients and ask:
  - *What makes up the greatest part of the 5 gram sachet?* (sucrose)
  - *What fraction of a milligram are the ingredients rounded to?* (one tenth)
  - *What does it mean if an amount has .0 as the decimal?* (It means the figure is rounded.)
  - *What fraction of a gram are the ingredients sucrose and lactose rounded to?* (hundredths of a gram)
- Discuss why they use different units. (It could be because, for example, 2.82 g would be 2820 mg and consumers would realise that there is almost 6 times as much sugar as vitamin C.)
- Use the beans in a tomato sauce label in the same way. Tell the children that the label gives the information for 100 g of beans in sauce but that one serving is 80 g. Let them discuss how much of each ingredient there would be in an 80 g serving. Allow them to use calculators to work this out.

Decimal place value

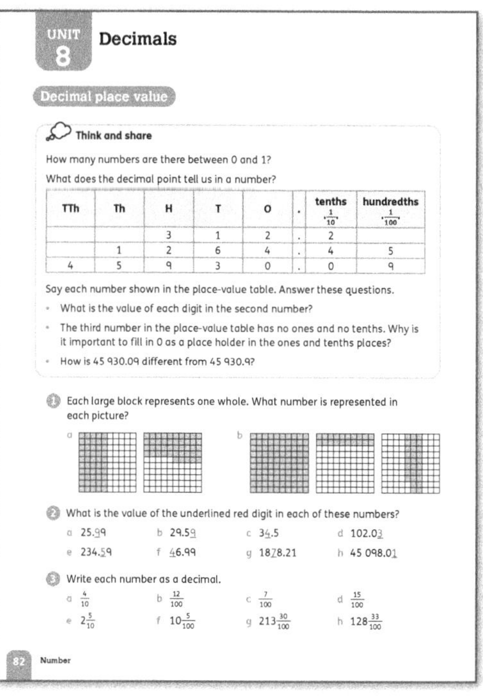

Materials
0–9 digit cards.

Warm-up
- As a mental starter for this lesson, draw a row of six boxes on the board, putting a decimal point before the last two boxes. Ask the children to make a copy of this.
- Explain that you are going to pick a card from a set of 0–9 digit cards and then return it to the pack and shuffle the cards before picking again.
- For each number you pick, the children must select a box to put the number in. Once positioned a digit cannot be moved.
- The aim is to create the largest number when all six boxes have been filled.
- After one round, discuss with the children the strategies they used to make decisions about where to place numbers.
- Repeat a few times, and then do the same activity trying to make the smallest number. Discuss how the strategies change when you have to make a small number.
- Next, ask the children to define a decimal and refine their suggestions to get a definition that works for the class.

Focus
- Think and share: Turn to **Pupil book page 82.** Use the Think and share activity to have a number talk (pages 17–18). Make sure the children realise that there is an infinite number of numbers between 0 and 1. Let them share their ideas about place value and then answer the questions as a class. (In the second number, the digits have place value 1 thousand, 2 hundreds, 6 tens, 4 ones, 4 tenths and 5 hundredths. If the third number did not have 0 as a place holder in the ones and tenths places, you would think the the 9 represented 9 ones instead of 9 hundredths. The number 45 930.09 has 0 tenths and 9 hundredths whereas the number 45 930.9 has 9 tenths and 0 hundredths.)
- Let the children complete questions 1–3 independently or in pairs. Remind the children that this is not new work as they have worked with decimal fractions in previous years.

Interesting mistakes
- There are many common misconceptions surrounding the concept of decimals. Some children see the numbers after the decimal point as a mirror of those before, and thus think of 34.56 as 'thirty-four point fifty-six'.
- This leads to incorrect comparison of numbers so that, for example, if the children were to compare 34.56 with 34.7, they would think that 34.56 is larger, as 'fifty-six is bigger than seven'.
- To avoid this, it is important to use place-value tables (page 21) or notation cards to emphasise the value of digits in a number and regularly practise how to say numbers correctly.

- Another common error relates to sequences of decimals, for example '0.6, 0.7, …'. When reaching 0.9 children often follow with 0.10, their line of thinking being 'nine tenths, then ten tenths'.
- Time needs to be spent on what is meant by ten hundredths; diagrams often help to clarify this. Using a place-value table can show how 0.1 and 0.10 are the same.

Answers for Pupil book page 82

1   a   0.5 and 0.45   b   1, 0.2 and 0.28

2   a   9 tenths   b   9 hundredths
    c   4 ones   d   3 hundredths
    e   5 tenths   f   4 tens
    g   7 tens   h   1 hundredth

3   a   0.4   b   0.12
    c   0.07   d   0.15
    e   2.5   f   10.05
    g   213.3   h   128.33

Thousandths

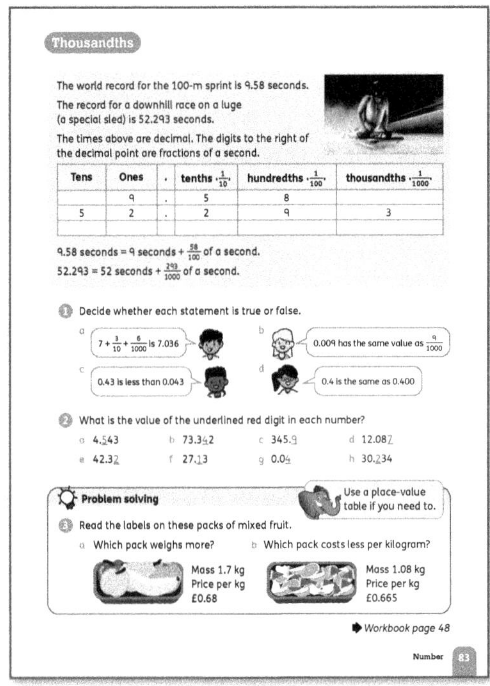

Material
Stopwatch or timer with split seconds; 1000 square (see below); number lines; place-value tables (page 21); counters.

Warm-up
Do some multiplication and division by 10 and 100 activities as a mental warm-up for this lesson. Include questions such as: *What is 10 times greater than … ? What is 100 times smaller than … ?*

Focus
- The children have already worked with decimals to thousandths. Use **Workbook page 48** as a baseline assessment task to see whether the children are able to decompose and represent decimals with 3 decimal places.
- Observe the children as they work to identify any areas that are problematic.
- If you feel that the children need to revise place value to thousandths show them a 1000 chart with some squares shaded. In this example $\frac{376}{1000}$ square are shaded.

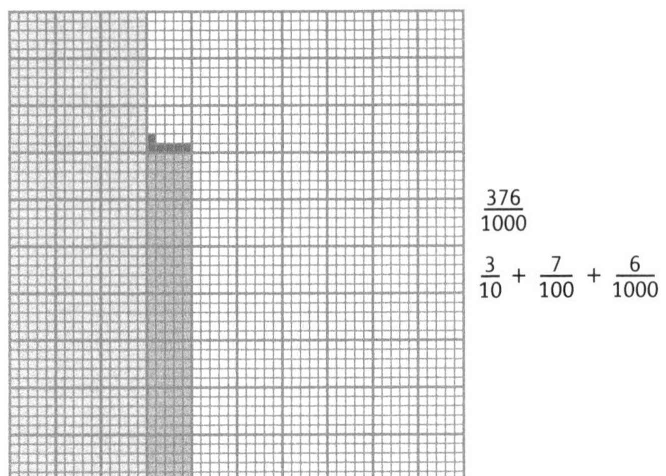

- Point out that $\frac{376}{1000} = \frac{3}{10}$ (the long strips) $+ \frac{7}{100}$ (the squares) and $\frac{6}{1000}$ (the smallest squares). We can write this as 0.376.
- Point to the grid and show that 0.376 is more than $\frac{3}{10}$ but less than $\frac{4}{10}$. This means it is between 0.3 and 0.4 on a number line.

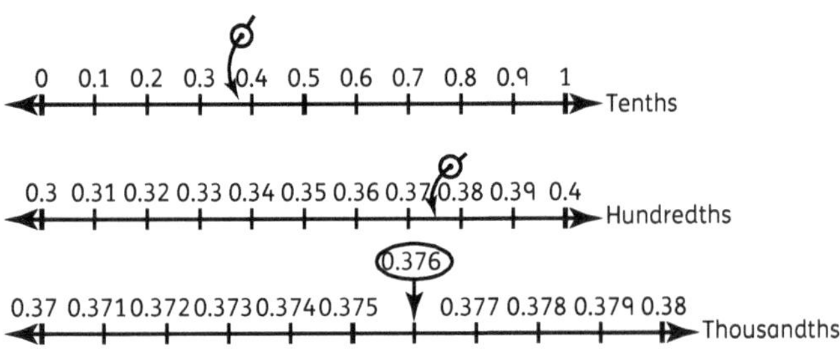

- If we zoom in, we get the second number line that shows the divisions between 0.3 and 0.4. Show the class that 0.376 is between 0.37 and 0.38.
- If we zoom in further, we get the third number line that shows the divisions between 0.37 and 0.38. We can show the exact position of 0.376 on this number line.
- Do a different example with the class if necessary and ask them to expand the number on the number lines.
- Turn to **Pupil book page 83**. Remind the class that we can use decimals to give times to a fraction of a second. Read aloud the times with decimal points and use a timer to measure 9.58 seconds and 52.293 seconds.
- Ask the children to suggest other ways of showing each number and let them come to the front of the class and show their ideas (for example, on a 1000 chart or number line).

- The children can do question 1 orally. Ask them to explain their decisions.
- The children should do question 2 independently in writing.
- <u>Problem solving:</u> In question 3, the children can use place-value tables or number lines to compare the two masses.

Support
- Let the children physically model the decimals using counters and place-value tables and use these to draw their representations.
- Decomposing and recomposing the numbers will help them to gain a better understanding of place value to thousandths.

Challenge
Give the children some pairs of decimals, for example: 3.4 and 3.5 or 10.25 and 10.26. Ask them to place these on a number and then show five decimals with 3 decimal places that would be between each pair.

Answers for Pupil book 83

1 a false   b true
    c false   d true

2 a 5 tenths   b 4 hundredths
    c 9 tenths   d 7 thousandths
    e 2 hundredths   f 1 tenth
    g 4 hundredths   h 2 tenths

3 a 1.7 kg   b £0.665

Answers for Workbook page 48

1 a 2 tens, 6 ones, 4 tenths, 5 hundredths, 6 thousandths
    b 3 tens, 7 ones, 2 tenths, 0 hundredths, 6 thousandths
    c 3 hundreds, 8 tens, 9 ones, 4 tenths
    d 4 tens, 2 ones, 3 tenths, 2 hundredths
    e 2 tens, 8 ones, 9 tenths
    f 3 hundreds, 4 tens, 5 ones, 2 tenths, 3 hundredths, 4 thousandths

2 a   b

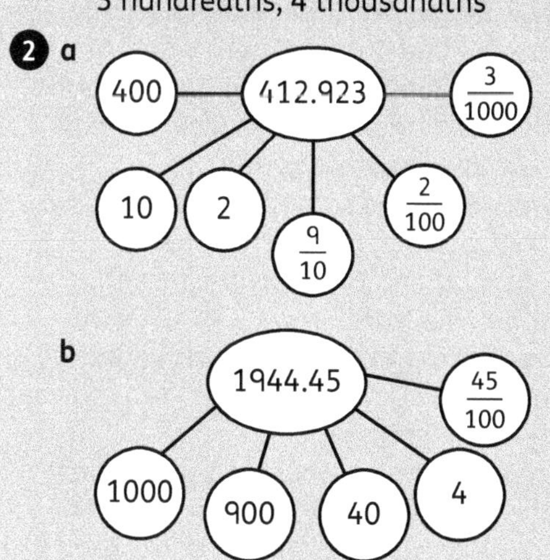

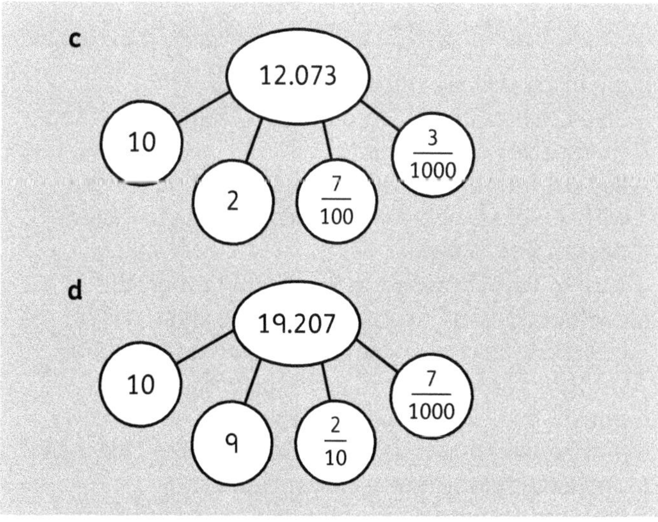

Compare and order decimals

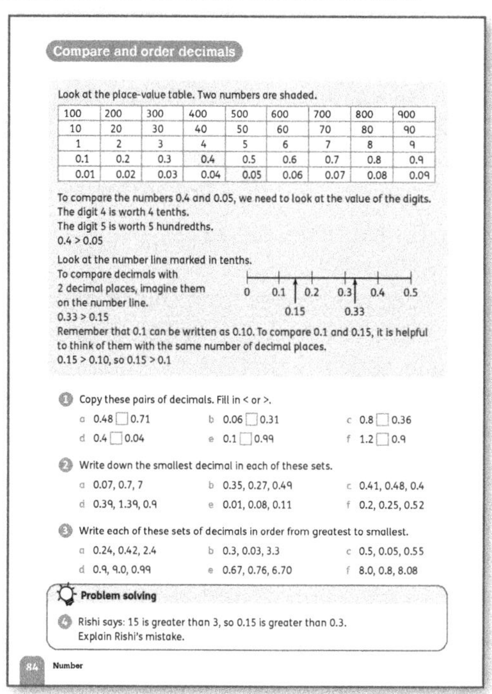

Materials
Flashcards, each showing a 2- or 3-place decimal.

Warm-up
Use 'Greater or smaller' (page 22) as a starter for this lesson. This will also help to revise the use of place value in comparing and ordering numbers.

Focus
- Remind the class that we compare numbers using the =, < and > symbols. Write statements on the board such as: '10.315 is more than 10.310'. Ask the children to say whether the statements are true or false and how they decided.
- Ask the class what it means to put numbers in ascending or descending order.
- Give each group of children a flashcard showing different decimals with 2 decimal places.

- Ask the other children to give the group instructions in order to put themselves in position so the cards are in ascending order.
- Give another child a card and ask them to position themselves on the line.
- Repeat with different decimals, including some with 3 places, and vary the instructions to include descending order.
- Turn to **Pupil book page 84**. Work through the explanation and examples with the class, asking questions to make sure the children understand.
- Let the children work in pairs to complete questions 1–3 but ask them to write the answers in their own books. Discuss the strategies they use and explore any interesting mistakes.
- <u>Problem solving:</u> Discuss the answer to question 4 as a class.

Interesting mistakes

- Children may misapply rules for working with whole numbers when they try to compare and order decimals. For example, they may say that 0.065 > 0.21 because 65 > 21 or that 3.06 > 3.5 because it has more digits.
- Similarly, they may misapply what they know about fractions. For example, because $\frac{1}{405} > \frac{1}{450}$, they may say that 0.405 > 0.450.
- Other mistakes may arise from thinking that zeros to the right of the decimal change its value, so, for example, they say that 0.3 < 0.300 because 3 < 300. Children may also mistakenly think that decimals with only tenths are greater than those with thousandths because 'tenths are bigger than thousandths'.
- If children make these mistakes have a number talk about them (pages 17–18). Ask the children to show visually how they arrived at their conclusions. The visual representation will often help them to see why they made the mistake.
- Shading the decimal part of a number on the 1000 grids will also help the children to understand these comparisons.

Answers for Pupil book page 84

1 a 0.48 < 0.71 b 0.06 < 0.31
  c 0.8 > 0.36 d 0.4 > 0.04
  e 0.1 < 0.99 f 1.2 > 0.9

2 a 0.07 b 0.27
  c 0.4 d 0.39
  e 0.01 f 0.2

3 a 2.4; 0.42; 0.24 b 3.3; 0.3; 0.03
  c 0.55; 0.5; 0.05 d 9.0; 0.99; 0.9
  e 6.70; 0.76; 0.67 f 8.08; 8.0; 0.8

4 $0.15 = \frac{15}{100}$; $0.3 = \frac{3}{10} = \frac{30}{100}$ so 0.3 > 0.15
  Rishi has not considered the place value of the digits.

Round decimals

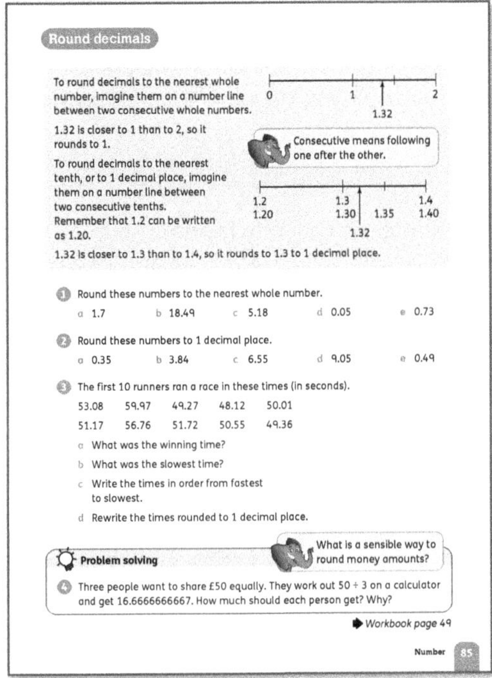

Materials
Place-value tables (page 21); highlighter pens.

Warm-up
Use any of the 'Rounding and estimating' activities (page 23) using whole numbers as a starter for this lesson.

Focus
- Revise rounding decimals with the class:
  - Display a number line from 0 to 5. Mark out the tenths between each whole number.
  - Remind the children about rounding to approximate numbers and how they know how to round whole numbers to the nearest 10 and 100. Explain that you are now going to look at rounding decimals in a similar way.
  - Mark 3.7 on the line. Ask the children which whole number this is nearest to. Tell them that this means that 3.7 rounded to the nearest whole number is 4. Repeat for other examples.
  - Mark 2.5 on the line and ask them what they think they should do. Remind the children that when we round whole numbers, if they fall halfway between two numbers, we round up.
  - Explain that we do the same with decimals, so 2.5 rounded to the nearest whole number is 3.
  - Now look at rounding to the nearest tenth. Use a number line from, say, 1.2 to 1.5, with the hundredths marked on. Ask the children to position numbers such as 1.38 and then say which tenth they are nearest to.
  - Emphasise that when rounding to the nearest tenth, if the hundredths digit is 5 or more, the tenth digit increases by 1, but if it is less than 5 the tenths digit stays the same.

- Ask the children to read through the explanation text and examples on **Pupil book page 85** to consolidate what they have just learnt. The children can work independently on questions 1 and 2.
- For question 3, read the problem with the class and ask the children how they would decide what the winning time was. Make sure they realise that the winner runs the fastest, so they run it in less time. In running terms, the slowest time is the highest decimal and not the lowest.
- <u>Problem solving:</u> Question 4 is a real-life problem that will require some discussion. The children may decide to split the amounts slightly differently, but they must recognise that you cannot give people fractions of a single penny (or cent).

Follow-up
- Use **Workbook page 49** to assess how well the children are able to round numbers to a specified number of places.
- Check the answers with the class and address any misconceptions.

Challenge
- Give the children a set of decimals and a set of clues for identifying one of them. Once they have found the decimal, let them write their own clues for three others. The children can exchange these to find each other's decimals. For example:

8.062	8.861	8.820	8.723
8.702	8.828	8.559	8.653

- **Clues**
  - *I have more than 5 thousandths.*
  - *I have an odd number of tenths.*
  - *If rounded to 2 decimal places, I round up*
  - *Rounded to the nearest tenth I will be 0.1 less than 8.7.*
  - *What number am I?*
  (8.559)

Support
- Encourage the children to use a highlighter to mark or underline the digit in the place they are rounding to. They can write the numbers on a place-value table if they need help to do this. Then focus them only on the digit to the right.
- Be aware that when using the term 'round down', some children may think they have to reduce the digit by 1. You may prefer to say *If the digit is 5 or more, round up, if not, round off the number.*

Interesting mistakes
- Some children might get confused when asked to round to tenths and round to the nearest 10 instead. For example, they may round 36.34 to 36 rather than 36.3. If children do this, continue to show numbers on a place-value table or encourage them to write the place value above the digits.

- Children may also round up to the nearest tenth correctly but write the digits to the right when they write the answer. So, for example, they round 3.467 to 3.567 instead of 3.5.
- Another interesting mistake is when children round back from the 'end' of the number and round to hundredths before they round to tenths. This may leave them with the right answer, but it can also make the answer incorrect. For example, 3.447 would round to 3.45 and then to 3.5 using this method (which is wrong). But 3.437 would round to 3.44 and then to 3.4.
- For any of these errors, hold a number talk (pages 17–18) about what the children did and why.
- Showing the decimals on 'zoomed' number lines can help the children to see where the number lies and what it is closest to.

Answers for Pupil book page 85

1. a 2   b 18   c 5
   d 0   e 1
2. a 0.4   b 3.8   c 6.6
   d 9.1   e 0.5
3. a 48.12   b 59.97
   c 48.12; 49.27; 49.369; 50.01; 50.55; 51.17; 51.72; 53.08; 56.76; 59.97
   d 53.1; 60.0; 49.3; 48.1; 50.0; 51.2; 56.8; 51.7; 50.6; 49.4
4. £16.66 with a remainder of £0.02. It could not be £16.67 as multiplied by 3 gives £50.01 which is more money that they have to share.

Answers for Workbook page 49

Number	Rounded to nearest hundredth	Rounded to nearest tenth	Rounded to nearest whole number
12.452	12.45	12.5	12
23.976	23.98	24.0	24
14.299	14.30	14.3	14
45.004	45.00	45.0	45
93.901	93.90	93.9	94
116.667	116.67	116.7	117
0.9823	0.98	1.0	1
1.0457	1.05	1.0	1

2. a 1   b 3   c 4
   d 13   e 26
3. a 99.999   b 98.765
   c 12.345   d 50.499

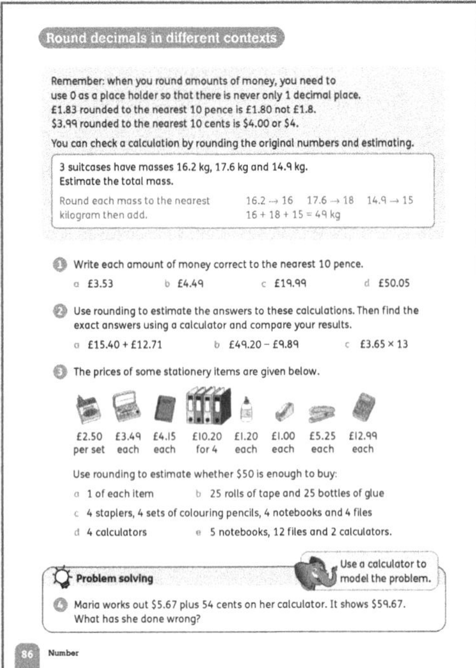

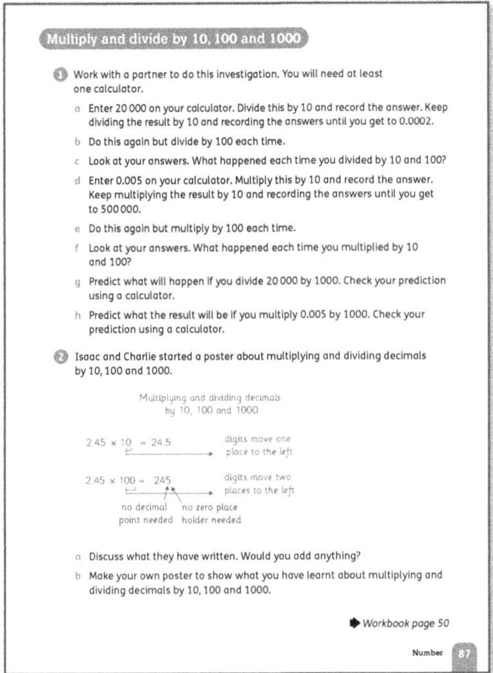

Round decimals in different contexts

Materials
Calculators.

Warm-up
Use any 'Rounding and estimating' activity using whole numbers (page 23) as a starter for this lesson.

Focus
- This lesson encourages the children to see how and where we use rounding of decimals to estimate in our everyday lives.
- Ask the children to turn to **Pupil book page 86**.
- Let them read through the information and example in pairs, and then ask them to work through questions 1–3 independently.
- <u>Problem solving:</u> The children can do question 4 orally in pairs or small groups.

Answers for Pupil book page 86
1. a £3.50 b £4.50
   c £20.00 d £50.10
2. a £28, £28.11 b £40, £39.31 c £52, £47.45
3. a $40. So, $50 is enough.
   b $50. This is an underestimate so $50 is not enough.
   c $58. So, $50 is not enough.
   d $52. So, $50 is not enough.
   e $76. So, $50 is not enough.
4. She added $5.67 and $54 instead of $5.67 and $0.54 which gives $6.21.

Multiply and divide by 10, 100 and 1000

Materials
A set of cards with decimals (including 1-, 2- and 3-place decimals); calculators; large sheets of paper; coloured pens; place-value tables (page 21).

Warm-up
Write a whole number on the board. Ask a child to multiply it by 10, 100 or 1000. Go round the class, asking children to multiply or divide the previous result by 10, 100 or 1000 (whole number answers only).

Focus
- Invite a child to pick a decimal card and show it to the class. The children must find the product of the decimal and 10/100/1000.
- Compare answers and, if there is disagreement, ask the children to explain how they found the answer and why they think it is correct. Through this discussion establish the correct solution.
- Repeat for some division activities.
- Use **Pupil book page 87** to investigate multiplying and dividing decimals by 10, 100 and 1000.
- For question 1, let the children complete the calculator investigation in pairs. Talk about their results as a class.
- Encourage the children to explain the effects of these with reference to place value.
- For question 2, the children can work in pairs to make the poster.

Follow-up
- Let the children complete the practice sets in **Workbook page 50** over five days to consolidate their ideas and build their confidence.
- Display the completed posters in the classroom.

Challenge

- Ask the children to make a list of ten multiplications and/or divisions by powers of 10 with some sneaky mistakes in them. Ask them to swap their list with a partner who has to try to find the mistakes.

Support

- Give the children a set of numbers, such as 1800, 95, 210.
- Ask them to work in pairs to write two multiplications and two divisions by 10, 100 or 1000 that will give each answer.
- Start with whole numbers and work to a decimal with one place, for example:
  - $1.8 = 180 \div 100$, $1.8 = 18 \div 10$, $1.8 = 0.18 \times 10$, $1.8 = 0.018 \times 100$
- The children can discuss the problem, use place-value tables and check the answers using a calculator.

Answers for Pupil book page 87

 **1 a** 20 000; 2000; 200; 20; 2; 0.2; 0.02; 0.002; 0.0002

b 20 000; 200; 2; 0.02; 0.0002

c When dividing by 10 the digits move one place to the right; when dividing by 100 the digits move two places to the right.

d 0.005; 0.05; 0.5; 5; 50; 500; 5000; 50 000; 500 000

e 0.005; 0.5; 50; 5000; 500 000

f When multiplying by 10 the digits move one place to the left; when multiplying by 100 the digits move two places to the left.

g 20   **h** 5

2 a, b Individual answers.

Answers for Workbook page 50

1

One-week mental practice				
Day 1	Day 2	Day 3	Day 4	Day 5
34	35.44	23	13	27
2	2.544	45.5	1300	37
122	2544	93.32	26	370
142	2450	9810	260	34.5
2.4	24.5	100.1	0.5	1 965 000
20.4	2450	230	6	0.23
50	0.365	4500	1.25	0.23
79	36.5	9007	1.25	0.38
3	3.65	1230	12.5	127
0.02	36.5	210	4	19 340
123	365	240	400	250
424	26	15.46	0.012	1.3
0.7	5.25	0.0342	45	0.25
0.004	52.5	455.5	34.5	24.6
23.95	4300	4.52	123.45	9
12	4300	10.001	1.97	929
12.5	43	2.3	0.176	3260
1.25	4.3	4.56	1.23	3.42
24.5	0.43	2.345	1.945	34.2
245	0.0043	23.45	2.05	342 000

2 Individual answers.

More operations with 10, 100 and 1000

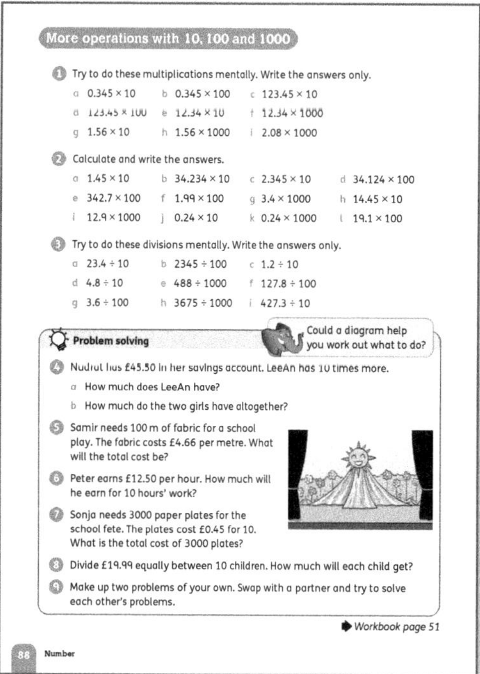

More operations with 10, 100 and 1000

1 Try to do these multiplications mentally. Write the answers only.

- a 0.345×10 b 0.345×100 c 123.45×10
- d 123.45×100 e 12.34×10 f 12.34×1000
- g 1.56×10 h 1.56×1000 i 2.08×1000

2 Calculate and write the answers.

- a 1.45×10 b 34.234×10 c 2.345×10 d 34.124×100
- e 342.7×100 f 1.99×100 g 3.4×1000 h 14.45×10
- i 12.9×1000 j 0.24×10 k 0.24×1000 l 19.1×100

3 Try to do these divisions mentally. Write the answers only.

- a $23.4 \div 10$ b $2345 \div 100$ c $1.2 \div 10$
- d $4.8 \div 10$ e $488 \div 1000$ f $127.8 \div 100$
- g $3.6 \div 100$ h $3675 \div 1000$ i $427.3 \div 10$

Problem solving

4 Nadirul has £45.50 in her savings account. LeeAn has 10 times more.
- a How much does LeeAn have?
- b How much do the two girls have altogether?

5 Samir needs 100 m of fabric for a school play. The fabric costs £4.66 per metre. What will the total cost be?

6 Peter earns £12.50 per hour. How much will he earn for 10 hours' work?

7 Sonja needs 3000 paper plates for the school fete. The plates cost £0.45 for 10. What is the total cost of 3000 plates?

8 Divide £19.99 equally between 10 children. How much will each child get?

9 Make up two problems of your own. Swap with a partner and try to solve each other's problems.

➜ *Workbook page 51*

88 Number

Materials

Calculators.

Warm-up

Use the activities on **Workbook page 51** as a mental starter for this lesson. Check the answers before moving on.

Focus

- Turn to **Pupil book page 88** and ask the children to complete questions 1–3 to consolidate mental multiplication and division by powers of 10.
- Check the answers with the class.
- <u>Problem solving:</u> Read through the word-problem questions 4–9 with the class. Ask the children what they need to do to solve each one and let them suggest whether a diagram would be helpful (and if so, let them draw one for the class).
- After you have discussed the problems, let the children work out the solutions independently.
- Encourage them to self-check and correct using a calculator as necessary.

Interesting mistakes

- When children have to multiply or divide decimals by a power of 10, they sometimes treat the whole number and decimal part separately, and multiply or divide them both separately. For example, they might say $4.3 \times 10 = 40.30$ or $80.4 \div 10 = 40.2$.
- If they do this, you will need to revisit place value and use a place-value table (page 21) to move digits to the left or right to show that the number is treated as one thing when you do these operations.

Answers for Pupil book page 88

Answers for Pupil book page 88

1 a 3.45   b 34.5   c 1234.5
d 12 345   e 123.4   f 12 340
g 15.6   h 1560   i 2080

2 a 14.5   b 342.34   c 23.45
d 3412.4   e 34 270   f 199
g 3400   h 144.5   i 12 900
j 2.4   k 240   l 1910

3 a 2.34   b 23.45   c 0.12
d 0.48   e 0.488   f 1.278
g 0.036   h 3.675   i 42.73

4 a £455   b £500.50

5 £466   **6** £125

7 £135   **8** £1.99

9 Individual answers.

Answers for Workbook page 51

1 220; 2820; 14 350; 147; 11 870; 2310; 80
2.2; 28.2; 143.5; 1.47; 118.7; 23.1; 0.8
220; 28.2; 14.35; 147; 11.87; 2310; 0.8
2200; 282; 143.5; 1470; 118.7; 23 100; 8
20.7; 3.1; 0.7; 1.9; 14.7; 208.1; 210
2.07; 0.31; 0.07; 0.19; 1.47; 20.81; 21

2 a × 10, ÷ 10, × 100, × 100, ÷ 100
b × 10, ÷ 100, × 100, ÷ 100, × 10
c ÷ 10, × 100, ÷ 100, × 100, ÷ 10
d Individual answers.

Multiply and divide decimals

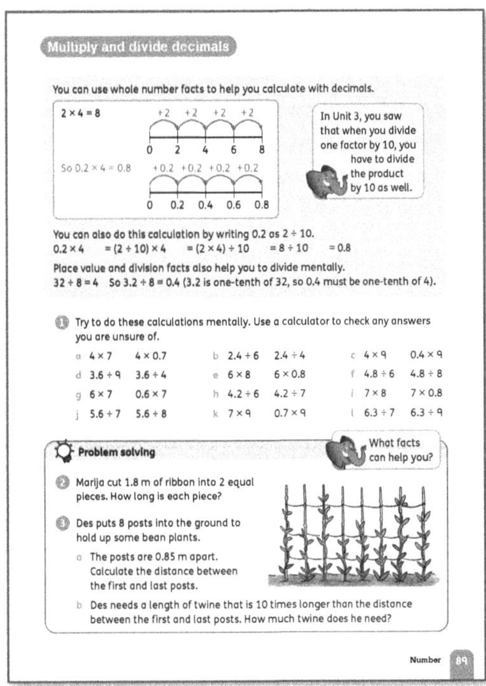

Materials

Calculators; drawing equipment.

Warm-up

Use any suitable 'Place value and number sense' activity (pages 22–23) as a starter for this lesson.

Focus

- Work through the example on **Pupil book page 89** to revise multiplying a decimal by a 1-digit number.
- Work through the first few calculations in question 1 as a class to reinforce the ideas. Let the children work independently to solve the rest of the calculations.
- <u>Problem solving:</u> For question 2, a simple bar model (page 21) will show the children what to do.
- <u>Problem solving:</u> For question 3, encourage the children to draw a sketch so they can see that even though there are 8 posts, there are only 7 spaces between them.

Interesting mistakes

- Children may make mistakes in calculations due to mental arithmetic errors, for example incorrect times tables or place-value errors, such as not putting the decimal point in the correct place.
- Encourage the children to estimate first and then to check their work to decide whether their answers are reasonable.

Answers for Pupil book page 89

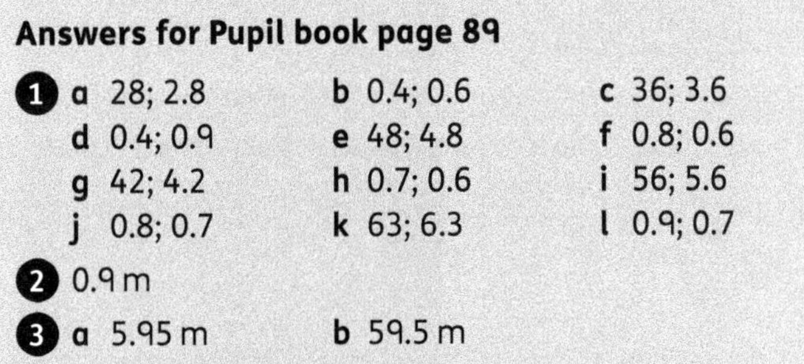

1 a 28; 2.8   b 0.4; 0.6   c 36; 3.6
d 0.4; 0.9   e 48; 4.8   f 0.8; 0.6
g 42; 4.2   h 0.7; 0.6   i 56; 5.6
j 0.8; 0.7   k 63; 6.3   l 0.9; 0.7

2 0.9 m

3 a 5.95 m   b 59.5 m

Fractions and decimal equivalents

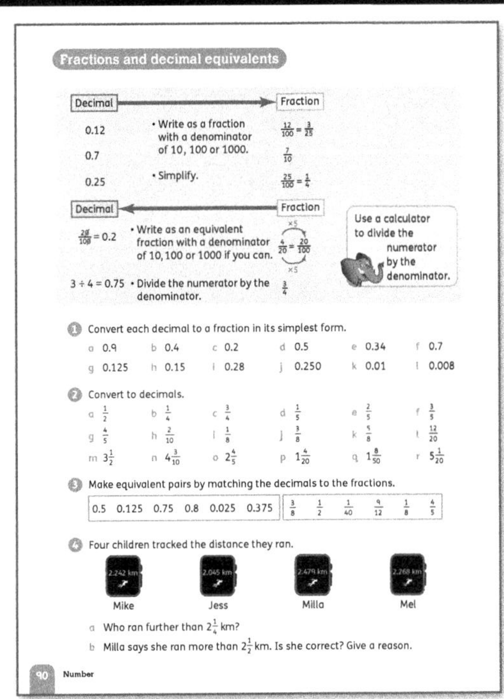

Materials

A set of ten objects with some the same (for example, 3 pencils, 2 markers and 5 pens); flashcards showing fractions and decimals less than 1; sets of cards with equivalent common fractions and decimals; calculators.

Warm-up

Write a fraction on the board and ask the children to work in groups to find 5 equivalent fractions. Share their answers – each one that no other group has found gains a point. Repeat with other fractions.

Focus

- Show the class the ten objects you have collected. Ask questions to relate the number of items to fractions and decimals, for example:
  - *There are 3 pencils, what fraction of objects are pencils?* $\left(\frac{3}{10}\right)$
  - *What is this as a decimal?* (0.3) *How do you know?*
  - *Is it correct to say that 0.2 of the objects are markers?* (Yes) *How do you know?*
  - *Five of the ten objects are pens. Is this the same as $\frac{1}{2}$ the objects?* (Yes) *How do you know?*
  - *What is this as a decimal?* (0.5)
- Hand out a mixed set of fraction and decimal flashcards to each group and ask them to order the cards from smallest to largest. Encourage the children to explain how they made their choices.
- Show the class a card $\left(\text{for example, } \frac{4}{20}\right)$ and ask them to express the amount on the card as a simplified fraction and a decimal.
- Work through the conversion examples on **Pupil book page 90** with the class.
- Once you are sure they understand the concepts, they can complete questions 1–4 independently.

Challenge

- Ask the children to develop their own game of dominoes using equivalent decimals and fractions. Give them time to discuss and plan their game, make the dominoes and play a game.
- Give groups a chance to explain their game to the class. Use these games to revise and consolidate the concepts taught.

Support

- Make sets of cards with equivalent common fractions and decimals. You will need one set per group of children. Use these to play a number of matching games such as 'Snap' where the children each get a number of cards which they play in turns. If the card played is equivalent to the one on the top of the pile, the child who says 'Snap' first is allowed to take all the cards on the pile.
- Alternatively, spread all the cards face down and the children take turns to turn over two cards. If the cards are the same, the child who turned them over may take the cards. If not, they turn them back over and the next child has a turn to find a matching pair.

Interesting mistakes

- Children need to realise that fractions can be changed to decimal fractions. To teach this, we rely on their understanding of equivalent fractions and division by 10 and 100. So, for example, to convert $\frac{1}{4}$ to a decimal, children can rewrite the fraction as $\frac{25}{100}$ and then divide by 100 to get 0.25.
- Once the children have understood this, they can move on to working with a calculator to do the conversions.
- In turn, this may raise issues of rounding because some fractions will produce recurring (and non-terminating) decimal places. A class discussion around this will help the children to make sense of their answers and decide what to do with them.

Answers for Pupil book page 90

1. a $\frac{9}{10}$ b $\frac{2}{5}$ c $\frac{1}{5}$
   d $\frac{1}{2}$ e $\frac{17}{50}$ f $\frac{7}{10}$
   g $\frac{1}{8}$ h $\frac{3}{20}$ i $\frac{7}{25}$
   j $\frac{1}{4}$ k $\frac{1}{100}$ l $\frac{1}{125}$
2. a 0.5 b 0.25 c 0.75
   d 0.2 e 0.4 f 0.6
   g 0.8 h 0.2 i 0.125
   j 0.375 k 0.625 l 0.6
   m 3.5 n 4.3 o 2.8
   p 1.2 q 1.16 r 5.05
3. $0.5 = \frac{1}{2}$; $0.25 = \frac{1}{4}$; $0.8 = \frac{4}{5}$; $0.025 = \frac{1}{40}$; $0.325 = \frac{3}{8}$
4. a Milla and Mel b No; 2.479 < 2.5

Add and subtract decimals

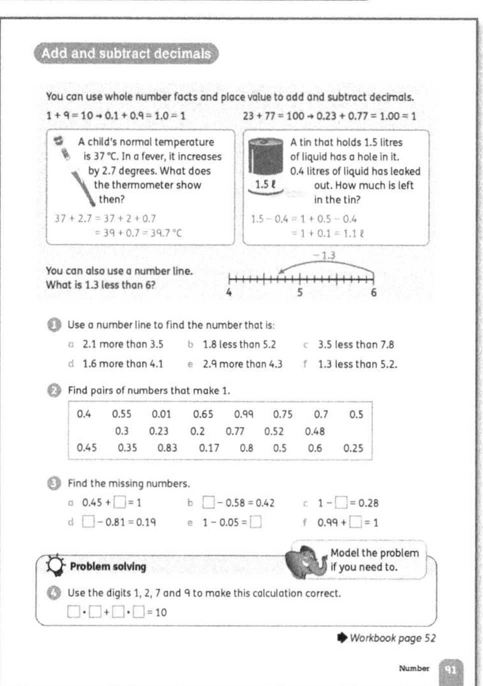

Materials

Measuring tapes; 10 by 10 squares drawn on squared paper; counters.

Warm-up

- The children should already know how to combine pairs of decimals to make 1 and 100 but use **Workbook page 52** to check that they remember.
- Let them compare and check each other's work.

Focus

- Use a measuring tape to demonstrate adding and subtracting any decimal amounts, for example:
  - Show a measurement of 0.8 m, and say: *I want a length that is 0.4 metres longer than this. What will that be?* (1.2 m)
  - Say: *This is 1.3 metres. What is 0.5 metres less than this?* (0.8 m)
- Relate your question to work with a number line, and let the children explain how they can find sums and differences of decimal amounts.
- Work through the different methods of adding and subtracting on **Pupil book page 91** with the class.
- The example box at the top of the page asks: What is 1.3 less than 6? (4.7)
- Discuss each method and encourage the children to suggest any other methods they think will work.
- Let the children use whichever methods they feel most comfortable with to answer questions 1–3.
- <u>Problem solving:</u> The children can make their own digit cards to help them solve question 4.

Support

Give each pair of children a 10 by 10 square drawn on squared paper and counters. Let one child cover some small squares with the counters while the other works out what fraction of the grid is covered and what fraction is uncovered. This will help them consolidate adding and subtracting hundredths.

Answers for Pupil book page 91

1   a   5.6   b   3.4   c   4.3
    d   5.7   e   7.2   f   3.9

2   0.4 and 0.6, 0.55 and 0.45, 0.01 and 0.99, 0.65 and 0.35, 0.75 and 0.25, 0.83 and 0.17, 0.5 and 0.5

3   a   0.55   b   1   c   0.72
    d   1   e   0.95   f   0.01

4   7.9 + 2.1 = 10

Answers for Workbook page 52

1   0.88 and 0.12; 0.65 and 0.35; 0.7 and 0.3; 0.1 and 0.9; 0.4 and 0.6; 0.99 and 0.01; 0.8 and 0.2; 0.72 and 0.28; 0.75 and 0.25; 0.45 and 0.55; 0.37 and 0.63

2   The missing numbers are:
    a   14, 14.8, 15, 15.4, 16, 16.3
    b   3, 0.7, 0.2, 0.7, 1.2, 0.7

3   0.7 + 0.8 = 1.5

4

7.5	+	8	→15.5
+		+	
2.9	+	1.6	→4.5

↓ 10.4   ↓ 9.6

7.4	+	8.1	→15.5
+		+	
3	+	1.5	→4.5

↓ 10.4   ↓ 9.6

7.7	+	7.8	→15.5
+		+	
2.7	+	1.8	→4.5

↓ 10.4   ↓ 9.6

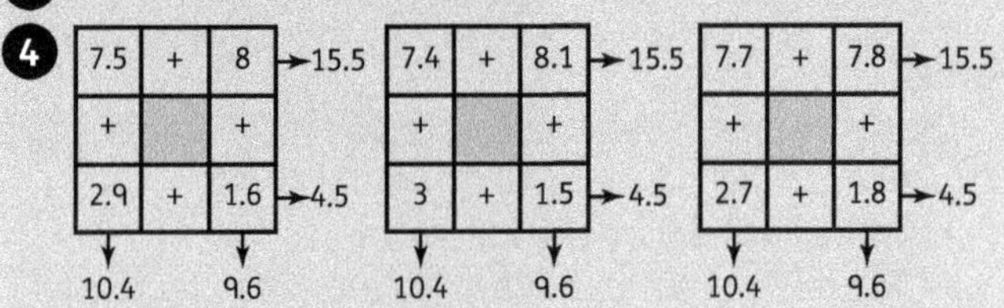

Written addition and subtraction with decimals

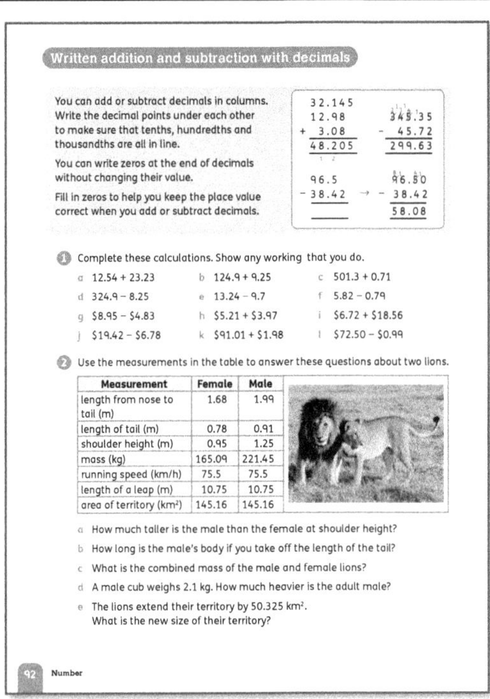

Written addition and subtraction with decimals

You can add or subtract decimals in columns. Write the decimal points under each other to make sure that tenths, hundredths and thousandths are all in line.

$$\begin{array}{r} 32.145 \\ 12.98 \\ +\ \ 3.08 \\ \hline 48.205 \end{array}$$

$$\begin{array}{r} 345.35 \\ -\ \ 45.72 \\ \hline 299.63 \end{array}$$

You can write zeros at the end of decimals without changing their value.

Fill in zeros to help you keep the place value correct when you add or subtract decimals.

$$\begin{array}{r} 96.5 \\ -\ 38.42 \\ \hline \end{array} \quad \rightarrow \quad \begin{array}{r} 96.50 \\ -\ 38.42 \\ \hline 58.08 \end{array}$$

1   Complete these calculations. Show any working that you do.

   a   12.54 + 23.23   b   124.9 + 9.25   c   501.3 + 0.71

   d   324.9 − 8.25   e   13.24 − 9.7   f   5.82 − 0.79

   g $8.95 − $4.83   h $5.21 + $3.97   i $6.72 + $18.56

   j $19.42 − $6.78   k $91.01 + $1.98   l $72.50 − $0.99

2   Use the measurements in the table to answer these questions about two lions.

Measurement	Female	Male
length from nose to tail (m)	1.68	1.99
length of tail (m)	0.78	0.91
shoulder height (m)	0.95	1.25
mass (kg)	165.09	221.45
running speed (km/h)	75.5	75.5
length of a leap (m)	10.75	10.75
area of territory (km²)	145.16	145.16

   a   How much taller is the male than the female at shoulder height?

   b   How long is the male's body if you take off the length of the tail?

   c   What is the combined mass of the male and female lions?

   d   A male cub weighs 2.1 kg. How much heavier is the adult male?

   e   The lions extend their territory by 50.325 km².
      What is the new size of their territory?

92   Number

Materials

A set of 'magic squares' (page 25); squared paper.

Warm-up

Use 'Magic squares' using whole numbers (page 25) as a starter for this lesson.

Focus

- Write some additions and subtractions of decimals horizontally on the board and ask the children to try to solve them mentally or by using pictures.
- Use their ideas to assess their understanding of the concepts of addition, subtraction and place value.
- Discuss methods with the children and explain that you can use written column addition to add and subtract decimals.
- Ask the class how they think this would be the same as whole number addition and how it might be different. (They would need to add in columns using place value, take the decimal point into consideration and use 0 as a placeholder for decimals with different numbers of places.)
- Turn to **Pupil book page 92**. Work on the examples with the class, talking through the steps. For example, say:
  - *I'm going to line up the numbers so the same places are below each other.*
  - *I will leave a separate column for the decimal point.*
  - *There isn't a digit in this place* (at the end of a decimal) *so I'm going to fill in 0.*
  - *Remember that 0.5 is equivalent to 0.50, so writing the zero at the end does not change the value.*
  - *Next, I'll add from right to left. I'll carry/exchange here …*
- If necessary, do a few more examples with the class to make sure there are no misunderstandings.
- Let the children do the calculations in question 1. Do not force children to use column addition and subtraction if they find another method more efficient at this stage. Using squared paper may help them to line up digits correctly in calculations.
- For question 2, give the children time to read the table and to discuss what information it gives them. Talk through the questions before asking them to work out their own answers.

Challenge

The children make up three addition and/or subtraction problems of their own. They calculate the answers.

Interesting mistakes

- Children may add or subtract decimals starting from the right, without thinking about place value. If they do this, encourage them to decompose or partition each number and then say how they could add or subtract them.
- Place-value tables (page 21) may also help clarify the concept, as will working on squared paper, with a column dedicated to the decimal point.
- Some children are likely to make mistakes carrying out column addition and subtraction. It is important to identify whether or not this is due to a misunderstanding of place value. One way of doing this is to give the children examples of calculations that have been answered incorrectly and ask them to identify where mistakes have been made.

Problems involving decimals

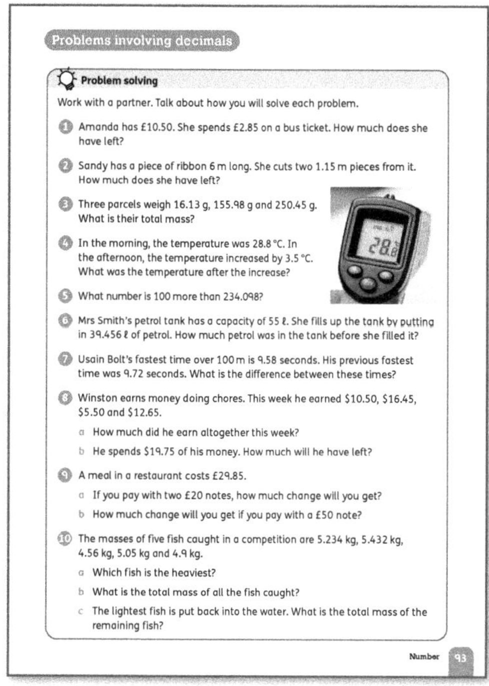

Warm-up

Use any suitable 'Mental problem-solving' activity (pages 23–24) as a starter for this lesson.

Focus

- <u>Problem solving:</u> No new skills are taught in this lesson. The children think about what they have learnt and choose the best methods to solve questions 1–10 on **Pupil book page 93**.
- You could allocate different problems from **Pupil book page 93** to different pairs or groups of children to solve and then ask the groups to explain the problems they worked on and how they solved them.
- Alternatively, set a few of the problems as homework.

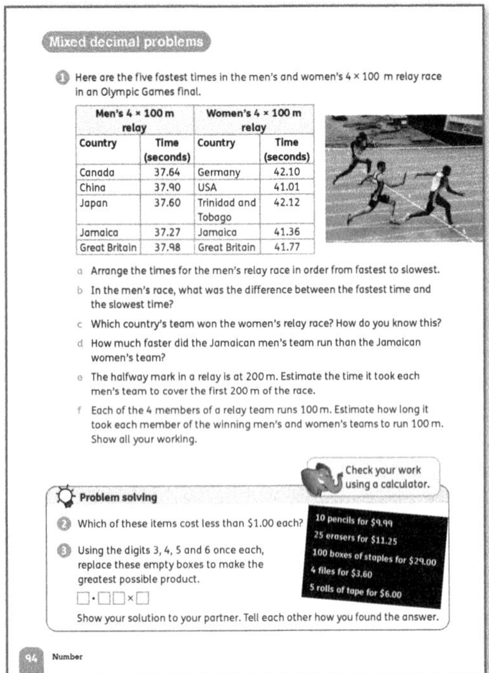

Materials
Calculators.

Warm-up
Use any suitable 'Mental problem-solving' activity (pages 23–24) as a starter for this lesson.

Focus
- Spend some time discussing the questions on **Pupil book page 94** with the class, to identify what is being asked and what strategies the children would use to solve the problems.
- Let the children work in pairs to complete question 1.
- Problem solving: In question 2, encourage the children to use estimation or reasoning to decide which items cost less than $1 each. For example, 25 erasers at $1 each would cost $25, so when 25 erasers cost less than this ($11.25), each must cost less than $1. The children can make digit cards to help them answer question 3.

Challenge
- Write an addition or subtraction calculation on the board involving decimals.
- Ask the children to write a word problem that matches the calculation.
- Discuss the children's word problems. Identify whether they make sense and which the children think are good problems and why.

Support
Ask the children to work in pairs to make up some word problems of their own using the data from the table. Encourage them to think about the wording and how they can make their problems challenging.

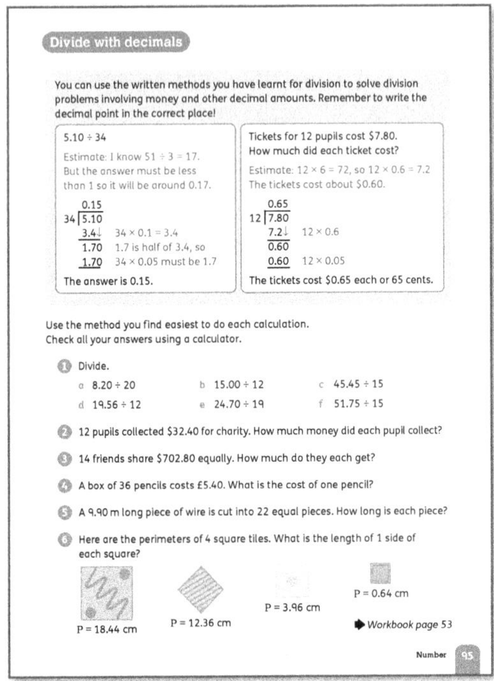

Materials
Calculators.

Warm-up
Use 'Halving' with whole numbers and decimals (page 24) as a starter for this lesson.

Focus
- Use **Workbook page 53** to introduce the concept of dividing a decimal by a whole number (that is not a multiple of 10). In this case, the children are essentially halving decimals.
- Discuss their strategies as a class, making sure they realise that halving a number is the same as dividing by 2.
- Ask the children to suggest how they could show dividing a decimal by 2 using a written method. Share their ideas as a class.
- Turn to **Pupil book page 95**. Explain that the methods the children have already learnt for whole number division work for decimals as well. Remind them that they have already accepted this for addition, subtraction and multiplication and that division is no different.

- Work through the examples with the class. Focus on the estimation and how important this is when you work with decimals because it's so easy to misplace the decimal point.
- If necessary, do the related whole number division alongside to make the point:

$$
\begin{array}{r}
15 \\
34\overline{)510} \\
34 \\
\hline
170 \\
170
\end{array}
\qquad
\begin{array}{l}
1 \times 34 = 34 \\
\\
5 \times 34 = 170
\end{array}
\qquad
\begin{array}{r}
34 \\
68 \\
102 \\
136 \\
170 \\
204
\end{array}
$$

- Talk about how to check the work with a calculator and focus on how a calculator displays decimal amounts. Remind the children that money amounts need 2 decimal places. Ask them what it means when a calculator gives the result 4.5 in a calculation involving money (this is 4 dollars and 0.5 of a dollar, which is equivalent to 4 dollars and 50 cents: we write this as $4.50).
- Next, let the children work on their own to complete questions 1–6.

Interesting mistakes

- Assuming that the children can confidently multiply and divide numbers, the biggest difficulty they are likely to encounter will relate to place value.
- It is important to not simply teach the children rules for multiplying and dividing with decimals, but to ensure that they are clear about why the rules work. For example, when multiplying decimals by 10, although the children may see it as 'moving the decimal point one place', they need to be aware that it is actually the digits that are each becoming ten times greater and thus are 'moving' up a place. This can be demonstrated using a place-value table (page 21).

Answers for Pupil book page 95

1 a 0.41 b 1.25 c 3.03
d 1.63 e 1.3 f 3.45

2 $2.70

3 £50.20

4 15p

5 0.45 m

6 4.61 cm; 3.09 cm; 0.99 cm; 0.16 cm

Answers for Workbook page 53

1 a 29, 14.5 [Provided as an example], 7.25, 3.625
b 42.4, 21.2, 10.6, 5.3, 2.65, 1.325
c 13.7, 6.85, 3.425
d 15.2, 7.6, 3.8

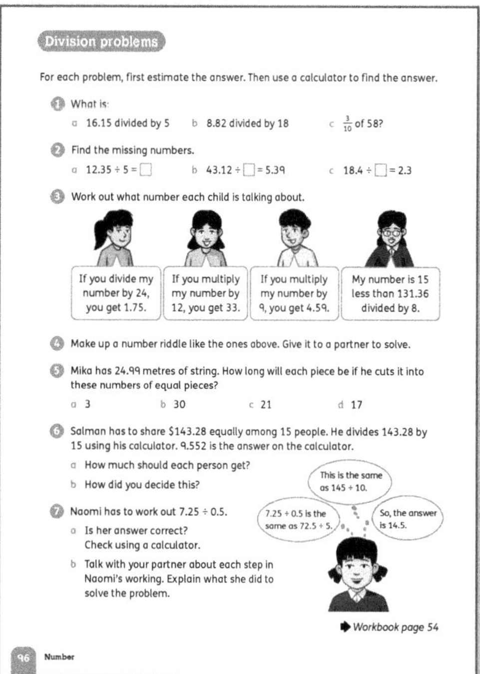

Materials

Rulers; calculators.

Warm-up

Start this lesson by asking the children to complete the measuring and dividing activities on **Workbook page 54**. They can work in pairs to decide how best to solve the problems.

Focus

- Turn to **Pupil book page 96**. Let the children work in pairs to complete questions 1–7. Note that the focus is on estimating (and thus making sense of the division) and that the actual calculations can be done with a calculator.
- Check answers and discuss how close they were to the estimates. Discuss the strategies the children used.

Answers for Pupil book page 96

1 a 3.23 b 0.49 c 17.4

2 a 12.35 ÷ 5 = 2.47   b 43.12 ÷ 8 = 5.39
c 18.4 ÷ 8 = 2.3

3 42; 2.75; 0.51; 1.42

4 Individual answers.

5 a 8.33 m b 0.833 m
c 1.19 m d 1.47 m

6 a $9.55
b The digit in the thousandths cannot be considered as there are no coins worth less than cents. When you round the hundredths digit you need to remember that the total cannot be greater than the money you have to share.

7 a Yes
b She multiplied both the numerator and denominator by 10 so that she could divide by a whole number, which is easier.

Answers for Workbook page 54

1 25.5 cm; each part is 2.55 cm

2 12.5 cm; each part 2.5 cm

3 10.5 cm; each part 3.5 cm

4 5.8 cm; each part 2.9 cm

Mixed work with decimals

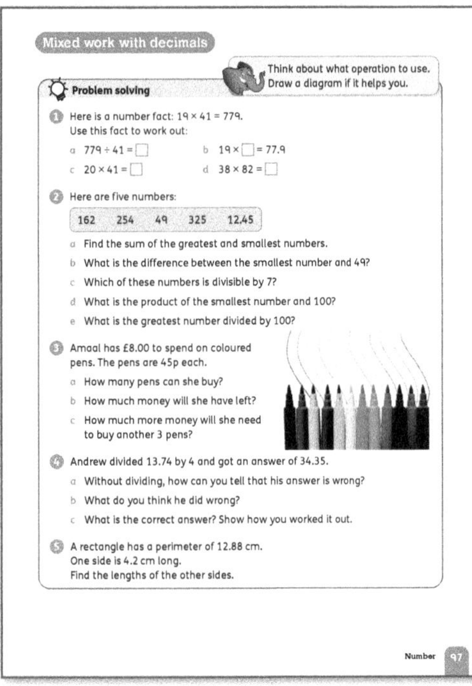

Materials

Rulers; calculators.

Warm-up

Use 'Round decimals' (page 23) as a starter for this lesson.

Focus

- Problem solving: Treat the questions on **Pupil book page 97** as a test of the concepts covered in this unit.
- Let the children work through the questions on their own, but encourage them to look back to examples from previous lessons if they need to (treat it like an open book test).
- Go through the answers with the class and let the children mark their own work and rate their understanding and mastery of the concepts.
- Ask them to write down anything that they think they need to revise or revisit and make time for them to help each other to go over anything they are unsure of.

Answers for Pupil book page 97

1 a 19   b 4.1   c 820   d 3166

2 a 337.45   b 36.55   c 49
   d 1245   e 3.25

3 a 17   b 35p   c £1

4 a 34.35 is greater than 13.74.
   b He divided 137.4 by 4.   c 3.435

5 2.24 cm

End-of-unit check

Use all or some of these questions and problems to assess how well the children have understood the concepts in this unit.

- *What is this digit in (give a decimal) worth?*
- *What is the value of 4 in the number 23.184? Give your answer as a decimal.* (0.004, 4 thousandths)
- *How do you say (give a decimal)?*
- *What is 0.75 as a fraction?* $\left(\frac{3}{4}\right)$
- Provide a set of decimals. *Put these decimals in order, starting with the smallest. When ordering the decimals, what did you look at first? Then what did you do?* (For example: write the numbers in a column with the decimal points lined up. Compare digits, starting from the left. Stop at the first digit that is different and compare these.)
- Provide a set of decimals. *Which numbers do you find hardest to order? Why?*
- *A child says '3.009 is greater than 3.1 because 9 is greater than 1'. How would you show this child that this is incorrect?* (For example: show both numbers in a place-value table or on a number line.)
- *Give a number that lies between 3.25 and 3.35. How many answers do you think there are to this question?* (For example: 3.26, 3.348; there is an infinite number of answers, as you can always create another number by going to the next decimal place.)
- *I have 25 cm of rope measured to the nearest centimetre. What is the longest the rope could be if measured to two decimal places? Explain why.* (25.004 cm – assuming that the child only writes numbers to thousandths, otherwise 2.004999 … cm. This is the greatest length that will round to 25.00 cm, as 25.005 cm will round to 25.01 cm.)
- *How could you give instructions to someone to explain how to round decimal fractions to the nearest whole number?* (Look at the digit in the tenths place; if it is less than 5 round to the previous whole number; if it is 5 or more, round to the next whole number.) *Now tell someone how to round to the nearest tenth or hundredth.* (Nearest tenth: look at the number in the hundredths place; nearest hundredth: look at the number in the thousandths place.)
- Provide a mixed set of fractions and decimals. *What is the largest number? What is the smallest number? Place the numbers in ascending order.*
- *What do you need to add to (give a decimal) to make 10?*
- Give a length such as 0.87 m. *What is the length of the piece required to make 1.5 metres?* (0.63 m)
- *How much change will you get from $10 if you buy something that costs $3.96?* ($6.04)
- *What is 1.34 + 0.25?* (1.59)
- *How much do you have to take from 1.6 kilograms to get 0.85 kilograms?* (0.75 kg)
- *What is double 0.92?* (1.84)
- *What is half of 4.86 litres?* (2.43 litres)
- *I want to cut a length of rope that is 5.86 m long into 4 equal pieces. How long should each piece be?* (1.465 m)
- *What is 3.2 divided by 4?* (0.8) *What is 3.2 divided by 0.4?* (8)
- Give a problem involving decimals. *What did you do to solve this problem? Describe each step.*
- *My calculator gives 12.9. What does that mean in money terms?* (£12.90, or other currency)
- *Sally says, '(give a decimal) divided by 10/100/1000 is (give wrong answer). Is she correct? How could you check?*

Percentages

Learning objectives
- Recognise percentages of shapes and numbers.
- Recall and use equivalences between simple fractions, decimals and percentages, including in different contexts.
- Solve problems involving the calculation of percentages and the use of percentages for comparison.

Key words
percentage   discount

Unit introduction

Teaching guidance
- Statistics in the media are often given as percentages. This makes media reports easier to understand, but percentages can also be misleading. For example, 60% of dentists recommending a brand of toothpaste could mean a lot of dentists recommended that brand, but it could also mean five dentists were asked and three recommended it.
- Similarly, when the media wants a number to look impressive, a percentage might not be reported. For example, during the Covid-19 pandemic, the media would report how many people were infected worldwide using numbers in the millions rather than saying that 1% of the world's population was infected. Hold a class discussion about how percentages are used in daily life, what they are intended to suggest to us and what they could mean.
- Ask the children to find examples used in advertisements and news articles, and share them with the class. Talk about what each percentage means.
- Use a simple pie chart to illustrate the fact that percentages do not give us actual amounts or numbers. For example, this pie chart shows the proportion of ingredients in a cake.

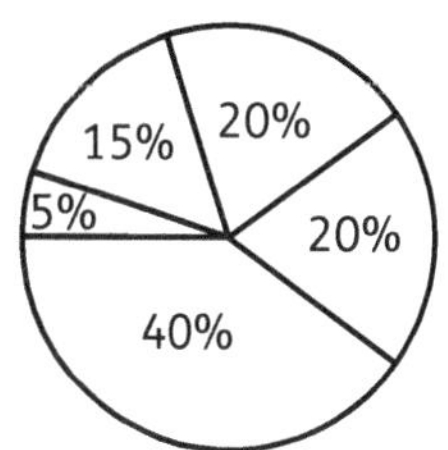

| Flour |
| Sugar |
| Eggs |
| Chocolate |
| Other |

- Give the children information about the ingredients and ask them to work out what proportion of the cake is represented by each sector on the graph.

For example, give them these statements:
- $\frac{4}{10}$ of the ingredients are flour.
- There is about the same proportion of eggs as there is sugar in the cake mix.
- If the ingredients have a mass of 800 grams, then 120 grams are chocolate.
- Butter, salt and baking powder are the smallest proportion of the ingredients.

Revisit percentages

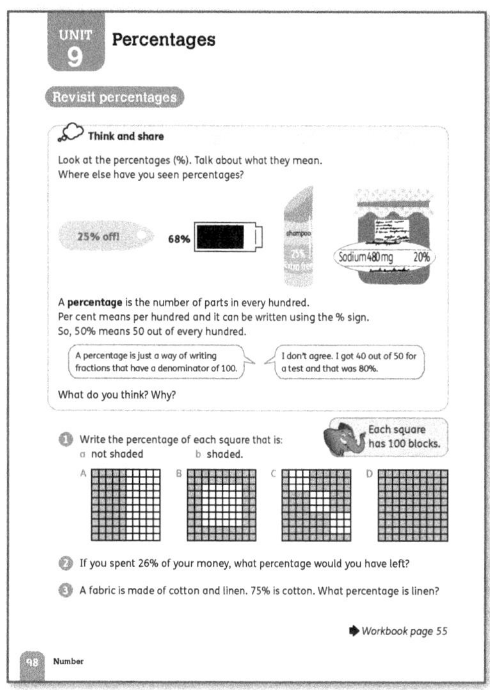

Materials
1–100 table.

Warm-up
Flip a counter on to a 1–100 table. Ask a child to subtract this number from 100. Get the child to flip the counter for the next number. Go round the class until everyone has had a turn.

Focus
- <u>Think and share:</u> Turn to **Pupil book page 98**. Let the children work in small groups to discuss the pictures and answer the Think and share questions. Discuss with the children where they have seen percentages. The question asks the children to say what they think about the statements and to explain their answers. (Both statements are correct: a percentage is a way of writing fractions that have a denominator of 100, but all other fractions can be written as percentages too if they are first written as equivalent fractions with a denominator of 100. For example, 40 out of 50 is equivalent to 80 out of 100 which is 80%.)
- Let the children complete questions 1–3 independently.

Follow-up

- Use **Workbook page 55** to check that the children are able to interpret the numbers and use them to work out the percentage of the squares that are shaded or unshaded.
- They can check each other's work when they have finished.

Support

Let the children create a poster showing percentages used in everyday life.

Answers for Pupil book page 98

1 a A = 50%; B = 36%; C = 27%; D = 0%

   b A = 50%; B = 64%; C = 73%; D = 100%

2 74%

3 25%

Answers for Workbook page 55

1 a 40% coloured, 60% not coloured

   b 80% coloured, 20% not coloured

   c 90% coloured, 10% not coloured

   d 50% coloured, 50% not coloured

   e 36% coloured, 64% not coloured

   f 6% coloured, 94% not coloured

   g 60% coloured, 40% not coloured

   h 25% coloured, 75% not coloured

Percentages, fractions and decimals

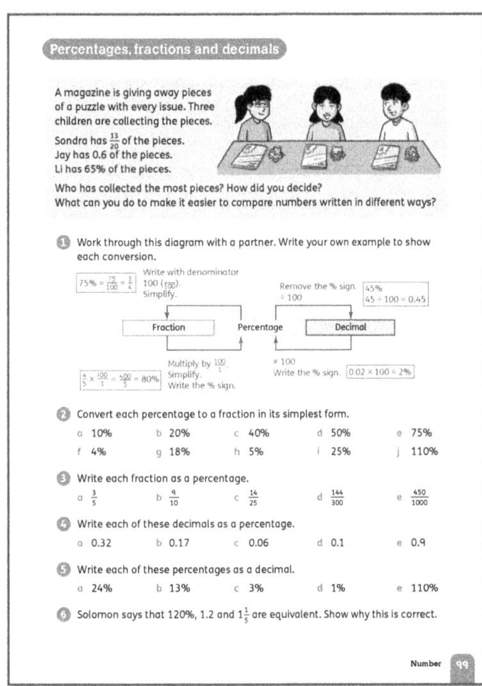

Materials

A set of cards showing mix of percentages, fractions and decimals; sets of cards with equivalent percentages, fractions (for example, half, quarters, tenths and fifths) and decimals.

Warm-up

Revisit work on fractions, decimals and percentages from Level 5 by writing fractions with denominator 100 on the board, and ask the children to write each one as a percentage, a decimal and a fraction in its lowest terms.

Focus

- Turn to **Pupil book page 99**. Read the problem with the class and let the children discuss the question.
- The explanation box at the top of the page asks: Who has collected the most pieces? (Sandra and Li) How did you decide? (For example, colour in a 100 chart to represent the different proportions, use a double number line marked in percentages and 20ths.) What can you do to make it easier to compare numbers written in different ways? (Convert them all to the same type of number.)
- Let them share the ways they worked to answer them. Make a list of their suggestions for making it easier to compare numbers that are expressed in different ways.
- Hand out cards with a mix of percentages, decimals and fractions to the children and ask them to order the numbers from smallest to largest.
- Encourage the children to explain how they made their choices, especially how they compared fractions or decimals with percentages.
- Show the children a card with either a percentage, fraction or decimal and ask them to express the amount on the card in the two other forms.
- Use the example $\frac{24}{40}$ to show that you can find an equivalent fraction with a denominator of 10 first $\left(\frac{6}{10}\right)$ and use that to find a fraction with a denominator of 100 $\left(\frac{60}{100}\right)$.
- Next, show the class that $\frac{24}{40} \times 100 = \frac{2400}{40} = 60\%$.
- Compare the two methods to show how they give the same answer.
- The children work in pairs for question 1. Give them time to discuss the diagram and to work out how to convert between different forms of the same number. Let the children write their own examples for each conversion. Share some of these with the class, checking that they are correct as you do so.
- The children can work independently on questions 2–6. For question 6, explain that they should show some calculations to show why the statement is correct.

Challenge

- Ask the children to make their own game of dominoes using equivalent percentages, decimals and fractions. Give them time to discuss and plan their game, make the dominoes and play a game.
- Give groups a chance to explain their game to the class. Use these games to revise and consolidate the concepts taught.

Support

- Give each group a set of cards with equivalent percentages, fractions (for example, half, quarters, tenths and fifths) and decimals.
- Use these to play matching games such as 'Snap' where the children each get a number of cards which they play in turns. If the card played is equivalent to the one on the top of the pile, the child who says 'Snap!' first is allowed to take all the cards on the pile.
- Alternatively, spread all the cards face down and ask the children to take turns to turn over two cards. If the cards are the same, the child who turned them over may take the cards. If not, they turn them back over and the next child has a turn to find a matching pair.

Answers for Pupil book page 99

1 Individual answers.

2
- a $\frac{1}{10}$
- b $\frac{1}{5}$
- c $\frac{2}{5}$
- d $\frac{1}{2}$
- e $\frac{3}{4}$
- f $\frac{1}{25}$
- g $\frac{9}{50}$
- h $\frac{1}{20}$
- i $\frac{1}{4}$
- j $1\frac{1}{10}$

3
- a 60%
- b 90%
- c 56%
- d 48%
- e 45%

4
- a 32%
- b 17%
- c 6%
- d 10%
- e 90%

5
- a 0.24
- b 0.13
- c 0.03
- d 0.01
- e 1.1

6 $1.2 \times 100 = 120\%$ and $1\frac{1}{5} \times 100 = 120\%$

Equivalent percentages, fractions and decimals

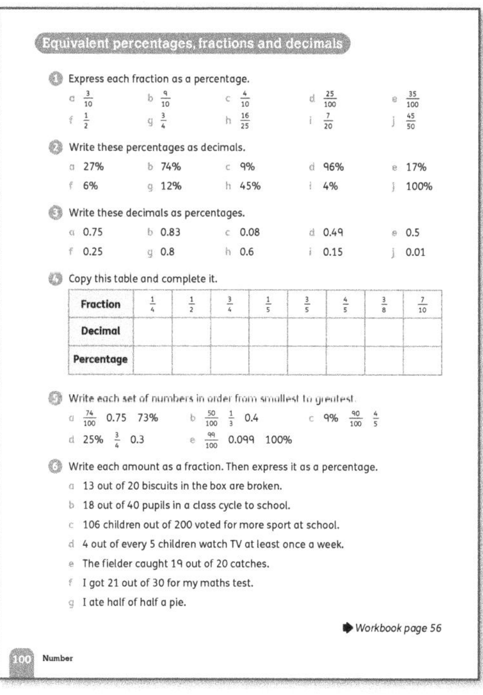

Warm-up

As a mental starter for this lesson let the children work independently to complete the matching activity on **Workbook page 56**.

Focus

- Remind the class that the ability to convert between fractions, decimals and percentages is important because in real life we see all three. For example, we may hear that 18 out of 40 people like football and 23 out of 48 people like basketball. It is difficult to say which group is larger using these two fractions, but easy if you consider $\frac{18}{40} \times 100 = 45\%$ and $\frac{23}{48} \times 100 = 48\%$.
- Let the children work on their own or in pairs to complete questions 1–6 on **Pupil book page 100** to consolidate their understanding.

Challenge

- When the children understand the basic equivalences, you may like to teach them how to use the % key on their calculators to find percentages such as $\frac{23}{48}$.
- To do this they would enter $23 \div 48$ and then press the % key to get an answer of 47.91666.
- They can round this to the nearest whole number using what they already know about rounding decimal fractions and get 48%.

Answers for Pupil book page 100

1
- a 30%
- b 90%
- c 40%
- d 25%
- e 35%
- f 50%
- g 75%
- h 64%
- i 35%
- j 90%

2
- a 0.27
- b 0.74
- c 0.09
- d 0.96
- e 0.17
- f 0.06
- g 0.12
- h 0.45
- i 0.04
- j 1

3
- a 75%
- b 83%
- c 8%
- d 49%
- e 50%
- f 25%
- g 80%
- h 60%
- i 15%
- j 1%

4

Fraction	$\frac{1}{4}$	$\frac{1}{2}$	$\frac{3}{4}$	$\frac{1}{5}$	$\frac{3}{5}$	$\frac{4}{5}$	$\frac{3}{8}$	$\frac{7}{10}$
Decimal	0.25	0.5	0.75	0.2	0.6	0.8	0.375	0.7
Percentage	25%	50%	75%	20%	60%	80%	37.5%	70%

5
- a 73%; $\frac{74}{100}$; 0.75
- b $\frac{1}{3}$; 0.4; $\frac{50}{100}$
- c 9%; $\frac{4}{5}$; $\frac{90}{100}$
- d 25%; 0.3; $\frac{3}{4}$
- e 0.099; $\frac{99}{100}$; 100%

6
- a $\frac{13}{20}$; 65%
- b $\frac{18}{40}$; 45%
- c $\frac{106}{200}$; 53%
- d $\frac{4}{5}$; 80%
- e $\frac{19}{20}$; 95%
- f $\frac{21}{30}$; 70%
- g $\frac{1}{2}$; 50%

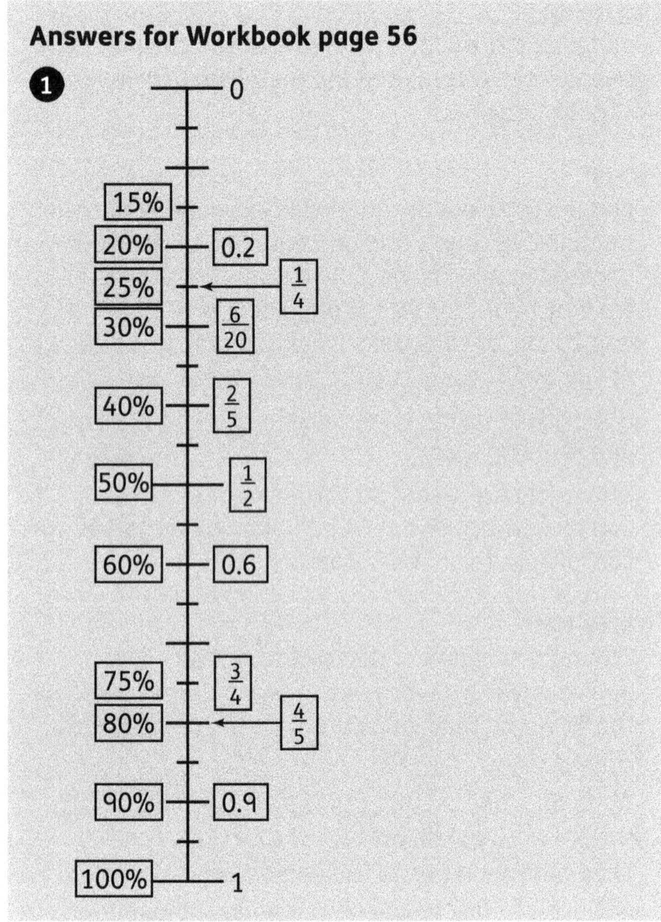

Percentages of amounts

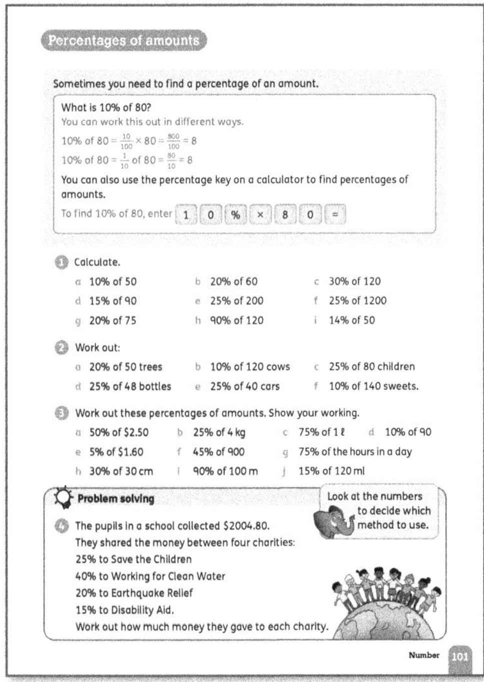

Materials
Calculators.

Warm-up
- As a starter for this lesson, write a number such as 320 in a circle in the middle of the board. Draw 'arms' coming from the circle and label each with a percentage: 10%, 25%, 5%, 45%, etc.
- Invite the children in turn to work out the correct percentage.
- Encourage the children to find their answers mentally and discuss the strategies they used. For example, they might say: 'I found 10% by dividing by 10 and I knew that 5% would be half of this'.

Focus
- Remind the children how to find percentages of quantities by converting the percentage into a fraction then finding the fraction of the quantity.
- Where appropriate, encourage the children to use mental methods. For example, to find 60%, they find 10% by dividing by 10 and then multiply that by 6.
- Turn to **Pupil book page 101**. Work through the example with the class, discussing the methods and letting the children say which they find easier and why.
- If necessary, do a few more examples with the class before asking them to complete questions 1–3 on their own.
- <u>Problem solving:</u> For question 4, prompt the children to check their answers by checking that the amounts they calculate add up to $2004.80.

Challenge
- Ask the children to design a poster explaining how to find percentages of amounts.
- Display the posters in the classroom for reference.

Support
- Make a list of percentages of prices, such as 25% of £80, 30% of £10, 50% of £125.
- Let the children work out how to order the amounts from least to greatest. Give them time to discuss how to do this and let them share the strategies they used with the class.

Interesting mistakes
If the children are just taught a rule for finding percentages of quantities, they are unlikely to remember it or use it correctly. It is important to discuss with the class why the method works. This can be done by continually thinking about what we mean by percentage and building rules up together, and by using diagrams to illustrate what is happening.

Answers for Pupil book page 101

1 **a** 5   **b** 12   **c** 36
   d 13.5   **e** 50   **f** 300
   g 15   **h** 108   **i** 7

2 **a** 10 trees   **b** 12 cows
   c 20 children   **d** 12 bottles
   e 10 cars   **f** 14 sweets

3 **a** $1.25   **b** 1 kg   **c** 750 ml
   d 9   **e** £0.08   **f** 405
   g 18 hours   **h** 9 cm   **i** 90 m
   j 18 ml

4 Save the Children $501.20; Working for Clean Water $801.92; Earthquake Relief $400.96; Disability Aid $300.72

Calculate percentages

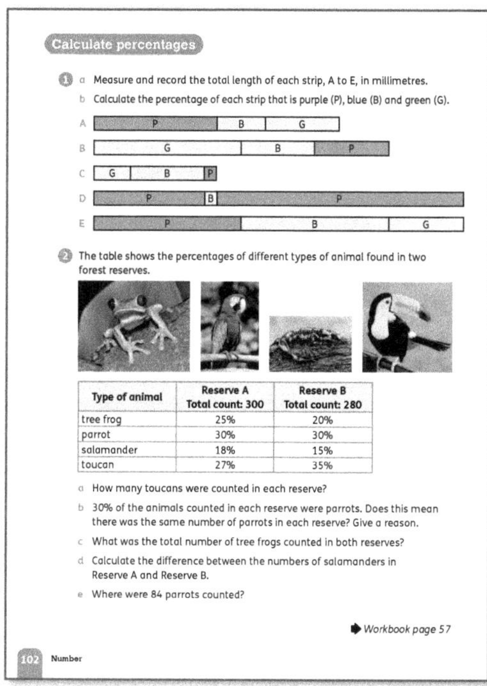

Materials
Rulers; calculators.

Warm-up
Use any of the 'Rounding and estimating' activities (page 23) as a starter for this lesson.

Focus
- Remind the children how to write one number as a percentage of another. For example, 3 out 10 children are left-handed. Write the fraction $\frac{3}{10}$, then convert this to a percentage $\left(\frac{30}{100} = 30\%\right)$. Repeat for more examples, using calculators to convert more difficult fractions, such as $\frac{5}{11}$, to a decimal and then a percentage.

- Ask the children to turn to **Pupil book page 102**.
- For question 1, the children should measure as accurately as they can to the nearest millimetre. Let them compare and check their measurements before they calculate the percentages. They can use calculators to do this.
- For question 2, remind the children that percentages do not give us actual numbers. Discuss what the table shows and make sure the children can read and interpret it.

Follow-up
- Use **Workbook page 57** to consolidate measuring and calculating percentages.
- Let the children check each other's work.

Answers for Pupil book page 102

1 **a** All measurements are in millimetres.
     A: P = 50, B = 20, C = 30;
     B: G = 60, B = 30, P = 30;
     C: G = 15, B = 30, P = 5;
     D: P = 45, B = 5, P = 100;
     E: P = 60, B = 60, G = 30
   b A: P = 50%, B = 20%, C = 30%;
     B: G = 50%, B = 25%, P = 25%;
     C: G = 30%, B = 60%, P = 10%;
     D: P = 30%, B = 3.3%, P = 66.7%;
     E: P = 40%, B = 40%, G = 20%

2 **a** Reserve A: 81 toucans; Reserve B: 98 toucans
   b No. In reserve A there are 0.3 × 300 = 90 parrots and in reserve B there are 0.3 × 280 = 84 parrots.
   c 131   **d** 12   **e** in reserve B

Answers for Workbook page 57

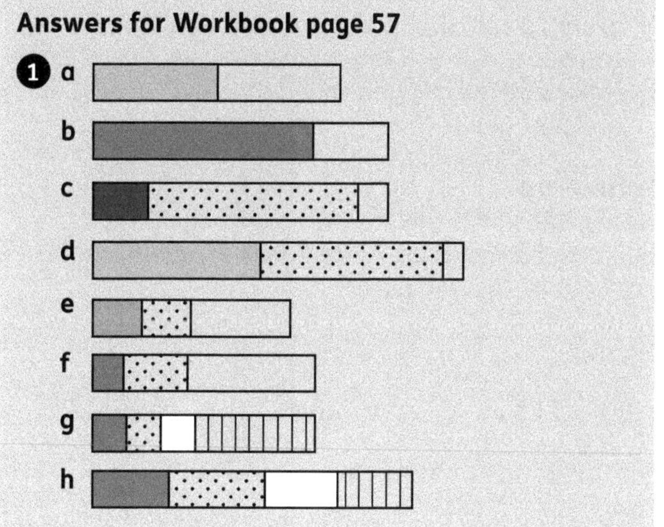

Solve problems involving percentages

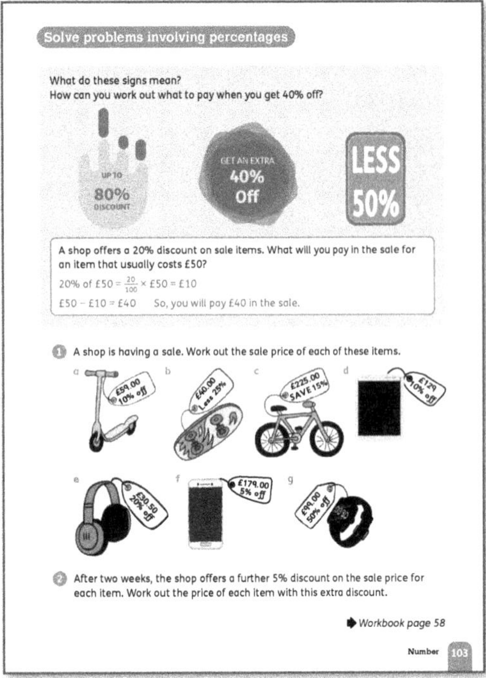

Materials
Calculators; adverts showing percentage discount.

Warm-up
Ask the children to work out $\frac{1}{2}$, $\frac{1}{4}$ and $\frac{3}{4}$ of different quantities as a mental starter for this lesson.

Focus
- Show the children real examples of 'discounts' shown in percentages (for example, in newspaper adverts, on shop posters). Demonstrate how to find sale prices if we know the original price and the discount.
- Turn to **Pupil book page 103**. Work through the example with the children, discussing each step as you go.
- Make sure the children understand the concept of discount before they answer questions 1 and 2.

Follow-up
Let the children complete **Workbook page 58** to assess that they can calculate percentages of amounts and work out discounted prices.

Answers for Pupil book page 103
1 a £53.10   b £30.00   c £191.25
   d £116.10   e £24.40   f £170.05
   g £49.50

2 a £50.45   b £28.50   c £181.69
   d £110.30   e £23.18   f £161.55
   g £47.03

Answers for Workbook page 58

1

Original price	Percentage discount	Money off	Sale price
£75	10%	£7.50	£67.50
£40	5%	£2.00	£38.00
£52	10%	£5.20	£46.80
£125	20%	£25.00	£100
£60	50%	£30.00	£30.00
£245	20%	£49.00	£196.00
£339	30%	£101.70	£237.30
£850	25%	£212.50	£637.50
£1260	15%	£189.00	£1071.00
£2612	5%	£130.60	£2481.40

2 a £16.00   b £11.60   c £44.00
   d £15.99   e £96.00   f £60.79

3 a 50%   b 75%   c 70%   d 80%

End-of-unit check

Use some or all of these questions and problems to assess how well the children have understood the concepts in this unit.

- Provide a mixed set of fractions, decimals and percentages. Ask the children to write them in ascending order.
- *What percentages can you easily work out in your head?* (Give some examples, such as 10%, 25%, 50%.)
- *What is (give a percentage) as a fraction? What is it as a decimal?*
- *If you have $\frac{4}{5}$ of an amount, what percentage is that?* (80%)
- *'To calculate 10% of a quantity you divide by 10, so to find 30% of a quantity you divide by 30.' Is this statement true? Why not?* (The statement is false. 30% is three times as much as 10%, so you divide by 10 and then multiply by 3.)
- *A teacher has 100 sheets of card. The class uses $\frac{1}{2}$ of the sheets to make nets of cubes and $\frac{1}{5}$ of the sheets to make vocabulary flashcards. What percentage of the sheets is left?* (30%)
- *How can you find a percentage of a quantity?* (Multiply by the percentage and divide by 100.)
- *Explain how to find out what percentage 24 is of 50?* (Write 24 out of 50 as a fraction $\left(\frac{24}{50}\right)$ and convert it to an equivalent fraction with denominator 100 $\left(\frac{48}{100}\right)$. The percentage is the numerator (48%).)
- *Which is larger, (give a fraction) or (give a percentage)? How do you know?*
- Give a decimal and an incorrect conversion of the decimal into a percentage. *I have converted this decimal into this percentage. Am I right? Why not?*
- *I want to buy a shirt in a sale offering 25% off. The original price is $15. What is the sale price?* ($11.25)
- *20% of the people on a cruise say that they cannot swim. There are 4300 people on the cruise. How many of them can swim?* (3440)

Measures and money

Learning objectives

- Solve problems involving the calculation and conversion of units of measure, using decimal notation up to 3 decimal places where appropriate.

- Use read, write and convert between standard units, converting measurements of length, mass, volume and time from a smaller unit of measure to a larger unit and vice versa using decimal notation to up to 3 decimal places.

- Convert between miles and kilometres.

- Convert between time intervals expressed as a decimal and in mixed units.

Key words

metric   convert   kilometre   mile   line graph

Unit introduction

Materials

Internet access.

Teaching guidance

- Ask the class to find out about the history of measuring systems and to prepare short reports on both the 'metric' and imperial systems of measurement. Their reports should include units used and where these come from.
- Ask the class to think of examples where it is important to 'convert' accurately between systems. Here are two examples:
  - You know your weight in pounds and a doctor needs it in kilograms to decide how much medicine to give you.
  - For an engineer a millimetre might mean the difference between safety and the collapse of a structure.
- The children can find mistakes caused by inaccurate conversions by searching for 'unit conversion disasters' online. An example is the loss of a $125 million space probe in 1999 because one team used metric measures and the other used imperial.
- Investigate online conversion apps and programs with the class. This is a fun way of exposing the children to the different units and their equivalents.

Metric units

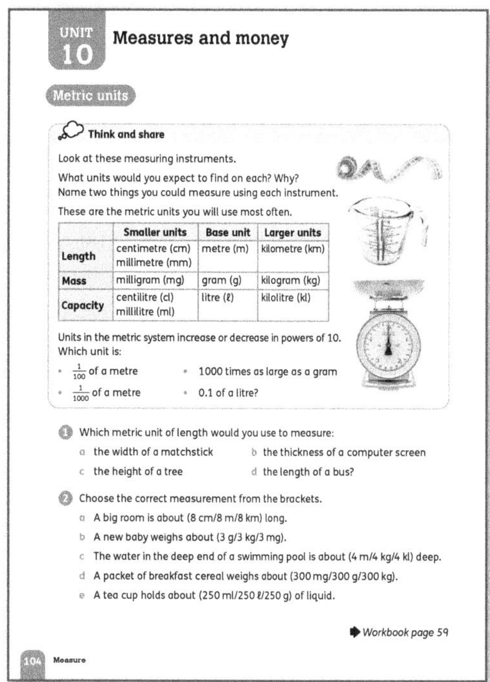

Materials

Examples (or pictures) of different measuring instruments.

Warm-up

Use the 'Multiplying and dividing by 10'/'by 100 and 100' activities (page 27) as a starter for this lesson.

Focus

- Show the class different measuring instruments. Ask them what each one is used for and look closely at the scale to identify the units and the intervals between them.
- Ask the children to give examples of other measuring instruments they know of. Ask them what is measured, what units are used and how the scale is divided up.
- Ensure digital measuring devices are mentioned. For example, point out how many bathroom scales give mass in kilograms and tenths of kilograms (100-gram increments), so you can get a mass of 55.4 or 55.5 but you cannot get anything in between.
- <u>Think and share:</u> Turn to **Pupil book page 104**. Give the children time to read through and discuss the information. Discuss the answers as a class. The first question asks the children to say the units that they would expect to find on each of the measuring instruments shown and to name two things that could be measured with them. (Tape measure: centimetres (cm) or millimetres (mm); for example how tall you are or the dimensions of a table.

Measuring jug: millilitres (ml); for example water or milk used in a recipe. Scales: grams (g) or kilograms (kg); for example flour for a recipe or a parcel to work out the postage cost.) The second question asks the children to answer these questions: What unit is $\frac{1}{100}$ of a metre? (1 cm) What unit is 1000 times as large as a gram? (1 kg) What unit is $\frac{1}{1000}$ of a metre? (1 mm) What is 0.001 of a litre? (1 ml)

- Also check the children's understanding of 'kilometre'. Work through questions 1 and 2 orally with the class, asking different children to share their answers.

Follow-up

- Use **Workbook page 59** to revise reading measuring scales. Spend time making sure the children can read and make sense of different measuring scales.
- Use additional instruments, such as weighing scales, measuring jugs, tape measures and other marked scales, to demonstrate different measures.

Challenge

- Ask the children to investigate different and unusual measuring scales. They can photograph or print images of the scales they find and make a poster showing the different scales.
- Ask the children to label each scale to say what it measures, what units are shown and how they are divided up. They don't have to know the units to do this.
- Note that if the children find imperial measuring scales, then the lack of decimal divisions could lead to an interesting discussion in class.

Support

- Give the children a list of things to measure, for example: the length of a pencil, the distance between two towns, the mass of a brick, how much water a cup holds. Ask them to name a sensible unit for measuring these.
- Let them compare and check each other's work.

Interesting mistakes

- Children may find it difficult to read and make sense of scales, particularly when intervals are not labelled. Continue to read and make sense of real measuring scales to show the children how to interpret them.
- Help the children to work out what the intervals are by asking questions like: *What do these marks on the scale tell you? Why are some marks longer/thicker than others? How many marks are there between the marked units? How can this help you read the scale?*

Answers for Pupil book page 104

1 a mm   b mm
    c m   d m
2 a 8 m   b 3 kg
    c 4 m   d 300 g
    e 250 ml

Answers for Workbook page 59

1 4.5 cm; 100.5 cm; 2.4 kg; 62 kg; 2.4 ℓ; 12.5 °C; 14 °C
2 720 ml, 0.72 ℓ; 470 ml, 0.47 ℓ

Convert between metric units

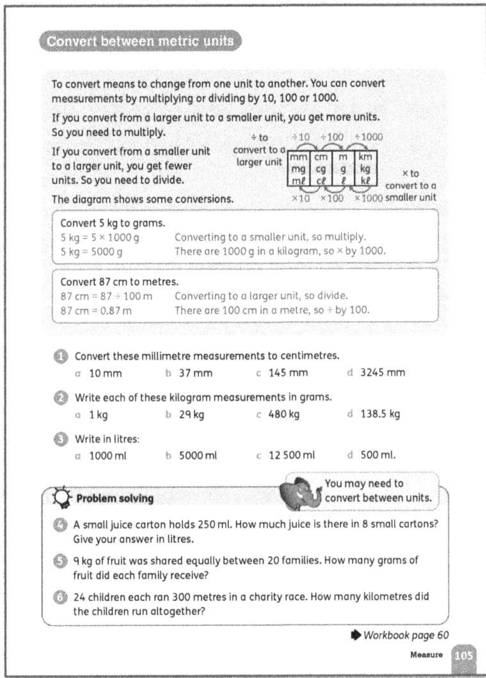

Materials

Calculators.

Warm-up

As a class, create a list of conversion facts for metric units, for example 1 m = 100 cm, 1 kg = 1000 g and so on, as a starter for this lesson.

Focus

- Previously, the children have converted from larger units to smaller units (by multiplying); for example, they have expressed 5 m in cm. Now they will apply their knowledge of decimals and place value to convert between any units of measurement with place values to thousandths (3 decimal places).
- Work through the examples on **Pupil book page 105** with the class. Revise place value as necessary for conversions.
- Do a few more conversions with the class before asking them to complete questions 1–3 on their own.
- Problem solving: Encourage the children to use bar models (page 21) or other diagrams to help them solve questions 4–6. Check that they have given their answers in the units stated in the question.

Follow-up

- Ask the children to complete **Workbook page 60** on their own.
- Let the children check their own work using a calculator.

Interesting mistakes

- Some children need additional support with both directions of conversion (larger to smaller and smaller to larger).
- Some children may find it easier to convert lengths than other measures, including capacity, perhaps because lengths are more familiar and easier to visualise.
- Therefore, it important to link the metric system to all the units you are working with, so that the children see that there is a pattern and a method that works for all measures, not just lengths.

Answers for Pupil book page 105

1   **a**   1 cm   **b**   3.7 cm
    c   14.5 cm   **d**   324.5 cm

2   **a**   1000 g   **b**   29 000 g
    c   480 000 g   **d**   138 500 g

3   **a**   1 litre   **b**   5 litres
    c   12.5 litres   **d**   0.5 litres

4   2 litres

5   450 g

6   7.2 km

Answers for Workbook page 60

1   **a**

Grams	4000	8000	9500	12 500	65 000	125 000
Kilograms	4	8	9.5	12.5	65	125

b

Metres	1500	3050	4750	9460	10 500	13 250
Kilometres	1.5	3.05	4.75	9.46	10.5	13.25

c

Kilometres	4	7.5	12.75	88.69	90.09	120.50
Metres	4000	7500	12 750	88 690	90 090	120 500

d

Millilitres	3500	5540	12 760	34 090	49 909	83 149
Litres	3.5	5.54	12.76	34.09	49.909	83.149

e

km	m	cm	mm
0.001	1	100	1000
0.004	4	400	4000
0.05	50	5000	50 000
0.2	200	20 000	200 000

f

kl	ℓ	cl	ml
1	1000	100 000	1 000 000
0.002	2	200	2000
0.0004	0.4	40	400
0.0005	0.5	50	500

Units of length

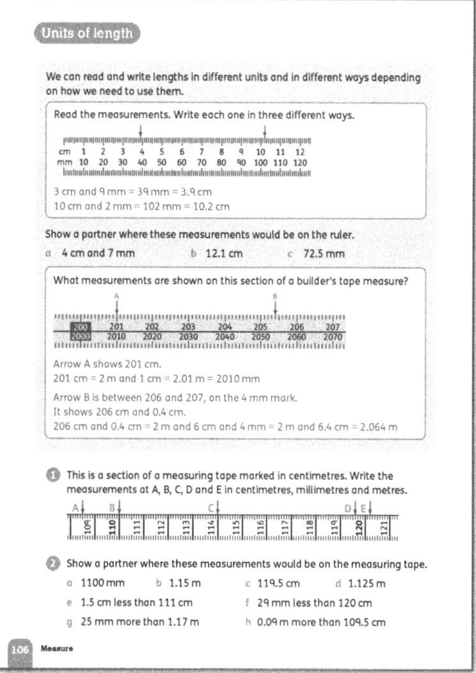

Materials

Rulers; measuring tapes; calculators.

Warm-up

As most conversions of units of length in this lesson involve units that are 10 or 100 times greater or smaller, use any activity that involves multiplying and dividing by 10 or 100 as a starter for this lesson.

Focus

- Ask the children how many centimetres are in a metre. Ask: *There are 10 mm in every centimetre. How many millimetres must there be in a metre?* (1000)
- Having established the answer, ask the children what fraction of a metre a millimetre must be. $\left(\frac{1}{1000}\right)$
- Draw a place-value table going from metres to millimetres on the board. Use this to practise writing millimetres as metres and vice versa.
- Write 5 cm 6 mm on the board. Ask the children how many millimetres this is in total. Encourage the children to explain how they worked it out, for example: 'There are 50 mm in 5 cm so it is 50 + 6, which is 56 mm.'
- Repeat for other examples. Ask the children to measure various lines in millimetres. Demonstrate the need to line up the ruler correctly in order to get an accurate measurement.
- Turn to **Pupil book page 106**. Either work through the examples with the class, or let the children work in groups.
- The measurements in question 2 refer to the diagram in the **Pupil book** but you could ask the children to locate them on a real measuring tape. Note that they have to do some calculation to work out the measurements before locating them.

Challenge

- People who do a lot of measuring in their work (contractors, architects, flooring professionals and others) often use electronic measuring devices (also called laser tape measures).
- Let the children investigate electronic measuring devices and ask them to explain how they work, why they are useful when you have to do lots of measuring and what other advantages (or disadvantages) there are in using them.

Interesting mistakes

Diagrams that ask the children to convert from millimetres to centimetres to metres to kilometres, for example, require them to work with a wide range of place values. Remind the children to work from one unit to the next unit up, for example from millimetres to centimetres before converting to metres and then kilometres.

Answers for Pupil book page 106

1 A: 108.5 cm, 1085 mm and 1.085 m
B: 110 cm, 1100 mm and 1.1 m
C: 114 cm, 1140 mm and 1.4 m
D: 119.6 cm, 1196 mm and 1.196 m
E: 120.2 cm, 1202 mm and 1.202 m

2

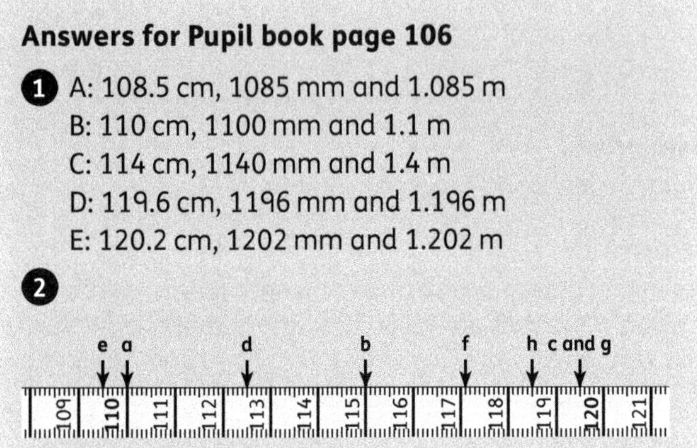

More conversions

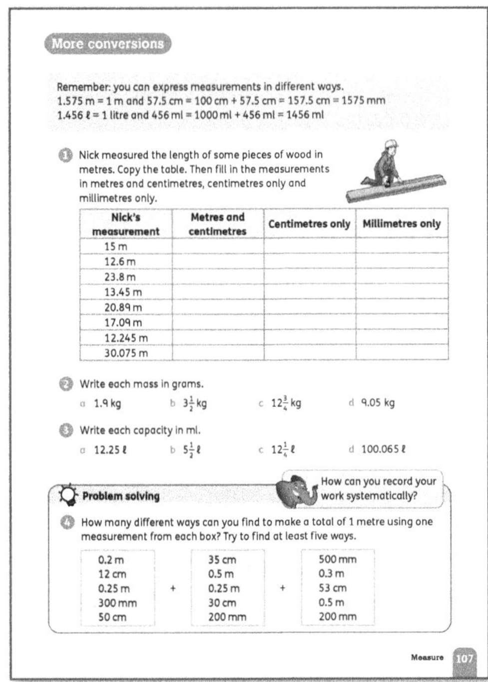

Warm-up

Use any activity that involves multiplying whole numbers and decimals by 1000, and dividing whole numbers by 1000, as a starter for this lesson.

Focus

- Let the children complete **Pupil book page 107** independently.
- <u>Problem solving:</u> Discuss their answers to question 4.

Challenge

Ask the children to draw a scale plan of the classroom. They will need to measure the dimensions of the room and the key pieces of furniture, agree a scale and then draw their plan, measuring and drawing accurately with a ruler. Provide graph or squared paper for them to work on.

Answers for Pupil book page 107

1

Nick's measurement	Metres and centimetres	Centimetres only	Millimetres only
15 m	15 m	1500 cm	15 000 mm
12.6 m	12 m and 60 cm	1260 cm	12 600 mm
23.8 m	23 m and 80 cm	2380 cm	23 800 mm
13.45 m	13 m and 45 cm	1345 cm	13 450 mm
20.89 m	20 m and 89 cm	2089 cm	20 890 mm
17.09 m	17 m and 9 cm	1709 cm	17 090 mm
12.245 m	12 m and 24.5 cm	1224.5 cm	12 345 mm
30.075 m	30 m and 7.5 cm	3007.5 cm	30 075 mm

2 a 1900 g   b 3500 g   c 12 750 g
d 9050 g

3 a 12 250 ml   b 5500 ml   c 12 250 ml
d 100 065 ml

4 0.2 m + 0.5 m + 0.3 m = 1 m, 0.2 m + 30 cm + 0.5 m = 1 m, 12 cm + 35 cm + 53 cm = 1 m, 0.25 m + 0.25 m + 0.5 m = 1 m, 300 mm + 0.5 m + 200 mm = 1 m

Kilometres

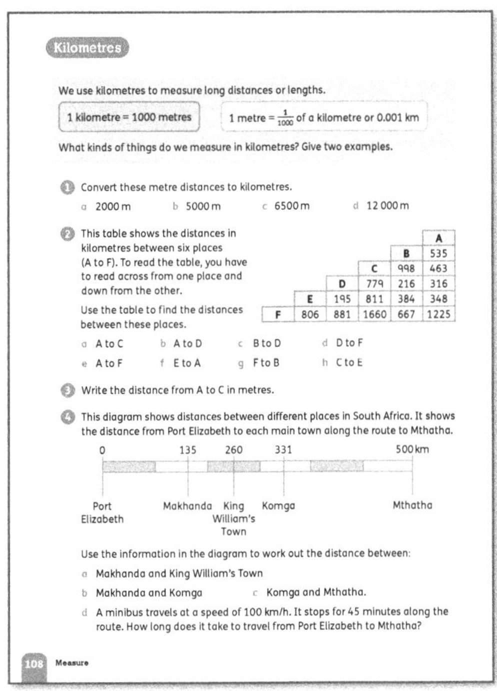

Warm-up

Give children a whole number and ask them to give you the number that is 1000 times smaller, or $\frac{1}{1000}$ of it. Repeat with whole numbers and decimals and ask for the number 1000 times larger.

Focus

- In this lesson, the children extend their knowledge of units of length to include kilometres. As a more abstract concept it is useful to relate it to distances in metres so that the children can get a sense of how far (how long) a kilometre is. For example, a full-size football field is generally about 100 m long. (The standard length is 105 m.) Tell the children that they would have to run up and down the field ten times to cover 1 kilometre.
- On the board write: kilometres, metres, centimetres and millimetres.
- For each, ask the children to think of things that would be best measured in this unit. Record suggestions, encouraging the children to explain how they made their decisions.
- Ask the children how many metres there are in a kilometre (1000).
- Having established the answer, ask the children what fraction of a kilometre a metre must be.
- Draw a place-value table going from kilometres to metres on the board. Use this to practise writing metres as kilometres and vice versa.
- Turn to **Pupil book page 108**. The explanation box at the top of the page asks: What kind of thing do we measure in kilometres? Give two examples. (We use kilometres for long distances, for example, the distance between two cities, the length of the equator, the height of a space rocket, the length of a river.)
- For questions 2 and 3, spend some time talking about the distance table and how to read it. The children need to work out that they read down from one place and across from the other. The distance between the places is in the block where the up and across movements meet.
- In question 4, the distance indicator is cumulative; it starts from 0 and shows how far from 0 the places along the route are. The total distance is 500 km.

Answers for Pupil book page 108

1   **a** 2 km   **b** 5 km   **c** 6.5 km
    d 12 km

2   **a** 463 km   **b** 316 km   **c** 216 km
    d 881 km   **e** 1225 km   **f** 348 km
    g 667 km   **h** 811 km

3   463 000 m

4   **a** 125 km   **b** 196 km
    c 169 km   **d** 5 hours and 45 minutes

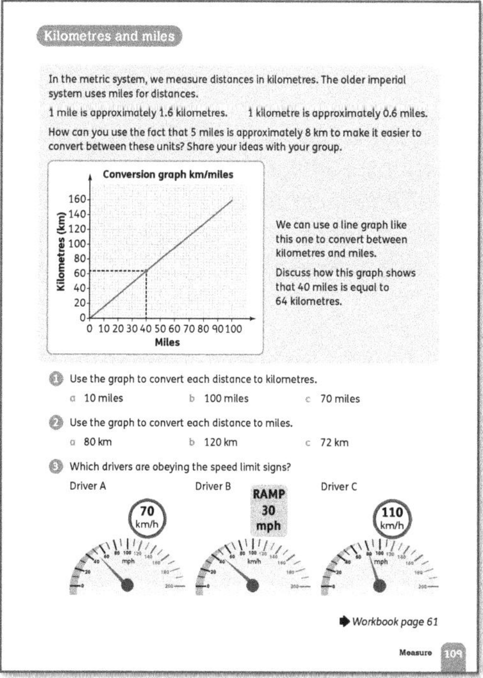

Workbook page 61

Materials

Calculators; rulers; online conversion apps, if possible.

Warm-up

Revise 5 and 8 times tables facts (both division and multiplication) as a starter for this lesson.

Focus

- Remind the class that almost all of the countries in the world have adopted the metric system of measurement. The USA, Liberia and Myanmar are the only three countries to still officially use the imperial system, but they use metric for many scientific and technical measurements. Although the imperial system is not the official system in other countries, people still use some imperial measurements, for example:
  - People may know their height in feet. A person who is 6 foot tall is about 1.8 m tall.
  - People in England use miles to talk about longer distances, and in many places speed limits are given in miles per hour. A speed limit of 30 miles per hour is about 50 km per hour.
  - People may talk about $\frac{1}{2}$-litre bottles of milk or other liquids as being 'about a pint'.
- Turn to **Pupil book page 109**. Read through the information about approximate conversion factors for kilometres and miles. The explanation box at the top of the page asks: How can you use the fact that 5 miles is approximately 8 km to make it easier to convert between these units? (Scale up, for example 10 miles is 16 km; divide the number of miles by 5 and multiply by 8 to convert to km; divide the number of km by 8, then multiply by 5, to convert to miles.) The children should be familiar with the term 'line graph'. They should remember how to read and

interpret a conversion graph, but if not, revise how to do this.

- Let the children work on their own to complete questions 1–3.

Follow-up

- Use **Workbook page 61** for additional practice and to check that the children can apply what they have learnt in different contexts.
- Remind the children to use 5 miles = 8 km to help them convert between units.

Support

Some children may need additional support to read and make sense of conversion graphs. Show them how to draw lines to find the matching units (or to place a ruler on one unit and read across or up from the other).

Interesting mistakes

- Children sometimes get the impression that imperial or customary measures are somehow less accurate than metric ones. Address this misconception directly. Use measuring instruments (such as tape measures and car speedometers) where both units are shown to demonstrate how you can measure just as accurately using either scale.

Answers for Pupil book page 109

1 a 16 km   b 160 km   c 112 km

2 a 50 miles   b 75 miles   c 45 miles

3 driver B

Answers for Workbook page 61

1

Jaffna	392	496	312	360	
245	Colombo	145.6	121.6	318.4	Kilometres
Miles	319	92	Galle	219.2	368
195	76	137	Kandy	200	
225	199	230	125	Batticaloa	

2 a 318.4 km   b Naresh
c Galle to Kandy

3 a

b around 11 km
c 8849 metres; 8.849 km; on troposphere.

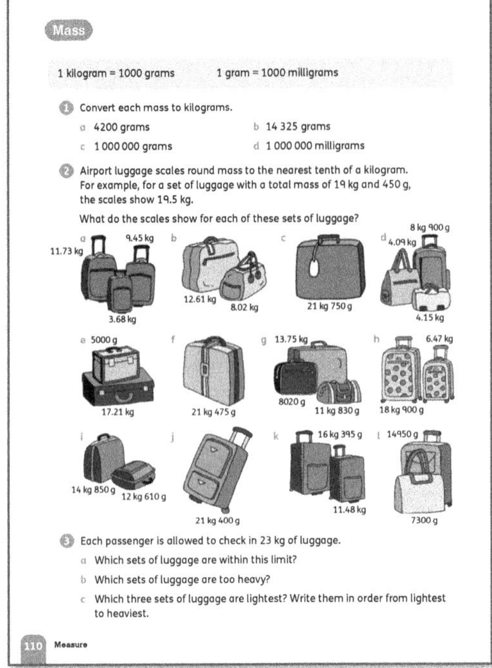

Mass

1 kilogram = 1000 grams   1 gram = 1000 milligrams

1 Convert each mass to kilograms.
  a 4200 grams   b 14 325 grams
  c 1 000 000 grams   d 1 000 000 milligrams

2 Airport luggage scales round mass to the nearest tenth of a kilogram. For example, for a set of luggage with a total mass of 19 kg and 450 g, the scales show 19.5 kg.

What do the scales show for each of these sets of luggage?

a 11.73 kg, 9.45 kg, 3.68 kg
b 12.61 kg, 8.02 kg
c 21 kg 750 g
d 8 kg 900 g, 4.09 kg, 4.15 kg
e 5000 g, 17.21 kg
f 21 kg 475 g
g 13.75 kg, 8020 g, 11 kg 830 g
h 6.47 kg, 18 kg 900 g
i 14 kg 850 g, 12 kg 610 g
j 21 kg 400 g
k 16 kg 395 g, 11.48 kg
l 14950 g, 7300 g

3 Each passenger is allowed to check in 23 kg of luggage.
  a Which sets of luggage are within this limit?
  b Which sets of luggage are too heavy?
  c Which three sets of luggage are lightest? Write them in order from lightest to heaviest.

110 Measure

Materials

Place-value tables (page 21); weighing scales marked in metric units.

Warm-up

Write 'missing number' calculations involving multiplying or dividing by 1000 on the board, and ask the children to solve them in groups, for example:
$2.6 \times \square = 2600$, $\square \div 1000 = 0.65$.

Focus

- Have a number talk (pages 17–18) in which you ask the class how they can use what they know about converting lengths to convert units of mass in the same way. Discuss what we measure in milligrams, for example using food labels.
- Let the children share their ideas and, if necessary, show the conversions on place-value tables (kilograms, grams and milligrams) as you did with units of length.
- Once the children understand that the conversions work in the same way, let them complete questions 1–3 on **Pupil book page 110** independently.

Challenge

- Ask the children to find information about the cost of sending parcels by post or courier in your country. (They can get this information from a post office or courier company, or online.) Use these to generate questions and additional activities.

Answers for Pupil book page 110

1 a 4.2 kg   b 14.325 kg   c 1000 kg
   d 1 kg

2 a 24.9 kg   b 20.6 kg   c 21.8 kg
   d 17.2 kg   e 22.2 kg   f 21.5 kg
   g 33.6 kg   h 25.4 kg   i 27.5 kg
   j 21.4 kg   k 27.9 kg   l 22.3 kg

3 a 20.6 kg; 21.8 kg; 17.2 kg; 22.2 kg; 21.5 kg;
     21.4 kg; 22.3 kg
   b 24.8 kg; 33.6 kg; 25.4 kg; 27.5 kg; 27.9 kg
   c 17.2 kg; 20.6 kg; 21.4 kg

Units of time

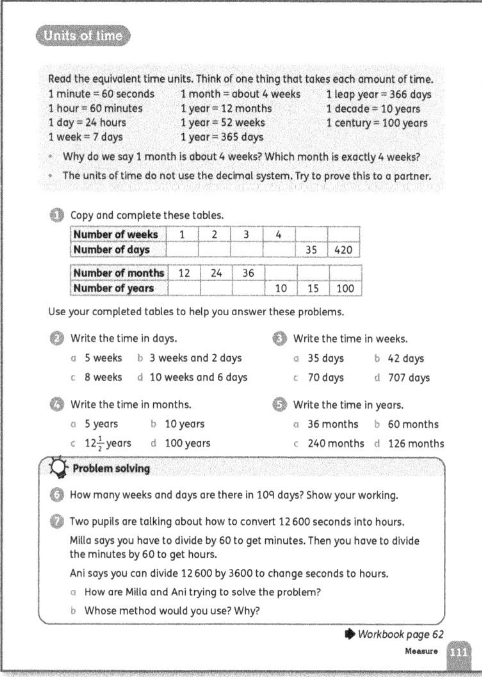

Materials
Flashcards showing the names of time units.

Warm-up
Use one of the 'Calendars and time' activities (pages 26–27) as a starter for this lesson.

Focus
- Start by revising the names of units of time with the class. Display the name of a unit, for example 'seconds', and ask the children to say what it refers to. For example, they might say that it's small unit of time; there are 60 seconds in one minute; a second is $\frac{1}{60}$ of a minute; a second is about as long as it takes you to say 'one Mississippi'.
- Turn to **Pupil book page 111**. The explanation box at the top of the page says: Why do we say 1 month is about 4 weeks? (Except in a leap year, months have either 28, 30 or 31 days. These are all closer to 4 weeks than 5 weeks. It is useful to use a round number of weeks for estimating time periods.) Which month is exactly 4 weeks? (February). The units of time do not use the decimal system. Try to prove

this to a partner. (In the decimal system, when you have 10 in one column of the place-value table you replace it with 1 in the column to the left. If you make a place-value table for weeks and days, when you have 7 in the days column you replace it with 1 in the weeks column.)
- For question 1, ask the children to complete the tables, then check them and ask different children to explain how they found the answers.
- Make sure the tables are all correct before moving on to questions 2–5, as the children may need to refer back to them.
- <u>Problem solving:</u> For question 6, spend time as a class solving and then discussing strategies to find the answer. Show the class how to use a number line to solve problems involving conversion of units of time. For example, they can use the number fact that there are 35 days in 5 weeks like this:

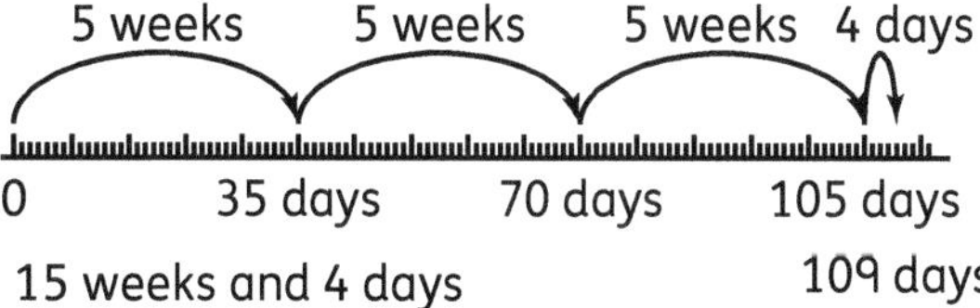

- This is an important strategy because time is not a decimal system and remainders need to be given in units of time, rather than as decimal fractions of the units. If the children do this problem on a calculator, they will get an answer of 15.57 and may mistakenly think this is 15 weeks and 57 days rather than 15 weeks and $\frac{57}{100}$ weeks.
- <u>Problem solving:</u> For question 7, ask the children to read and think about the problem before asking them to share their ideas.

Follow-up
- Use **Workbook page 62** to assess how well the children understand and can convert between units of time.
- Let the children complete the work independently and then mark the answers as a class.

Challenge
Let the children work in pairs to develop their own loop game activity using equivalent units of time. You may need to provide a set of blank cards or an example like the one below for them to organise the questions.

I have 42 days. Who has the equivalent number of weeks?	I have 6 weeks. Who has $1\frac{1}{2}$ years in months?	I have 18 months. Who has 3 centuries?
I have 300 years. Who has 10 decades?	I have one century. Who has 366 days?	I have a leap year. Who has …

Interesting mistakes

- Some children may not know or understand the terms 'decade' and 'century' in the context of time. Spend some time talking about the prefixes 'deca-' (meaning ten) and 'cent-' (meaning hundred).
- Encourage the class to find other words that relate to tens and/or hundreds. For example, a decagon is a ten-sided shape; a decathlete competes in ten different events; a century in cricket means 100 runs; a centimetre means $\frac{1}{100}$ of a metre, and a cent is $\frac{1}{100}$ of a dollar (or other currency).

Answers for Pupil book page 111

1

Number of weeks	1	2	3	4	5	6
Number of days	7	14	21	28	35	42

Number of months	12	24	36	120	180	1200
Number of years	1	2	3	10	15	100

2
- a 35 days
- b 23 days
- c 56 days
- d 76 days

3
- a 5 weeks
- b 6 weeks
- c 10 weeks
- d 101 weeks

4
- a 60 months
- b 120 months
- c 150 months
- d 1200 months

5
- a 3 years
- b 5 years
- c 20 years
- d 10.5 years

6 $109 \div 7 = 15.571...$
$15 \times 7 = 105$
$109 - 105 = 4$ days
109 days are 15 weeks and 4 days

7
- a Milla is converting seconds into minutes and then minutes into hours by dividing by 60 twice. Ani has worked out that there are $60 \times 60 = 3600$ seconds in 1 hour, so she is doing the conversion in one step.
- b Individual answers.

Answers for Workbook page 62

1
- a 60
- b 120
- c 4
- d 30
- e 15
- f 2
- g 60
- h 15

2
- a 365
- b 24
- c 366
- d 10
- e 1000
- f 1200

3
- a There are 240 hours in 10 days.
- b There are 48 months in 4 years.
- c There are 180 minutes in 3 hours/There are 180 seconds in 3 minutes.
- d There are 30 minutes in 0.5 hours.
- e There are 1440 minutes in 1 day.

4
- a 12
- b 90
- c 12.5
- d 1.025 hours
- e 420

5 Individual answers.

Calculate periods of time

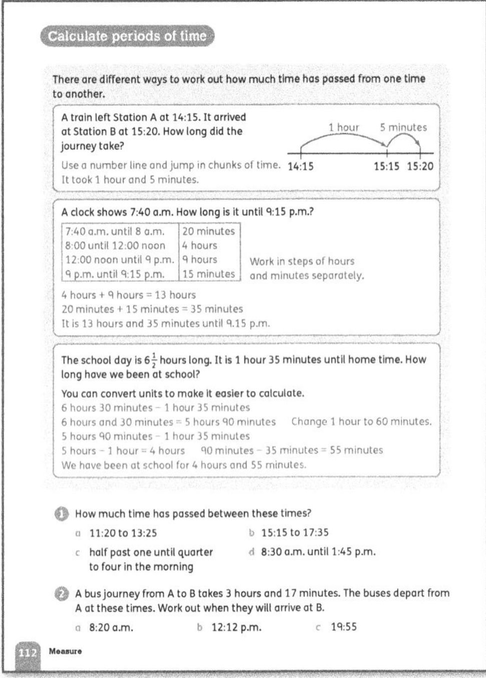

Materials

A large analogue clock or a model clockface with moveable hands.

Warm-up

Do some time unit conversions as a starter for this lesson.

Focus

- Use an analogue clock to demonstrate time intervals to the class before they attempt to calculate them:
  - Show a starting time on the clock.
  - Write the ending time on the board and then move the hands, counting and recording hours and minutes to find the interval.
- Then show the children that they can do this in writing using 'timeline' number lines.
- Play a game called 'How much longer?' in groups:
  - Let the children take turns to identify an event and a time in the school day and then pose a 'How much longer?' question for the others to answer. For example, say: *The start of assembly is 8.15. How much longer is it until first break?*
  - Let the children compare and discuss the strategies they use to work out the answers.
- Turn to **Pupil book page 112**. Work through the three examples with the class, talking about each step in the different methods. Make sure the children are able to draw and work with number lines to solve time problems.
- If necessary, work through some further examples to reinforce the concept.
- Let the children complete questions 1 and 2 in pairs.

Challenge

Give the children a time interval, for example $4\frac{1}{4}$ hours. Ask them to make up a multi-step word problem with that solution.

Support

Some children may find it very challenging to use the clockface. Allow them to physically model the time intervals before they proceed to a written method.

Answers for Pupil book page 112

1 a 2 hours and 5 minutes
b 2 hours and 20 minutes
c 2 hours and 15 minutes
d 5 hours and 15 minutes

2 a 11:37 b 15:29 c 23:12

Convert between time intervals

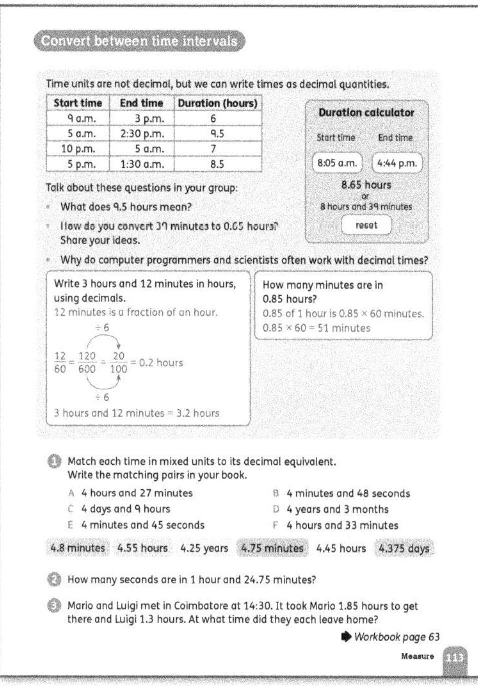

Materials

Calculators; examples of time duration calculator apps.

Warm-up

Revise writing one number as a fraction of another, and then as a decimal and percentage, as a starter for this lesson.

Focus

- Show the children how to use a time duration calculator app to find time intervals.
- Turn to **Pupil book page 113** to have a number talk (pages 17–18) about different ways of expressing time intervals.
- Give the children a chance to read the information and to talk through the questions. The example box at the top of the page asks: What does 9.5 hours mean? (9 and a half hours, or 9 hours 30 minutes) How do you convert 39 minutes to 0.65 hours? (Write 39 minutes as a fraction of one hour, simplify, then convert to a fraction with denominator 100: $\frac{39}{60} = \frac{13}{20} = \frac{65}{100} = 0.65$.) Why do computer programmers and scientists often work with decimal times? (It is easier to calculate with decimals than with fractions, and easier to type decimals than fractions.)

- If possible, let the children explore a different online app to see how it converts between time in hours and minutes and time as a decimal quantity.
- Discuss the answers with the class, focusing on the conversion of 39 minutes to 0.65 hours. Make sure the children can see that $\frac{39}{60} = 0.65$. Let them use calculators to do the division.
- Work through the two examples with the class.
- The children can try question 1 on their own and use an online app to check the answers if possible.
- Let the children use calculators for time interval conversions, such as in question 2.
- For question 3, remind the class that they can use timeline methods to solve the problem once they have done the conversion.

Follow-up

- Use **Workbook page 63** for additional practice.
- Spend some time discussing the problems that the children make up in question 2.

Interesting mistakes

- Children often get the wrong answers particularly when they use a calculator to solve problems, such as: *How many hours and minutes are there in 147 minutes?* If they enter 147 ÷ 60, they will get 2.45. Many children will incorrectly give the answer as 2 hours and 45 minutes, instead of working out that it is two hours and $\frac{45}{100}$ of an hour, which is only 27 minutes.
- Using a number line as on **Pupil book page 112** makes it clear that they have 27 minutes left over after subtracting two lots of 60 minutes.
- Once they understand, ask them to find the answer on a calculator and discuss why this is given in a different format.

Answers for Pupil book page 113

1 A = 4.45 hours; B = 4.8 minutes; C = 4.375 days; D = 4.25 years; E = 4.75 minutes; F = 4.55 hours

2 5085 seconds

3 Mario 12:39; Luigi 13:12

Answers for Workbook page 63

1

Time interval	Hours and minutes
45 minutes + 50 minutes	1 hour and 35 minutes
35 minutes + $\frac{1}{2}$ an hour + 0.25 days	7 hours and 5 minutes
$1\frac{3}{4}$ hours + 55 minutes + 0.5 days	14 hours and 40 minutes
$3\frac{1}{2}$ – 1 hour 10 minutes	2 hours and 20 minutes
$\frac{1}{2}$ a day – 4.5 hours	7 hours and 30 minutes
4 days – $25\frac{3}{4}$ hours	70 hours and 15 minutes

2 a 7 h 25 min; individual answers.
b 3 h 45 min; individual answers.
c 2 h 45 min; individual answers.

More time calculations

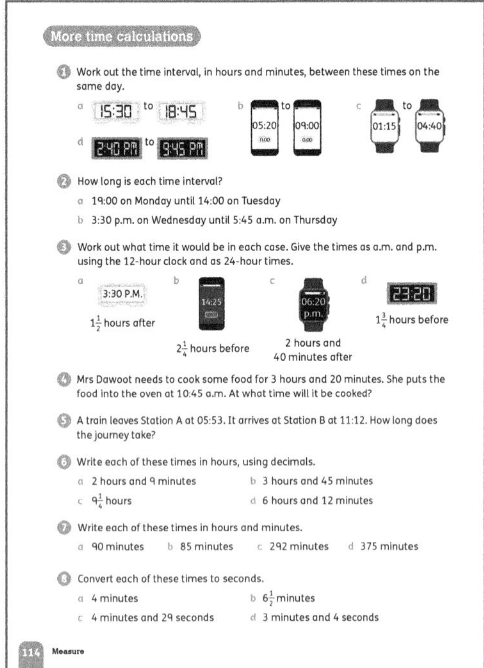

Warm-up
Use 'Finding times earlier and later than a given time' (page 27) as a starter for this lesson.

Focus
- There are no new concepts in this lesson. Questions 1–8 on **Pupil book page 114** provide additional practice and allow the children to apply what they have learnt about time.
- You could use it as a test to assess how well the children can work with time, or allocate some sections for the children to complete in class or as homework.

Support
If children continue to treat time as a decimal quantity (when it should not be), spend time revising units of time and continue to demonstrate how to work out time intervals using time lines and counting on and back to find the correct answers.

Answers for Pupil book page 114
1 a 3 hours and 15 minutes
 b 3 hours and 40 minutes
 c 3 hours and 25 minutes
 d 7 hours and 5 minutes
2 a 19 hours   b 14 hours and 15 minutes
3 a 5:00 p.m.   b 12:10
 c 09:00 p.m.   d 21:35
4 14:05
5 5 hours and 19 minutes
6 a 2.15 hours   b 3.75 hours
 c 9.25 hours   d 6.2 hours
7 a 1 hour and 30 minutes
 b 1 hour and 25 minutes
 c 4 hours and 52 minutes
 d 6 hours and 15 minutes
8 a 240 s   b 390 s   c 269 s   d 184 s

Work with money

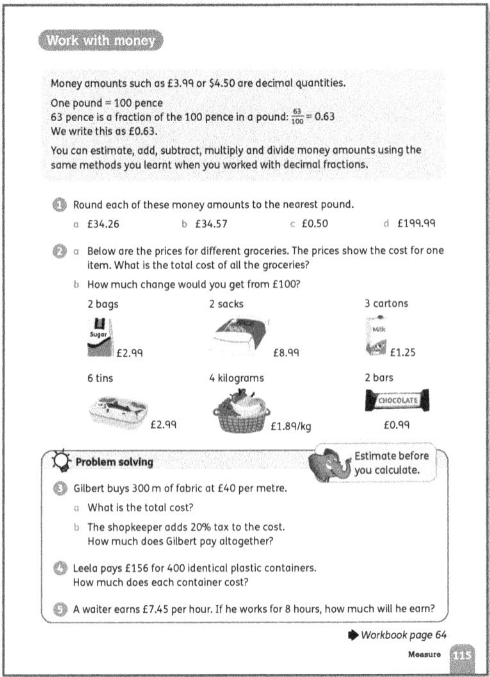

Materials
Calculators.

Warm-up
Use 'Magic squares' (page 26) with decimals as a starter for this lesson.

Focus
- No new concepts are introduced in this section. The children will need to apply what they already know about decimals and percentages to solve problems involving money amounts, change and tax charges.
- Turn to **Pupil book page 115**. Ask the children to share their ideas about how money amounts are similar to decimals.
- Spend some time discussing how to write amounts less than a pound or dollar (or any local units) as decimals. For example, write on the board 50p = £0.50.
- Stress that money amounts have 2 (and only 2) decimal places.
- Let the children work independently to complete questions 1 and 2. Check their answers as a class before moving on.
- <u>Problem solving:</u> In questions 3–5, encourage the children to use bar models (page 21) or other diagrams to help them write the calculation they need to do, and then use estimation to check that their calculated answer is sensible.

Follow-up
Let the children work in pairs with a calculator to check the calculations on **Workbook page 64**.

Interesting mistakes

- Children may get confused with money amounts on a calculator, as it will truncate (cut off any unnecessary zeros) the decimal and not include 2 decimal places for whole number answers and answers with no hundredths.
- Remind them that 0 is a place holder and that the convention for writing money is to include the zeros to the right of the decimal place, even if they are not strictly necessary.

Answers for Pupil book page 115

1 a £34 b £35 c £1 d £200

2 a £55.19 b £44.81

3 a £12 000 b £14 400

4 £0.39

5 £59.60

Answers for Workbook page 64

1 a List A £21; List B £75; List C £24; List D £28; List E £30

 b List A £79; List B £25; List C £76; List D £72; List E £70

 c List A £20.87; List B £75.80; List C £22.49; List D £30.03; List E £30.51

2 List A £79.13; List B £24.20; List C £77.51; List D £69.97; List E £69.49

Money calculations

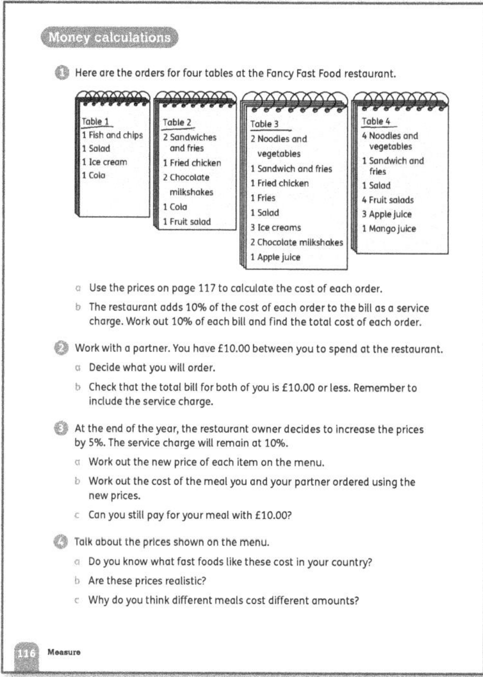

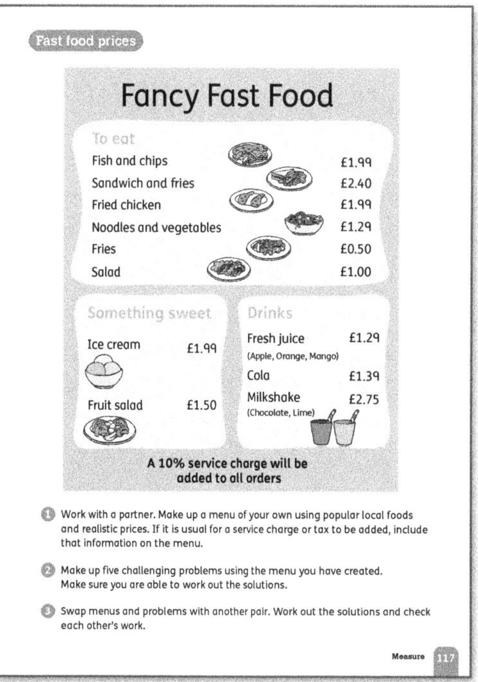

Materials

Calculators; menus from local fast food outlets with prices (or online versions).

Warm-up

Use any of the 'Rounding and estimating' activities (page 23) as a starter for this lesson.

Focus

- Turn to **Pupil book pages 116 and 117.**
- Discuss how you would work out the cost of each order with the class. Make sure the children understand that the price of each item is given on the menu on **Pupil book page 117.**
- Once you are confident that the children know how to do this, let them work in pairs through questions 1–3 on **Pupil book page 116.**
- For question 4 on **Pupil book page 116**, use the menus you have brought to class for a discussion about local pricing and how it differs from the menu in the book.
- The children should work in pairs on questions 1–3 on **Pupil book page 117.**

Answers for Pupil book page 116

1 **a** Table 1 £6.37; Table 2 £15.18; Table 3 £21.23; Table 4 £19.72

b Table 1 £0.64; Table 2 £1.52; Table 3 £2.12; Table 4 £1.97

Table 1: £7.01; Table 2: £16.70; Table 3: £23.35; Table 4: £21.69

2 **a, b** Individual answers.

3 **a** fish and chips £2.09; sandwich and fries £2.52; fried chicken £2.09; noodles and vegetables £1.35; fries £0.53; salad £1.05; ice cream £2.09; fruit salad £1.58; fresh juice £1.35; cola £1.46; milkshake £2.89

b, c Individual answers.

4 **a, b, c** Individual answers.

Answers for Pupil book page 117

1 Individual answers.

2 Individual answers.

3 Individual answers.

End-of-unit check

Use some or all of these questions and problems to assess how well the children have understood the concepts in this unit.

- Give the children something to measure. *What units would you use to measure this quantity? Why?*
- *Give an example of when we need to measure accurately using millimetres/milligrams/millilitres.* (for example: millimetres: sizes of building bricks; milligrams: masses of medicines; millilitres: when cooking)
- Give a measurement in centimetres and say: *Convert this measurement to metres.* (Repeat for other units.)
- *How many litres/kilograms/kilometres are there in 2000 ml/g/m?* (2 litres/2 kg/2 km)
- *How long is it from* (give a 12- or 24-hour time) *to* (give a 12- or 24-hour time)?
- *Which is longer: 3.25 hours or 200 minutes?* (200 minutes)
- *How many minutes are there in* (give a number) *hours?*
- *What is* (give a time in minutes and seconds or hours and minutes) *as a decimal?*
- *Which is longer, 2.45 hours or 2 hours and 45 minutes? How do you know?* (2 hours and 45 minutes: it is 2.75 hours, which is longer than 2.45 hours.)
- *How long does it take you to do your homework? Give your answer in hours and minutes, and minutes only.*

- *How could you check that you added the amounts in this list correctly?*
- *What would you pay for* (give various items and prices)?
- *Use estimates to work out whether you can afford these items* (give examples) *if you only have $50.*
- *A train leaves Kandy at 14:35 and arrives in Colombo 6.75 hours later. What is the arrival time?* (21.20)

Answers to Mixed practice 2

Answers for Pupil book pages 118–119

1 **a** $\frac{5}{8}$ of $64 = 40$ **b** correct
c correct **d** correct
e $0.01 = 1\%$ **f** $18\,\text{m} = 0.018\,\text{km}$
g $\frac{15}{18} > 0.8$ **h** correct

2 **a** $2\frac{3}{4}$ **b** $\frac{7}{25}$ **c** $\frac{7}{20}$
d $3\frac{13}{20}$ **e** $1\frac{1}{20}$ **f** $\frac{19}{20}$
g 8 **h** $5\frac{3}{4}$ **i** $\frac{1}{8}$
j $\frac{4}{5}$

3 **a** $\frac{1}{10}$ **b** $\frac{1}{4}$ **c** 37.5%
d 0.5 **e** 65% **f** 0.75
g 90% **h** $\frac{20}{20}$

4 $\frac{36}{8} = 4.5$

5 $\frac{7}{8}$

6 Beach; Gym; Trampolines

7 **a** $1\frac{1}{2} + \frac{25}{30} = \frac{3}{2} + \frac{25}{30} = \frac{45+20}{30} = \frac{70}{30} = \frac{7}{3} = 2\frac{1}{3}$

b $3\frac{11}{12} + 2\frac{1}{3} = \frac{47}{12} + \frac{7}{3} = \frac{47+28}{12} = \frac{75}{12} = \frac{25}{4} = 6\frac{1}{4}$

c $9\frac{14}{15} - 2\frac{1}{3} = \frac{149}{15} - \frac{7}{3} = \frac{149-35}{15} = \frac{114}{15} = \frac{38}{5} = 7\frac{3}{5}$

d $\frac{6}{7} \times \frac{4}{5} = \frac{24}{35}$

e $1\frac{3}{4} \times 8 = \frac{7}{4} \times 8 = 14$

f $\frac{27}{10} \times 1\frac{6}{7} = \frac{27}{10} \times \frac{13}{7} = \frac{351}{70} = 5\frac{1}{70}$

g $3\frac{1}{3} \div 3 = \frac{10}{3} \div 3 = \frac{10}{3} \times \frac{1}{3} = \frac{10}{9} = 1\frac{1}{9}$

h $\frac{17}{19} \div 3 = \frac{17}{19} \times \frac{1}{3} = \frac{17}{57}$

8 3.5 years **9** £19.30

10 9 **11** 9 kg

12 **a** $A = (-5, 4)$; $B = (-2, 3)$; $C = (-2, -2)$; $D = (-5, -1)$
b a rectangle
c $P' = (-2, 4)$; $Q' = (-5, 5)$; $R' = (-4, 2)$; $S' = (-1, 1)$
d $A' = (-2, 0)$; $B' = (1, -1)$; $C' = (1, -6)$; $D' = (-2, -5)$
e $P'' = (0, -2)$; $Q'' = (-3, -3)$; $R'' = (-2, 0)$; $S'' = (1, 1)$

13 $\frac{1}{10}$

14 £1037.40

Learning objectives
- Record, organise and represent data using a variety of tables, diagrams and graphs.
- Interpret and construct pie charts and line graphs and use these to solve problems.
- Calculate and interpret the mean, median, mode and range.
- Plan and carry out a statistical investigation.

Key words
frequency table   grouped data   line graph
pie chart   sector   mode   range
median   mean

Unit introduction

Materials
Newspapers, leaflets and Internet access so the children can search for graphs from real-life sources; large sheets of paper; glue; coloured pens.

Teaching guidance
- Spend some time looking at unusual graphs to revise the basic concepts of displaying data. Show the class examples like these:

A Type of cookies in a box

B What is in our rubbish?

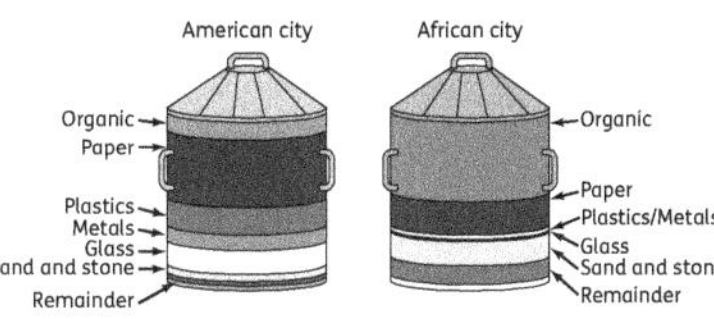

C The future of tourism

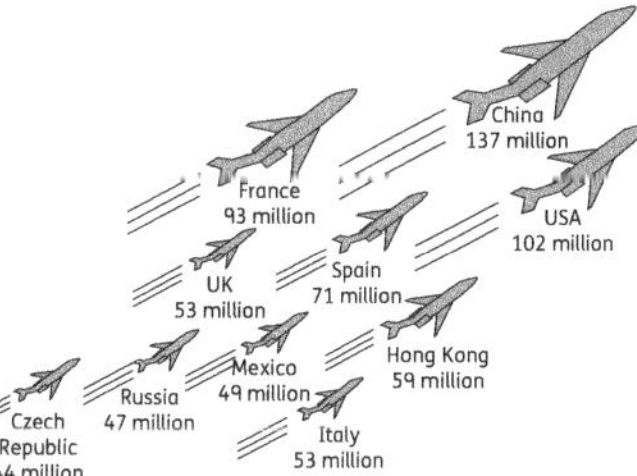

Where tourists will come from:

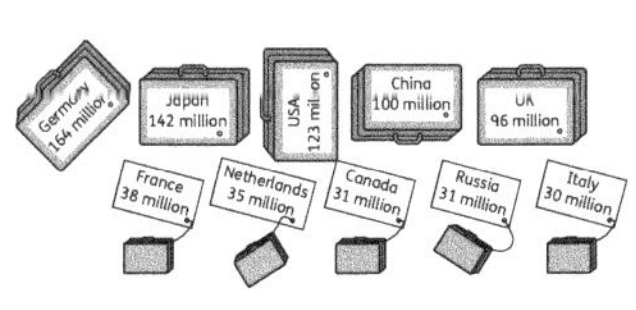

D % of replanted forests (5 years)

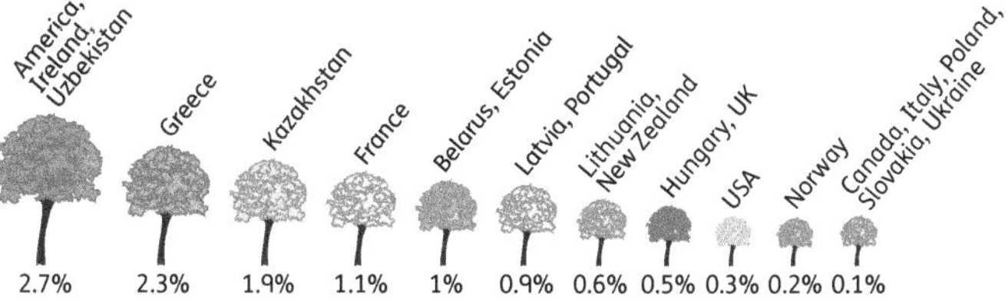

- Discuss the following questions with the class:
  - *What makes the graph unusual?*
  - *What information can you get from the graph?*
  - *Is the graph easy to understand?*
  - *How could you show this information on a graph in a more usual way?*
- Then display a different graph, such as:

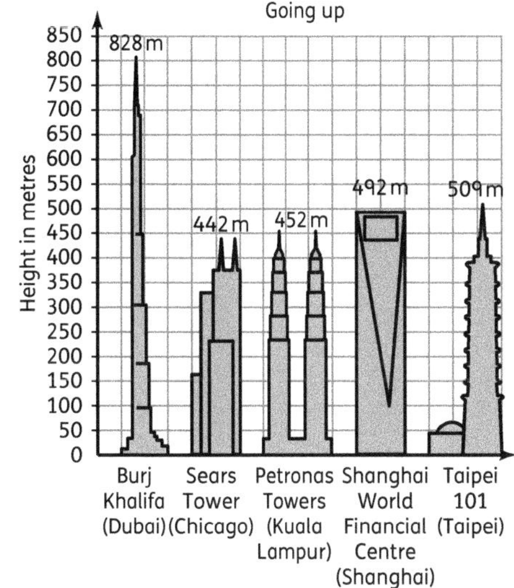

- Ask the children why they think it has been drawn like this and let them make up some questions of their own about the graph. Give different children a chance to ask their questions while the others answer them.
- Let the children work in groups to collect examples of unusual graphs from real-life sources and make a poster display of them.
- Encourage them to talk about what each graph shows and why it is unusual.
- In Year 6, the children will be expected to make sense of data in real-life contexts. This is an important skill as we are faced with many examples of graphs, tables and 'statistical facts' in the media on a daily basis.
- It is important that the children begin to think critically about how data is used and that they understand how data can be misleading if it is represented in particular ways.

Organise data

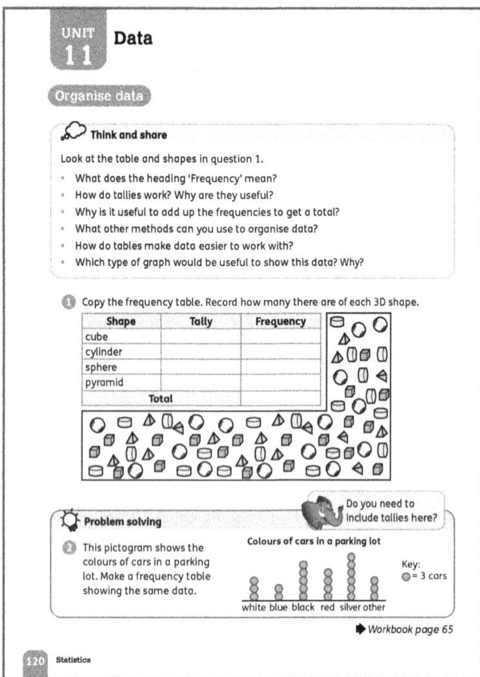

Warm-up

Use activities in which the children have to add sets of four or five small numbers as a mental starter for this lesson. This is a useful skill when working with frequency tables.

Focus

- Think and share: Turn to **Pupil book page 120**. Let the children work on the Think and share questions in pairs.
- Let different children share their answers. Allow different contributions, and resolve any misconceptions or mistakes. The questions ask the following about the table and shapes in question 1. What does the heading 'Frequency' mean? (The total number of each shape.) How do tally marks work? Why are they useful? (You can go through the data in the order in which it's given and draw a tally mark in the correct row for each item. 5 tally marks are shown by a diagonal line across the previous 4 marks. This makes it easier to find the total number of tally marks for each category.) Why is it useful to add up the frequencies to get a total? (You can then use the totals to draw a suitable chart or graph to show the data.) What other methods can you use to organise data? (Carroll diagrams) How do tables make data easier to work with? (They make it easier to spot patterns and compare the number of data values in each category.) Which type of graph would be useful to show this data? Why? (A bar chart or a 'pie chart' would be useful because the data is discrete and these allow you to compare the number of data values in each category.)
- You may need to check that the children remember the names of different types of 3D shapes before asking them to complete question 1.

- Problem solving: For question 2, remind the children to use the key to interpret the pictogram.
- Use **Workbook page 65** to revise the concept of a frequency table and to assess that the children can complete tables for ungrouped data before moving on to the next lesson about grouped data.

Answers for Pupil book page 120

1

Shape	Tally	Frequency
cube	IIII IIII IIII IIII	19
cylinder	IIII IIII IIII IIII	20
sphere	IIII IIII IIII	15
pyramid	IIII IIII IIII II	17
Total		71

2

Colour	Frequency
white	9
blue	6
black	15
red	12
silver	18
other	9

Answers for Workbook page 65

1

Number of people	Tally	Frequency
1	IIII	4
2	IIII	4
3	IIII	4
4	III	3
5	III	3
6	II	2

2 a

Number of times the pupil left the room	Tally	Frequency
0	IIII IIII II	12
1	IIII IIII	9
2	IIII III	8
3	IIII	4
4	II	2
5		0
6	I	1
Total		36

b 12   **c** 3

Grouped data

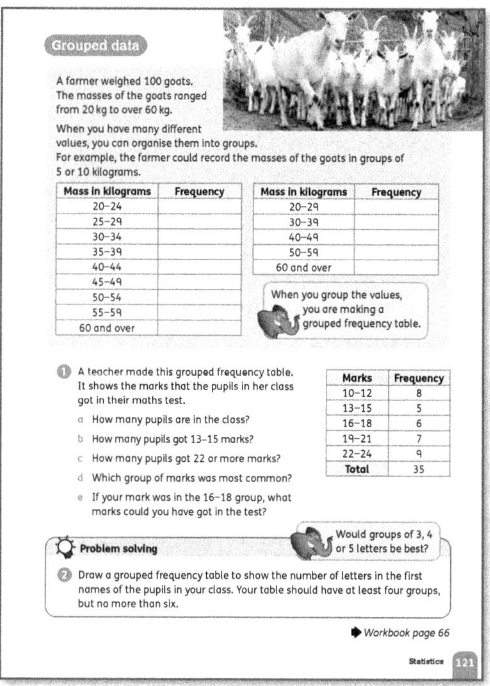

Mass in kilograms	Frequency
20–24	
25–29	
30–34	
35–39	
40–44	
45–49	
50–54	
55–59	
60 and over	

Mass in kilograms	Frequency
20–29	
30–39	
40–49	
50–59	
60 and over	

1. A teacher made this grouped frequency table. It shows the marks that the pupils in her class got in their maths test.

   a How many pupils are in the class?
   b How many pupils got 13–15 marks?
   c How many pupils got 22 or more marks?
   d Which group of marks was most common?
   e If your mark was in the 16–18 group, what marks could you have got in the test?

Marks	Frequency
10–12	8
13–15	5
16–18	6
19–21	7
22–24	9
Total	35

Problem solving

2. Draw a grouped frequency table to show the number of letters in the first names of the pupils in your class. Your table should have at least four groups, but no more than six.

Workbook page 66

Statistics 121

Materials
Cubes with faces numbered 1–6.

Warm-up
Continue to use activities in which the children add sets of four or five small numbers mentally as a starter for this lesson.

Focus
- Put the children into small groups. Let them take turns to roll two number cubes 50 times and to record how many times the possible scores (from 2 to 12) are rolled.
- Write the numbers from 2 to 12 on the board.
- Record how many times each score was rolled by the different groups. Use the format $n + n + n$ for the frequencies (for example, write 2 + 5 + 6).
- Total the number of times each score was rolled. Explain that you can make a 'frequency table' of the results by putting the results into groups.
- Draw a grouped frequency table on the board using the groups 1–3, 4–6, 7–9 and 10–12. Write in the frequencies, making sure the children realise these come from the totals you recorded earlier.
- Work as a class to draw a bar chart showing the data.
- Work through the example on **Pupil book page 121** with the class to consolidate the concept of grouped data.
- Stress that if you recorded the mass of each goat separately, you would have a table with so many rows it would not fit onto a page. Explain that you can make this simpler by grouping the masses, and that if you have more than 8 to 10 groups in a table it becomes very unwieldy. Usually, 4 to 6 groups

are sufficient. The groups should be of equal width (for example, 10–12, 13–15 and so on, not 10–15, 16–18, …) and should not overlap.
- Work through question 1 on **Pupil book page 121** with the class.
- Problem solving: For question 2, you could do the survey with the class. Ask each child to say how many letters are in their first name. Record these as they say them (for example, write 6, 3, 7, 8, 3, 9). The class can use the data to work out what groups to make and draw up their tables. Discuss how they grouped the data, and if anyone did it differently.
- After ensuring they have grasped the concepts, let the children work through **Workbook page 66** on their own. The groups are given for the data so it's easier to check that the children understand the concept. They should all get the same results.

Challenge
- Give each group a short section of text from a newspaper or magazine article.
- Ask them to count and tally the number of letters in each word in their article and then draw a grouped frequency table to show the data.
- Next, let them draw a bar chart to show the data.

Support
If the children need additional support with the grouped frequency concept, continue with practical activities such as letting them count the number of vowels in each child's first name or spinning a number spinner a number of times and recording the results.

Interesting mistakes
- Children may need additional support to choose appropriate and equal width groups for data. Help by telling them to look at the number range, for example 1–12, and then dividing it by a factor to make equal groups. Point out that you do not want too few or too many groups.
- Children may misplace numbers in grouped frequency tables or overlap the intervals. Spend some time talking about where a number on the margins should go, for example: *Where does a mass of 55 go?* This will help the children make sense of the groups and see that the heights at the ends of the intervals are included in the group.

Answers for Pupil book page 121

1. a 35   b 5   c 9
   d 22–24   e 16, 17, 18
2. Individual answers.

Answers for Workbook page 66

1

Mass in kilograms	Tally	Frequency
16–20	||	2
21–25	||	2
26–30	||||	4
31–35	ⅢⅠ |||	8
36–40	ⅢⅠ ||	7
41–45	ⅢⅠ ||	7
46–50		
	Total	30

2 a 30   b 8
c 0   d 31–35 kg
e 26 kg, 27 kg, 28 kg, 29 kg or 30 kg
f 8
g She would be in the group 31–35, because rounding 30.5 to 1 d.p. gives 31.

Line graphs

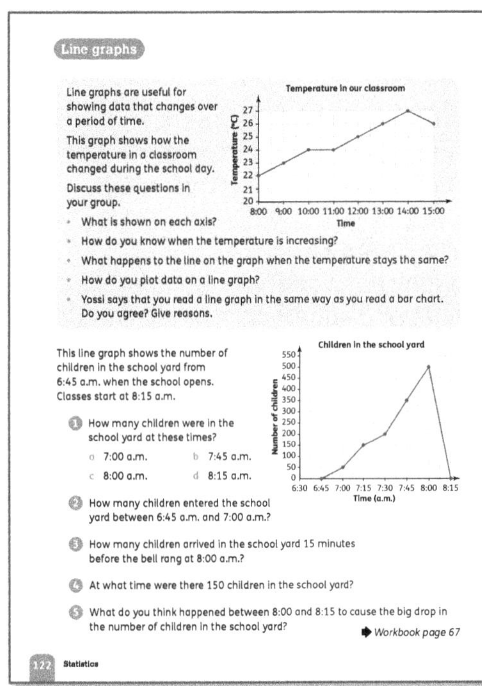

Materials
Rulers; videos or slide presentations showing how to draw line graphs (search online)

Warm-up
Use any suitable 'Mental problem-solving' activity (pages 23–24) or skip counting as a starter for this lesson.

Focus
- Use **Workbook page 67** as a baseline assessment activity to see how much the children remember about plotting and interpreting line graphs.
- Observe the children as they complete the work and help if necessary.

- Check the answers with the class and discuss any interesting mistakes.
- Turn to **Pupil book page 122**. Let the children work in groups to answer the questions about the temperature graph. The explanation box at the top of the page asks: What is shown on each axis? (Times from 8 a.m. to 3 p.m. on the horizontal axis; temperature in degrees Celsius on the vertical axis, from 20 ° to 27 °.) How do you know when the temperature is increasing? (The line slopes up from left to right.) What happens to the line on the graph when the temperature stays the same? (It is horizontal.) How do you plot data on a line graph? (Plot the temperature at each time, above the time on the horizontal axis. Use the scales on both axes.) Yossi says that you read a line graph in the same way as you read a bar chart. Do you agree? Give reasons. (Individual answers, for example: Yes, because you need to read the title and the information on both axes to understand the information.)
- Share the answers as a class to make sure that everyone can read and interpret the graph.
- The children can answer questions 1–4 orally in pairs or they can work on their own, writing the answers in their books. Take feedback from the class to check the answers.
- For question 5, note that this is a discussion and the children can only make suggestions based on what they know about how schools operate and the information they are given. So, for example, they may say that the bell rang at 8 a.m. and all the children went to assembly or to their home rooms. This is a logical reason, but there will be other acceptable answers.

Challenge
- Let the children investigate using computer programmes to draw line graphs. They can use familiar contexts such as temperature changes to do this.
- Encourage them to find tables of data online for their own region and then explore how to show the data on a graph.

Support
Let the children view some step-by-step videos or slide presentations showing how to draw a line graph. Select those most suited to your class and the contexts in which you work. This kind of visual presentation helps children who may not have grasped the basics.

Interesting mistakes
Children may need additional support to allocate data to the correct axis and to develop scales that are suitable and easy to read. You can address these mistakes by showing the class impractical and incorrect representations and discussing why they are incorrect and why they do not work.

Answers for Pupil book page 122

1. **a** 50 **b** 350 **c** 500 **d** 0
2. 50
3. 150
4. 7:15 a.m.
5. The children went to class.

Answers for Workbook page 67

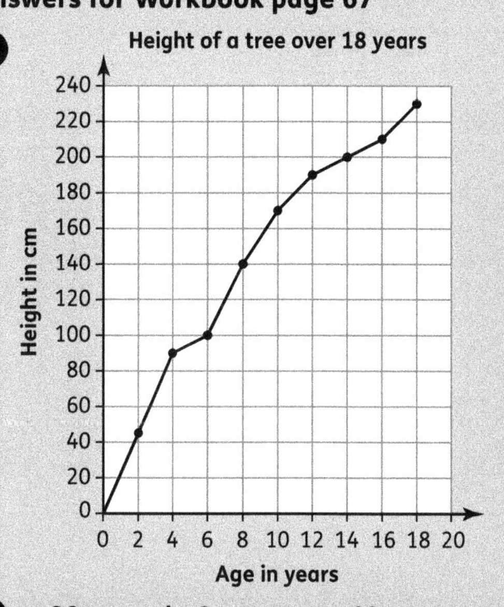

1.
Height of a tree over 18 years

2. **a** 20 cm **b** 2 years **c** 60 cm
3. **a** about 25 cm **b** about 95 cm
   c about 240 cm

More line graphs

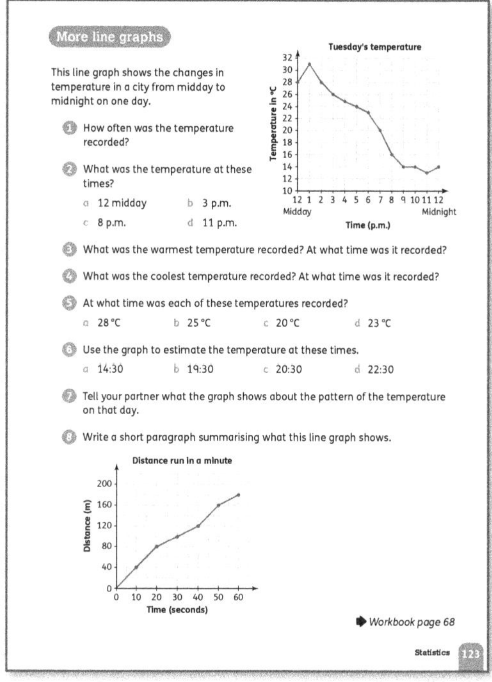

Materials

Rulers; two different-coloured pens.

Warm-up

Use any suitable 'Mental problem-solving' activity (pages 23–24) or skip counting as a starter for this lesson.

Focus

- Turn to **Pupil book page 123**. Let the children work independently or in pairs to complete questions 1–8.
- Check the answers as a class and make sure everyone can read and interpret the data.
- Use **Workbook page 68** to assess how well the children are able to plot and draw line graphs. Make sure they understand that they have to plot the two sets of data on the same axes so that they can compare them.
- Hold a class discussion about what they found challenging and check the answers.

Interesting mistakes

- Children may also see time and distance graphs as 'pictures' of the journey and ignore the mathematical aspect; they may say things like, 'He walked up a hill, went along flat for a bit and then walked down a less steep slope', when actually the person is travelling along at a given speed, then stopping for a while and then returning at a different speed.
- A number talk (pages 17–18) and discussions about the graphs and the data will help to address these misconceptions.

Answers for Pupil book page 123

1. every hour
2. **a** 28 °C **b** 26 °C **c** 16 °C **d** 13 °C
3. 31 °C; 1 p.m.
4. 13 °C; 11 p.m.
5. **a** midday and 2 p.m. **b** 4 p.m.
   c 7 p.m. **d** 6 p.m.
6. **a** 27 °C **b** 18 °C **c** 15 °C **d** 13.5 °C
7. It shows that it is hottest at 1 p.m. and then the temperature drops.
8. e.g. A pupil ran 180 metres in 60 seconds at an approximately constant speed.

1

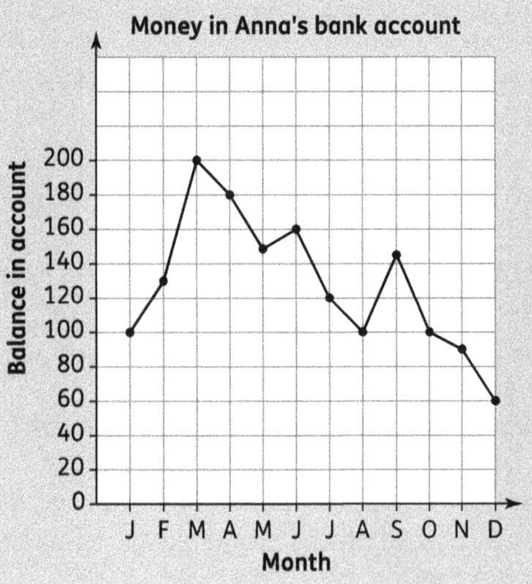

Money in Anna's bank account

(line graph: Balance in account vs Month)

2

Money in Anna's and Mae-Ling's bank account

(line graph: Balance in account vs Month, showing Anna and Mae-Ling)

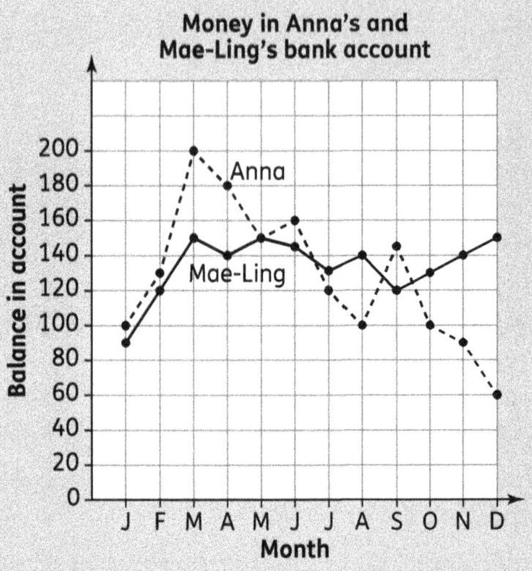

3 **a** May

b January; February; March; April; June; September

c It increased then decreased. It stayed between £100 and £200.

d It increased then decreased then increased again. It stayed between £120 and £150.

Make sense of line graphs

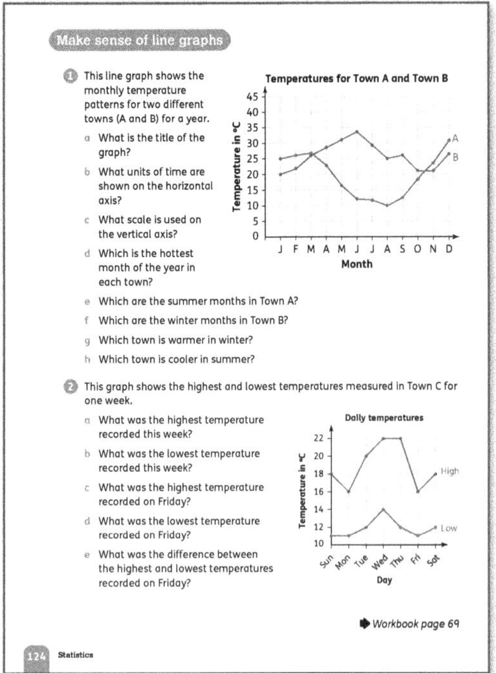

Materials
Rulers.

Warm-up
Revise counting in decimal intervals of one tenth as a mental starter for this lesson. Choose different starting points and have the children count forwards and backwards, including through whole numbers.

Focus
- This lesson builds on what the children have previously done and no new concepts are introduced.
- Turn to **Pupil book page 124**. Work through question 1 orally with the class. Give the children time to think about and answer the questions before taking contributions. Encourage the children to give reasons for their answers.
- The children can complete question 2 independently. Check their answers to make sure they can work with two sets of data on the same axes.
- Ask the class what makes a good line graph. Take suggestions and list them without initially commenting or ordering them.
- Ask the children how they could use these to make a checklist for assessing each other's line graphs.
- Develop this collaboratively with the class and leave it on display.

Follow-up

- Use **Workbook page 69** to consolidate work on accurately drawing line graphs.
- Spend time discussing the fact that the data contains decimals and that the scale you choose for the graph axes should make it easy to plot values in between whole numbers.
- Refer the children to the checklist and have them assess each other's line graphs. You can use a scale or a rating system that you agree with the children beforehand.

Interesting mistakes

Some children may make presentation errors when drawing a graph. For example, they might not give it a title; they might not label its axes. Discuss why it is important to fully label graphs. It is useful to display a checklist like the one developed in this lesson.

Answers for Pupil book page 124

1 **a** temperature figures for Town A and Town B
b months
c 1 square = 5 °C
d A December; B July
e January, February and March
f January, February and March
g Town B
h Town A

2 **a** 22 °C **b** 11 °C **c** 16 °C
d 11 °C **e** 5 °C

Answers for Workbook page 69

1 **a**

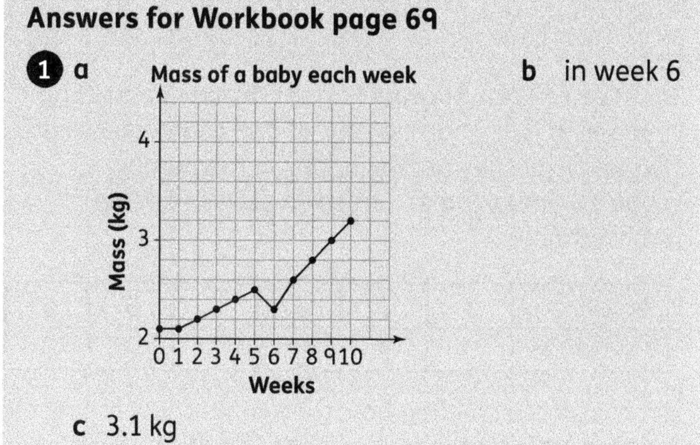

b in week 6

c 3.1 kg

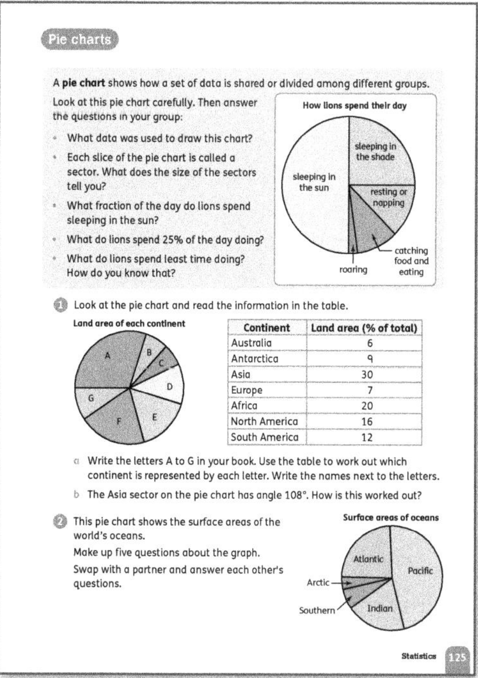

Materials

Circle template with 16 equal divisions.

Warm-up

Use any activity in which children give equivalent fractions, decimals and percentages as a mental warm-up for this lesson.

Focus

- This is the first time that the children are formally expected to interpret and construct pie charts.
- Generally, it is easier to read and make sense of pie charts than it is to construct them, so this lesson deals with interpreting the charts and making sense of the size of 'sectors'.
- Once children understand what the pie charts show and how they work (basically, they can be thought of as circular bar charts) they will find it easier to understand how to draw the charts.
- Turn to **Pupil book page 125**. Let the children work in groups to examine the pie chart about lions and answer the questions. The explanation box at the top of the page asks: What data was used to draw this pie chart? (time periods for lions' different activities in one day) What does the size of the sectors tell you? (The sectors show the proportion of a day/the times the lions spend on each activity. A larger sector represents a longer time.) What fraction of the day do lions spend sleeping in the sun? (one half) What do lions spend 25% of the day doing? (sleeping in the shade) What do lions spend least time doing? (roaring) How do you know that? (It is the smallest sector.)

- Take feedback from the groups. Focus on the fact that we can use what we already know about angles and amount of turn to work out what fraction and what percentage of the data is shown by different sectors.
- The children should understand that the size of each sector shows the data in each group, so, for this graph, half of the day is spent sleeping in the sun. This is by far the largest sector in the chart.
- Use question 1 to check that the children recognise that the largest percentage area will be shown by the largest sector, and that they can arrange the sectors by size to work out which continent is represented by each sector. Discuss the children's suggestions for how the angle for Asia is calculated.
- For question 2, the children should work in pairs to write and swap questions. You can share some of these with the class.

Challenge

Let the children make a collection of funny pie charts (bearing in mind that what they think is funny might vary). There are many available online and you may need to check that the children have selected appropriate examples.

Support

- A circle template with equal divisions can help the children compare the size of sectors. If you print these templates onto clear plastic the children can use them for any pie chart. Alternatively, print the circle smaller and have the children cut it out. When they place it on the pie chart, the chart will stick out around it.
- This one has 16 divisions, it allows the children to work out halves, quarters, eighths and sixteenths.

Answers for Pupil book page 125

1 **a** A = Asia; B = Europe; C = Australia ; D = South America; E = North America; F = Africa; G = Antarctica

b 360° ÷ 100 = 3.6° corresponds to 1% of land area. As Asia has 30% of land area, 0.3 × 360° = 108°

2 Individual answers.

Construct pie charts

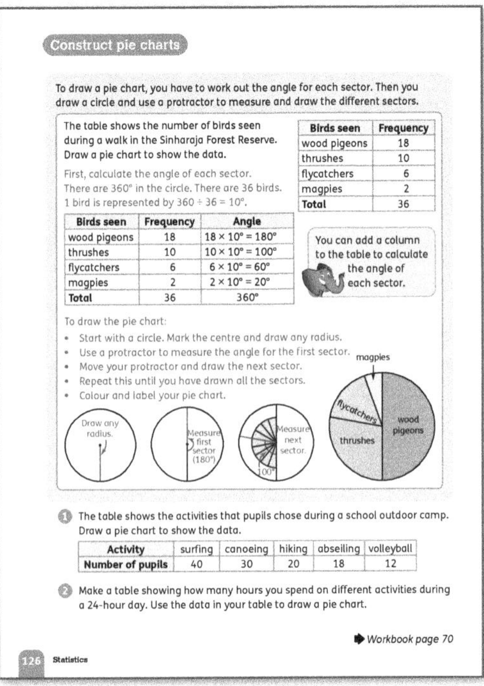

Materials

Circle template divided into 16 sectors (see previous lesson); cubes with faces numbered 1–6; compasses; rulers; protractors; video or slides on how to draw a pie chart (optional); calculators.

Warm-up

Use a calculation activity in which the children calculate percentages of amounts as a mental warm-up for this lesson. Include some simple percentages of 360.

Focus

- Ask the children to work in pairs. Give each pair a cube with faces numbered 1–6 and a circle template.
- Explain that they are going to roll the number cube 16 times and record the results in a table. For example:

Result	1	2	3	4	5	6
Frequency	2	1	6	1	2	4

- Next, ask the children how they could show the result on the circle template to make a pie chart. Guide them to see that a frequency of 2 would mean that they colour 2 sectors for that result. Let the children draw their own pie charts.
- Afterwards, ask: *How will people know what this pie chart shows?* Guide the children to recognise that the graph needs a title and that if they just coloured the sectors, they would need to make a key showing what each colour stands for.
- Explain that when we draw a pie chart, there will not always be a convenient number of pieces of data, so you have to work out what the angle of each sector is going to be.

- If possible, show the class a video or slide show that teaches how to draw a pie chart step by step.
- Turn to **Pupil book page 126**. Let the children work through the information and example in pairs to see how to calculate the angle for each sector and how to measure and draw these on the chart.
- For question 1, everyone needs a pair of compasses and a protractor. Allow them to use a calculator to check their calculations before they draw the chart.
- Let the children compare charts and discuss how they are the same or different. (The size of the circles might vary, but the sizes of the angles at the centre of each sector should be the same.)
- Discuss question 2 with the class. Talk about what categories you will use for the day. Let the children draw up a table of data before they start drawing the charts.

Follow-up
- Turn to **Workbook page 70**. Remind the class that we often use percentages instead of actual numbers and that we can draw pie charts to show the percentage of the data in each group.
- Explain that we have to work out the percentage of the whole circle which is 360°, for example: 10% of 360 = 36 degrees.
- Let the children work independently to complete the question. Observe them to make sure they are doing the correct calculations and help anyone who needs additional support to draw the sectors on the chart.

Interesting mistakes
Some children might need additional support to work out the size of the angles for each sector or to draw these once they have worked them out. Help the children by working with them to agree a 'recipe' for the calculation (for example, number of children/120 × 360°) and provide circles with equally spaced markings to help them draw the charts.

Answers for Pupil book page 126

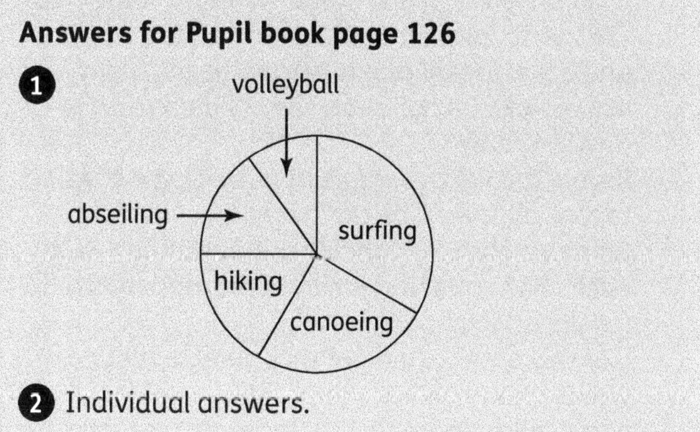

1

2 Individual answers.

Answers for Workbook page 70

1  a

Item	Proportion of money	Angle of sector
clothes	50%	180°
phone minutes/data	20%	72°
food	10%	36°
entertainment	15%	54°
transport	5%	18°

b

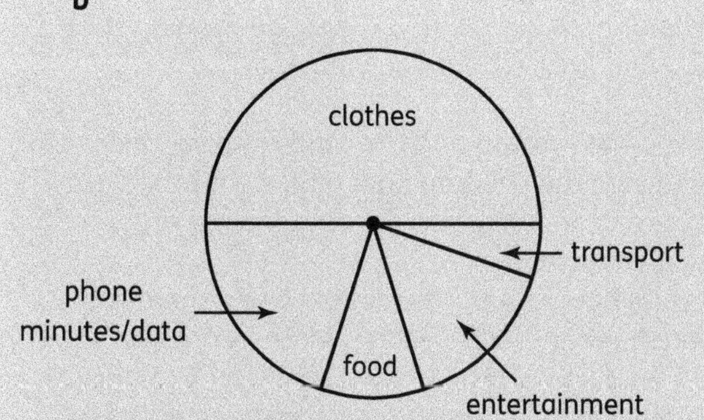

c clothes £45; phone minutes/data £18; food £9; entertainment £13.50; transport £4.50

Mode, median and range

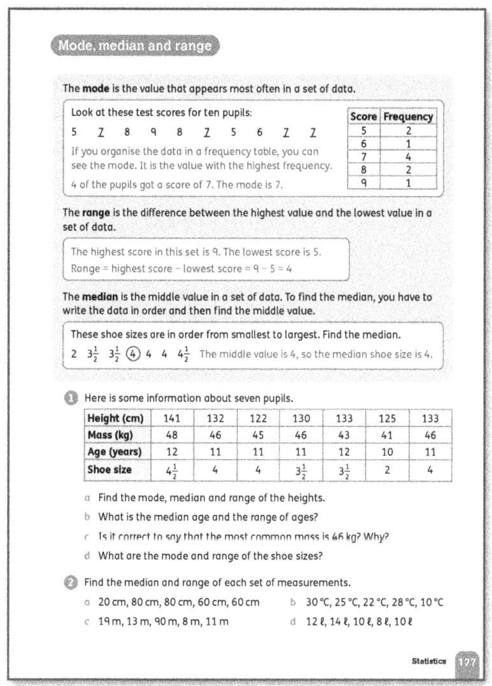

Materials

Calculators.

Warm-up

Choose any comparing and ordering activity as a mental warm-up for this lesson. Include questions asking for the number between two given numbers.

Focus

- The children have already worked with the mode, median and range of data sets.
- Revise the term 'mode' and work through the example on **Pupil book page 127** with the class to remind them how to find the mode from a set of data and from a frequency table.
- Work through the rest of the example to revise how to find the median and the range of a set of data. Make sure the children understand that the data have to be in ascending (or descending) order to find the median value.
- Discuss how to find the median when there is an even number of data values. Mathematically, you can add the two values and then divide by 2 to find the middle value (in other words, the mean of the two values) but the children can also use rough number lines to determine the midpoint between two values.
- Make sure the children understand that the median may be a number that is not in the data set.
- The children can work independently to complete the questions.

Challenge

Give the children opportunities to calculate the mode, median and range of data about the class, for example: the mode of shoe sizes of the children, the median number of people in different households or median amount of time spent on homework/studying each day, the range of numbers of pets.

Answers for Pupil book page 127

1 a mode = 133 cm; median 132 cm; range = 19 cm
  b median = 11 years; range = 2 years
  c Yes; it is the mode because it is the mass that appears the most times.
  d mode = 4; range = 2.5

2 a median = 60 cm; range = 60 cm
  b median = 25 °C; range = 20 °C
  c median = 13 m; range = 82 m
  d median = 10 litres; range = 6 litres

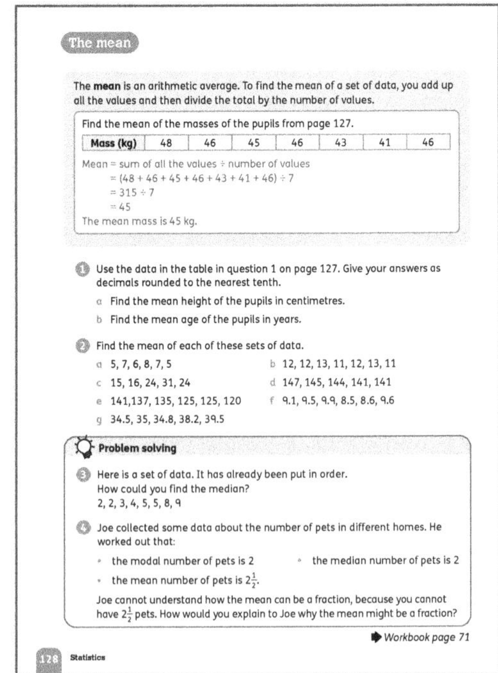

The mean

The **mean** is an arithmetic average. To find the mean of a set of data, you add up all the values and then divide the total by the number of values.

Find the mean of the masses of the pupils from page 127.

Mass (kg)	48	46	45	46	43	41	46

Mean = sum of all the values ÷ number of values
= (48 + 46 + 45 + 46 + 43 + 41 + 46) ÷ 7
= 315 ÷ 7
= 45
The mean mass is 45 kg.

1 Use the data in the table in question 1 on page 127. Give your answers as decimals rounded to the nearest tenth.
  a Find the mean height of the pupils in centimetres.
  b Find the mean age of the pupils in years.

2 Find the mean of each of these sets of data.
  a 5, 7, 6, 8, 7, 5   b 12, 12, 13, 11, 12, 13, 11
  c 15, 16, 24, 31, 24   d 147, 145, 144, 141, 141
  e 141, 137, 135, 125, 125, 120   f 9.1, 9.5, 9.9, 8.5, 8.6, 9.6
  g 34.5, 35, 34.8, 38.2, 39.5

Problem solving

3 Here is a set of data. It has already been put in order. How could you find the median?
2, 2, 3, 4, 5, 5, 8, 9

4 Joe collected some data about the number of pets in different homes. He worked out that:
  • the modal number of pets is 2   • the median number of pets is 2
  • the mean number of pets is $2\frac{1}{2}$.
Joe cannot understand how the mean can be a fraction, because you cannot have $2\frac{1}{2}$ pets. How would you explain to Joe why the mean might be a fraction?

➡ *Workbook page 71*

128 Statistics

Materials

Calculators.

Warm-up

Do some activities in which the children use a calculator to add sets of values as a warm-up for this lesson. Focus on working accurately with the calculator.

Focus

- Ask the children whether they have heard the word 'average' before and what they think it means. Build on the children's ideas to establish the correct meaning.
  ○ Without the children seeing, on the board write 30 everyday words that are familiar to the children. Cover them with a screen or large cloth.
  ○ Tell the children they will be shown the words and have one minute to remember as many as possible. Do not allow the children to write anything down.
  ○ Reveal the words for one minute for the children to look at; then hide them again.
  ○ Go round the class finding out how many words each child correctly remembered, and record the results on the board.
  ○ Use this data to find out the mean number of words remembered.
- Turn to **Pupil book page 128** and work through the example with the class to consolidate the concept.
- For question 1, make sure the children understand that they need to use the data from the previous lesson.
- For questions 1 and 2, the children should use a calculator as the focus is on finding the mean, rather than calculation skills.
- <u>Problem solving:</u> Give the children time to think about question 3 and then discuss it orally with the class.

- <u>Problem solving</u>: For question 4, the children think about their explanations and then ask different children to share their ideas with the class.

Follow-up

- Use **Workbook page 71** for additional practice and to consolidate working with different types of averages. Let the children complete the work independently, using a calculator if necessary.

Challenge

- Show the children a range of reading books aimed at different age ranges.
- Ask them how they might be able to identify which book was for which age range if they were not familiar with the stories.
- Use the children's ideas to write a hypothesis, for example: 'Books for older children have more long words in them.'
- Ask the class how they could test the hypothesis. Use their suggestions to set up a test of the books. For example, the children might find out the mode and mean length of words on a single page.
- Talk with the children about the importance of the size of a sample and pool their results to increase the sample size and gain a more accurate analysis.
- Discuss the children's findings and whether the hypothesis is true.

Interesting mistakes

- Children might not recognise that the mean does not have to be a value in the data.
- Also, they might not calculate the mean exactly. Help by encouraging them to include any remainders as part of the mean; they should think about the data and decide whether the remainder should be given as a number, a fraction or a decimal.

Answers for Pupil book page 128

1 a 130.9 cm   b 11.1 years
2 a 6.3   b 12   c 22   d 143.6
   e 130.5   f 9.2   g 36.4
3 You find the mean of the 4th and 5th pieces of data.
4 The mean does not need to be one of the data values; you divide by the number of values to find the mean, and that may give a fraction answer.

Answers for Workbook page 71

1 Pupil 1: mode 21; median 21; mean 21.6; mean of top three 22.7
   Pupil 2: mode 20; median 20; mean 20.8; mean of top three 21.7
   Pupil 3: mode 24; median 24; mean 23.4; mean of top three 24.3
   Pupil 4: mode 19; median 20 ; mean 20.4; mean of top three 21.3

Plan and carry out your own investigation

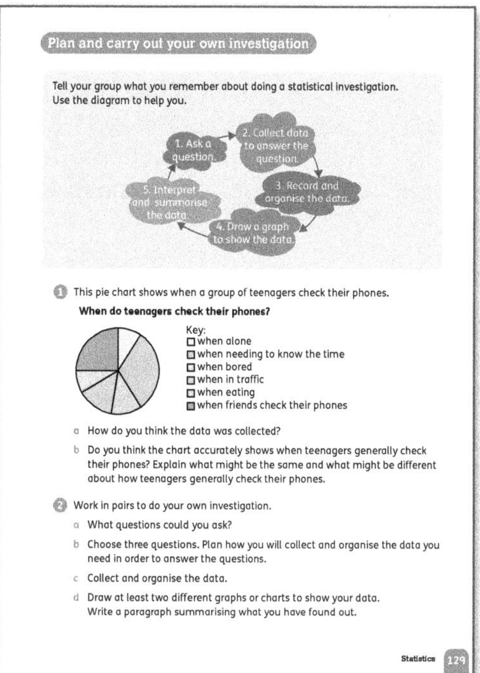

Materials

Squared paper; Internet access (optional).

Warm-up

- Spend time discussing the steps in a statistical investigation with the class by looking at **Pupil book page 129**.
- Pay attention to the first step 'Ask a question', and let the children suggest good questions for investigations. (It should be possible to collect or find data to answer the question.)
- Talk about different methods of collecting data, including surveys and research using secondary sources.

Focus

- Focus the class on the questions on **Pupil book page 129**.
- For question 1, discuss the pie chart. Let the children share their ideas about how the data could have been collected and then let them compare the data with their own (or friends' and family's) phone use.
- Be sensitive to the fact that some children might not have phone access of their own; if necessary, let the children work in groups, where at least one person has a phone.
- For question 2, work with the class to list a set of questions that they could investigate.
- Then let the children work in pairs to talk about the questions and decide how they could answer them by collecting, organising and displaying data.
- Work through the questions one by one, asking groups to share their plans.

- Explain that the children are going to plan and carry out their own investigation over the next few days. Let them write up a plan for collecting, organising and displaying the data. Check these are sensible and practical before they start work. Discuss whether finding any of the mode, median, mean or range would be sensible for the data they collect.
- Set some time aside over the next few lessons for the children to work on their investigations. Decide what they need to present to you or the class as a final product.

End-of-unit check

Use some or all of these questions and problems to assess how well the children have understood the concepts in this unit.

- Show an incomplete frequency table. *What do these tally marks add up to?*
- Use a grouped table like the one on **Pupil book page 121**. *How could you group this data to make a smaller table?* (Create groups that cover a greater range of masses.) *Why should the groups not overlap?* (Each value can only go into one group.)
- Show a grouped table of marks in a test. *How many children got* (give a group) *marks? How many pieces of data were collected altogether? How did you work that out?* (Add up all the frequencies.)
- Show a graph. *What are three facts you know from looking at this graph?*
- Show a graph. *What question could be answered using this graph?*
- *What tips would you give someone about drawing this type of graph?* (Pie chart: work out the total number of items and divide 360° by this number to find the angle for one item; multiply the frequency for each item by this angle; draw a circle and a radius; starting from the radius, use a protractor to measure the angle for each sector; colour and label the pie chart. Line graph: find the maximum value and choose a scale that will go up to this value with suitable intervals; draw and label the axes; plot the data points; join the data points with straight lines; label the graph.)

- Show some data. *What is the best way to represent this data? Why?*
- *How can the interval chosen for the scale make a line graph look different?* (It can make the changes look very small or very large.)
- Show a line graph. *Why does the line on this graph slope down to the right?* (As the values on the horizontal axis increase, the values on the vertical axis decrease.)
- Show a poorly drawn or poorly labelled graph. *What is wrong with this graph?*
- *Why is it important to label graphs clearly?* (So that anyone looking at the graph knows what it shows.)
- Give a data set. *What is the mode/median/range?*
- *How do you find the mean/mode of a set of data?* (Mean: add together all the values and divide by the number of values. Mode: find the value that occurs most often.)
- *I have calculated this mean of a set of data. What does this tell you?* (For example: the mean is a number that represents the data set.)
- *Why is the mean and mode of this set of data so different?* (For example: the mode may be the lowest or highest value.)
- *How do you find the median of a set of data?* (Put the values in order and find the middle value.)
- *When would it be useful to use the mean/mode/ median?* (For example: Mean: average height of children in a class. Mode: most common shoe size. Median: representative wage in a company, because it is not affected by very high wages for managers.)
- *What are the highest and lowest values in this set of data? How do they affect the mean/mode?*
- *What steps do you follow to do an investigation? Why?* (Ask a question. Collect data to answer the question. Record and organise the data. Draw a graph to show the data. Interpret and summarise the data. The steps form a cycle so that you can change the question according to what the data shows.)
- *Give an example of a good question for an investigation. What makes it a good question?* (For example: How much time do children spend playing sport each week? You can collect data on how much sport children do each week.)

UNIT 12 — Ratio and proportion

Learning objectives

- Understand the relationship between quantities when they are in direct proportion.

- Use knowledge of equivalence to understand and use equivalent ratios.

- Solve problems involving the relative size of two quantities where missing values can be found by using integer multiplication and division facts.

- Solve problems involving similar shapes where the scale factor is known or can be found.

- Solve problems involving unequal sharing and grouping using knowledge of fractions and multiples.

Key words

ratio   proportion   scale factor

Unit introduction

Materials

Coloured cubes or counters.

Teaching guidance

- The children worked with ratio and proportion in Level 5.
- Start this topic by putting the children into groups and presenting them with four different problems.
- Give each group a large sheet of paper to record how they work to solve each problem.
- Tell them that each member of the group must be prepared to tell the class how the group worked to solve their problem.
- Display the problems for the class:

  ○ **Problem A**
  Two angles, a and b, make a right angle. They are in the ratio 1 : 5. What is the size of each angle in degrees?

  ○ **Problem B**
  The ratio of cubes to cones in a set of shapes is 3 : 1.
  The ratio of cones to pyramids in the same set of shapes is 2 : 5.
  What is the ratio of cubes to pyramids in the set?

  ○ **Problem C**
  Jess is 10 years old and Naresh is 20 years old. They decide to share $48 in the ratio of their ages. How much will each person get?

  ○ **Problem D**
  This is the recipe for a fruit smoothie.

  Fruit smoothie (makes 3 glasses)
  3 frozen bananas
  2 fresh bananas, peeled and sliced
  475 ml oat or dairy milk
  25 g raspberries or blueberries

  How much of each ingredient will you need to make 6 glasses?
  How much will you need to make 5 glasses?

- Let the children work in groups to decide how to solve each problem and record their working. Encourage them to use coloured cubes or counters to model the problems, or to draw diagrams to help them.
- When they have finished, work through the problems one by one, asking different groups to share how they worked out their answers. Problems C and D involve sharing into unequal groups. Using diagrams or manipulatives to model these allows the children to see each 'share' as a fraction of the total.

(Problem A: $a = 15°$; $b = 75°$. Problem B: 6 : 5. Problem C: Jess gets $16 and Naresh gets $32. Problem D: for 6 glasses: 6 frozen bananas, 4 fresh bananas, 950 ml oat or dairy milk, 50 g raspberries/blueberries; for 5 glasses: 5 frozen bananas, 3 $\frac{1}{3}$ fresh bananas. 971$\frac{2}{3}$ ml oat or dairy milk, 41$\frac{2}{3}$ g raspberries/blueberries.)

Ratio

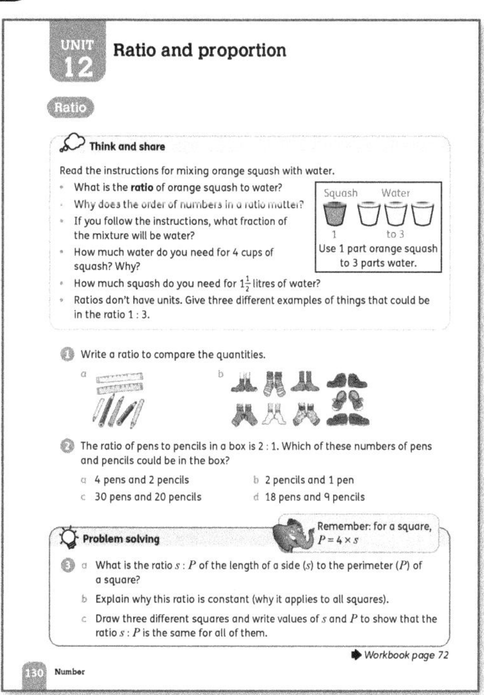

Materials

Blue, yellow and black colouring pencils; coloured beads or cubes.

Warm-up

Play 'Three or four in a row' (page 26) with equivalent fractions as a mental starter for this lesson.

Focus

- <u>Think and share:</u> Use the activity on **Pupil book page 130** to start a number talk (pages 17–18) with the class. The children are asked the following questions. What is the ratio of orange squash to water? (1 : 3) Why does the order of numbers in a ratio matter? (The order of the numbers in the ratio is the same as the order of the amounts when they are written in words. Changing the order would change the proportions.) If you follow the instructions, what fraction of the mixture will be water? $\left(\frac{3}{4}\right)$ How much water to do need for 4 cups of squash? Why? (12, because you need 4 times as much water as for 1 cup of squash.) How much squash do you need for $1\frac{1}{2}$ litres of water? $\left(\frac{1}{2} \text{ litre}\right)$ Ratios don't have units. Give three different examples of things that could be in the ratio 1 : 3. (For example, a recipe for jam asks for 1 cup of sugar for every 3 cups of fruit; in a blend of flour there is 3 times as much wheat flour as rice flour; in a garden there are 20 yellow flowers and 60 blue flowers.)
- Once you are confident that the children know what a ratio is and how to represent it, ask them to complete the colouring activity on **Workbook page 72**. They can compare and check each other's answers. Ask questions to make sure the children can see that the order or pattern in which you colour blue : yellow doesn't affect the ratio. If the ratio of blue : yellow is 1 : 2, all that matters is that for every one yellow square, there are two blue ones. The children could colour y, b, b, y, b, b, but they could also colour $\frac{1}{3}$ of the shape blue and $\frac{2}{3}$ yellow in solid colour blocks. Use the children's completed answers to revise simplifying ratios by dividing every number in the ratio by the same number.
- Turn back to **Pupil book page 130** and let the children work on questions 1 and 2 independently.
- For question 1, let the children discuss this, and in part b allow them to decide what they are comparing. For example, they could compare single socks to single shoes, pairs of socks to shoes, or feet to socks and shoes.
- <u>Problem solving:</u> The children could work on question 3 in pairs or hold a class discussion about how the ratio of side length to perimeter of a square $(s : P)$ will always be 1 : 4 because the perimeter of a square is always four times the length of one side.

Support

- If the children still seem unsure of the concept, let them work in pairs or small groups to make patterns with beads or cubes where the colours appear in given ratios. For example, the ratio of blue to yellow beads is 2 to 3.
- Ask questions to make sure they understand that the position of the beads in the pattern is irrelevant as long as there are two blue beads for every three yellow beads.

Challenge

Play some '5 in a row' games using equivalent ratios. Prepare a sheet like the one below and let the children work in pairs to take turns to choose and solve a problem. The first child to get five correct answers in a row is the winner.

5 : 15 = □ : 45	2 : 3 = 5 : □	7 : □ = 21 : 30	10 : 12 = 15 : □	1 : 8 = □ : 20
20 : 22 = □ : 11	This block is free – anyone can use it.	□ : 3 = 9 : □	3 : 1 = 1$\frac{1}{2}$: □	3 : 2 = □ : 40
□ : 12 = 5 : 15	16 : □ = 8 : 9	□ : 4 = 75 : 100	4 : 9 = □ : 27	12 : □ = 36 : 57
1 : 2 = □ : 9	10 : □ = □ : 54	7 : 8 = □ : 64	1 : 5 = □ : 180	2 : 3 : 1 = □ : 9 : □
2 : 3 = □ : 21	5 : □ = □ : 40	6 : 7 = □ : 63	□ : 10 = 9 : 45	125 : 1000 = 1 : □

Interesting mistakes

Children may not see the connection between ratio and the parts of the whole, so might not see that a ratio of 3 to 5 means that the whole is divided into two groups making $\frac{3}{8}$ and $\frac{5}{8}$. Lots of practical activities will help the children to develop the concept clearly.

Answers for Pupil book page 130

1 a rulers : pencils = 2 : 5
    b socks : shoes = 12 : 6 = 6 : 3 = 2 : 1

2 a, b and d

3 a 1 : 4
    b The perimeter is always four times the side length
    c Individual answers.

Answers for Workbook page 72

1 and **2**

a 16 blue shaded, 32 yellow shaded
yellow : blue = 8 : 4
blue : yellow = 4 : 8

b 12 blue shaded, 30 yellow shaded
yellow : blue = 5 : 1
blue : yellow = 1 : 5

c 12 blue shaded, 28 yellow shaded
yellow : blue = 7 : 3
blue : yellow = 3 : 7

d 24 blue shaded, 24 yellow shaded
yellow : blue = 3 : 3
blue : yellow = 3 : 3

e 8 blue shaded, 32 yellow shaded
yellow : blue = 4 : 1
blue : yellow = 1 : 4

f 27 blue shaded, 3 yellow shaded
yellow : blue = 1 : 9
blue : yellow = 9 : 1

g 28 blue shaded, 56 yellow shaded,
14 black shaded
yellow : blue = 8 : 4
blue : black = 4 : 2
black : yellow: blue = 2 : 8 : 4

h 5 blue shaded, 10 yellow shaded,
25 black shaded
blue : yellow = 1 : 2
black : blue = 5 : 1
black : yellow : blue = 5 : 2 : 1

Proportion

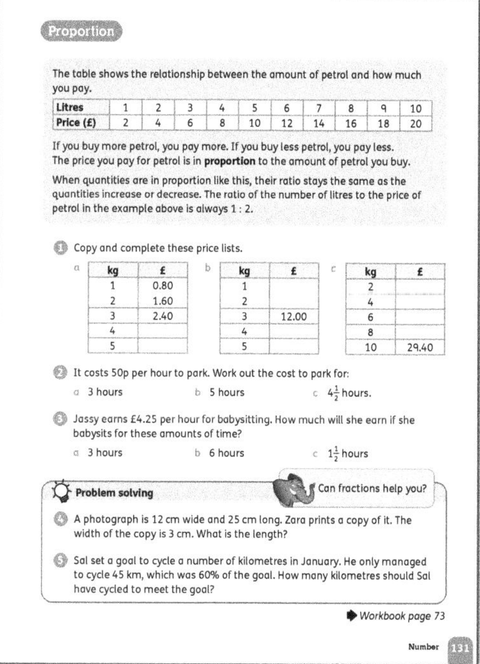

Materials

Colouring pencils.

Warm-up

Use any doubling or halving activities as a mental starter for this lesson. Include some decimals in the calculations.

Focus

- Use **Workbook page 73** to revise the concept of proportion. Let the children work on their own to complete the questions.
- Turn to **Pupil book page 131**. Read through the example with the class, making sure they can see the pattern in the number sequences and relate the amount of petrol to the price you pay.
- The children will know that you will pay more when you buy a number of things that cost the same. For example, one packet of chips might cost 50p, so 2 packets will cost £1 and 3 packets will cost £1.50. At this level, there is no need to discuss direct and indirect proportion. All the children need to understand is that amounts increase and decrease in proportion to each other and that they can use the relationship to solve simple problems.
- Use questions 1–3 to check that the children understand the basic concepts.
- In question 3c, the children will have to round their answer up to the nearest penny.
- <u>Problem solving:</u> Let the children discuss questions 4 and 5 in pairs before sharing their ideas with the class. For question 4 ensure the children understand that as the width of the copy is $\frac{1}{4}$ the original width, the length of the copy is $\frac{1}{4}$ the original length. For question 5, a bar model (page 21) showing that 60% $\left(\frac{6}{10}\right)$ is 45 km may help them to work out 10% and hence 100%.

Interesting mistakes

Children often need additional support with proportional reasoning when they have to solve problems involving ratio and proportion. Help them by providing practical activities in which you lay out items in specific ratios and multiply and/or divide them in the given proportion. A proportion can be given as a fraction, as a decimal or as a percentage. 'What proportion?' means 'What fraction?', or 'What decimal?', or 'What percentage?' For example:

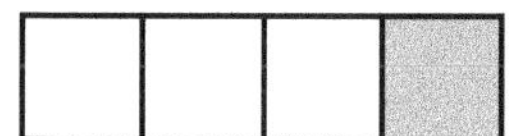

There are 4 squares altogether.

- 1 out of 4 squares is grey $\left(\frac{1}{4}, 0.25, 25\%\right)$.

- 3 out of 4 squares are white $\left(\frac{3}{4}, 0.75, 75\%\right)$.

We could say:

- The proportion of grey squares is 1 in every 4.
- The proportion of white squares is 3 in every 4.

Answers for Pupil book page 131

1 a

kg	£
1	0.80
2	1.60
3	2.40
4	3.20
5	4.00

b

kg	£
1	4.00
2	8.00
3	12.00
4	16.00
5	20.00

c

kg	£
2	5.88
4	11.76
6	17.64
8	23.52
10	29.40

2 a £1.50   b £2.50   c £2.25

3 a £12.75   b £25.50   c £6.38

4 6.25 cm

5 75 km

Answers for Workbook page 73

1 a $\frac{2}{5}$ b 1   c $\frac{2}{7}$ d $\frac{1}{5}$

e $\frac{1}{3}$ f $\frac{2}{3}$ g $\frac{1}{2}$ h $\frac{1}{4}$

2 Individual answers.

Scale

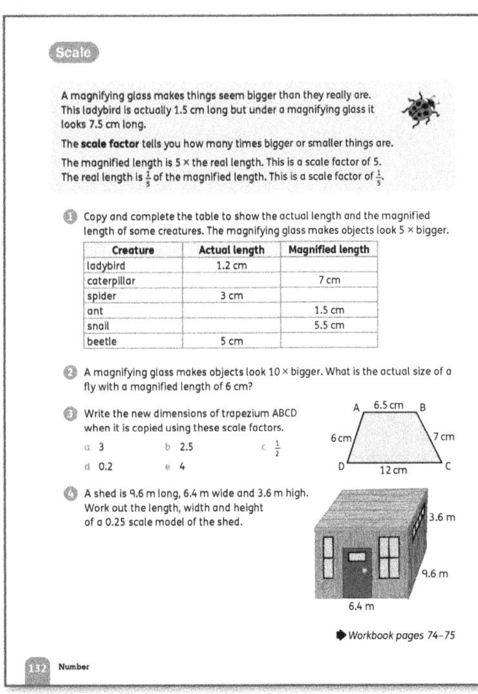

Materials
Phone camera, if possible; rulers; calculators.

Warm-up
Use doubling, tripling and halving activities with whole numbers, fractions and decimals as a mental starter for this lesson.

Focus
- The children have probably experienced enlarging or reducing images on a copier, screen, tablet or camera. Some cameras show you the 'scale factor' when you zoom in. Spend some time talking about what it means to enlarge or reduce an image.

- Guide the children to recognise that, when a picture is enlarged, every dimension on the picture is multiplied by the same amount. For example, if a picture is enlarged by scale factor 10 it means that it will be 10 × longer and 10 × wider, and that everything you can see in the picture will be 10 times the size of the original.
- Turn to **Pupil book page 132**. Use the ladybird example to talk about scale factors. A scale factor is really a ratio: if you enlarge by a scale factor of 10, the ratio of the original to the enlargement is 1 : 10; the original is $\frac{1}{10}$ of the size of enlargement.
- Let the children work through the questions 1–4 on their own. Then let them explain how they solved each problem.

Follow-up
- Put the children into groups and let them discuss the problems on **Workbook page 74**. Ask groups to share what they think they need to do to complete each question and how they will do this.
- Once the children have shared their ideas, let them work collaboratively to complete the questions.
- **Workbook page 75** involves making scale drawings. Ask the children to work independently to complete the questions.
- Let the children compare their drawings and check each other's work.

Challenge
The children can investigate map scales and how they work. They may have done some of this in geography lessons. Let them find different scales (line scales, word scales and ratios) on maps and explain what they mean.

Interesting mistakes
- Children may need additional support to increase or decrease amounts in proportion because they do not understand the concept of multiplicative reasoning. It is important to emphasise that doubling, halving, finding three times as much or a quarter as much, all involve multiplying or dividing.

Answers for Pupil book page 132

1

Creature	Actual length	Magnified length
ladybird	1.2 cm	6 cm
caterpillar	1.4 cm	7 cm
spider	3 cm	15 cm
ant	0.3 cm	1.5 cm
snail	1.1 cm	5.5 cm
beetle	5 cm	25 cm

2 6 mm

3 a 19.5 cm; 21 cm; 36 cm; 18 cm

b 16.25 cm; 17.5 cm; 30 cm; 15 cm

c 3.25 cm; 3.5 cm; 6 cm; 3 cm

d 1.3 cm; 1.4 cm; 2.4 cm; 1.2 cm

e 26 cm; 28 cm; 48 cm; 24 cm

4 2.4 m by 1.6 m by 0.9 m

Answers for Workbook page 74

1

Triangle	Length of sides (cm)	Scale factor
A	12	2
B	9	1.5
C	4.5	0.75
D	21	3.5
E	1.25	$\frac{5}{24}$

2

Shape	Perimeter (cm)	Scale factor 3	Scale factor 2.5	Scale factor $\frac{1}{4}$	Scale factor 0.75
A	44	132	110	11	33
B	6.8	20.4	17	1.7	5.1
C	625	1875	1562.5	156.25	468.75
D	12.6	37.8	31.5	3.15	9.45

3 a 0.1   **b** 1.5

Answers for Workbook page 75

1 a

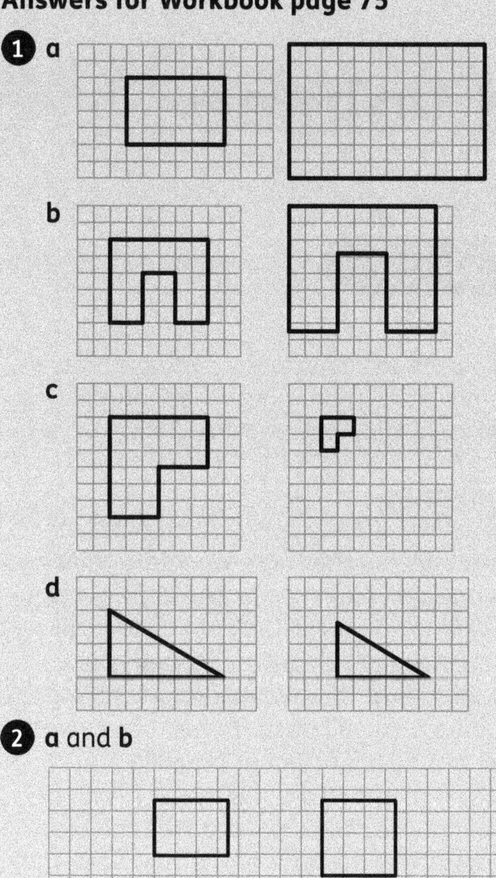

b

c

d

2 a and b

Ratio and proportion problems

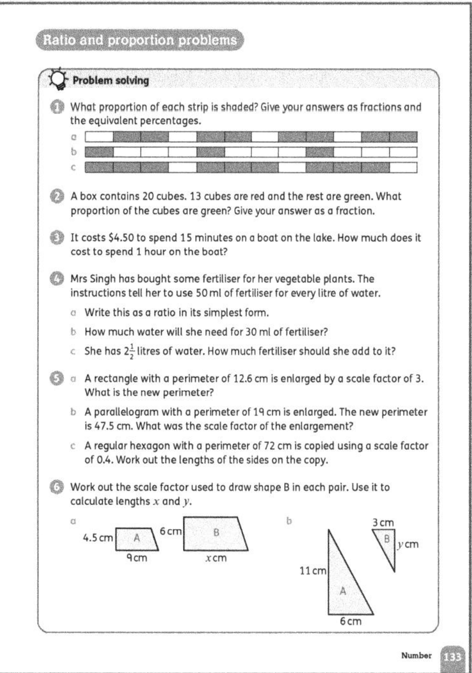

Warm-up

Use some different doubling, tripling and halving activities with whole numbers, fractions and decimals as a mental starter for this lesson.

Focus

Problem solving: There are no new concepts in this lesson. The children can work on their own or in pairs to solve questions 1–6 on **Pupil book page 133**. Encourage them to look back to the relevant examples in the **Pupil book pages 130–132** and **Workbook pages 72–74** if they are not sure what to do.

Answers for Pupil book page 133

1 a $\frac{8}{12}$; 66.7%   **b** $\frac{3}{12}$; 25%   **c** $\frac{9}{12}$; 75%

2 $\frac{7}{20}$

3 $18

4 a 1 : 20   **b** 600 ml   **c** 125 ml

5 a 37.8 cm   **b** 2.5   **c** 4.8 cm

6 a $x = 12$ cm   **b** $y = 5.5$ cm

End-of-unit check

Use some or all of these questions and problems to assess how well the children have understood the concepts in this unit.

- *There are 15 books and 5 pencils in a box. What sentence could you say to describe the items in the box using the word 'ratio'?* (For example: The ratio of books to pencils is 15 to 5 or 15 : 5 or three to one or 3 : 1.)
- *Write a question that has the answer 2 to 3.* (For example: In this box, there are 2 red pens and 3 blue pens. What is the ratio of red pens to blue pens?)

- *Write a word problem involving ratio that has the answer 30. (For example: the ratio of cows to sheep on a farm is 1 : 3. The farmer has 10 cows. How many sheep does he have?)*
- *When might you use ratio in real life?* (for example: cooking, mixing juice and water)
- *This recipe is for 6 people. What do you have to do to the ingredients if you only want to make enough for 3 people?* (Halve the amounts.) *What if you want to make enough for 2 or 9 people?* (2 people: divide the amounts by 3; 9 people: multiply the amounts by 1.5.)
- Show a necklace or picture of a beaded necklace. *What proportion of this necklace is (give a colour) beads?*

- *When or where do you think we use these ratios?*
  - *Use 1 litre to cover 5 square metres of wall.* (painting a wall)
  - *Mix 1 part flour to 1 part water.* (making glue or dough)
  - *Melt 100 g chocolate for every 200 ml of cream.* (making a chocolate sauce)
  - *You will need 5 ml of butter per slice of bread.* (making sandwiches)
  - *You need 1.5 m space for each person.* (working out how many people will fit in a room)
- Draw a set of shapes and colour them using three different colours. *What proportion of shapes is each colour?*
- Provide some problems involving ratio and proportion, for example:
  - *7 bags of dried fruit cost £17.50. What do 5 bags cost?* (£12.50)

UNIT 13 Perimeter, area and volume

Learning objectives
- Recognise that shapes with the same areas can have different perimeters and vice versa.
- Use simple formulae.
- Recognise when you can use a formula to find the area and volume of shapes.
- Use knowledge of the area of rectangles to estimate and calculate the area of right-angled triangles.
- Calculate the area of parallelograms and triangles.
- Understand the difference between volume and capacity.
- Calculate, estimate and compare the volume of cubes and cuboids using standard units, including cm^3 and m^3 and extending to other units (mm^3 and km^3).

Key words
perimeter   formula   composite shape   area
volume   cubic units   capacity

Unit introduction

Materials
Small irregularly shaped objects (for example, erasers, plastic toys, small balls or marbles, small containers); rulers; card; scissors; glue or tape; shoe box.

Teaching guidance
- The children have already learnt about perimeter, area and volume and they have worked with nets of cuboids and other 3D shapes. The activity below combines work on all these topics in a practical way.
- Give each child a small irregularly shaped object.
- Ask the children how they would design and make a net of a cuboid-shaped box to fit their object with as little space around the object as possible.
- Give the children time to discuss this. If necessary, show them that they can draw around the object to get a 'perimeter' for the base of the box. Next, they can use a ruler to work out the height of the tallest part to decide how tall the box needs to be.
- Give the children the stationery items and let them use the net they have designed to make a box. They place the object in the box to check that it fits.
- Let each group compare the boxes they have made. Ask the children to arrange them in order, from smallest to greatest in volume.

- Discuss how they could check that the order is correct (by calculating the volume of each cuboid using length × width × height; they learnt this in Level 5.)
- You can extend this to an area task, by asking how many boxes of a given size will fit next to each other on a shelf of given dimensions.
- You could also include more volume activities by asking how many of the small boxes could fit into a shoe box (provide a box or its dimensions).

Perimeter

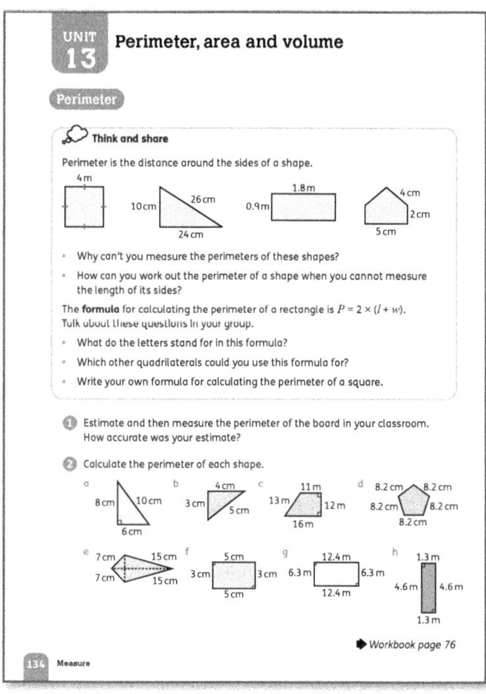

Plastic or cardboard shapes; rulers or measuring tapes.

Warm-up
- Use any suitable 'Mental problem-solving' activity (pages 23–24) as a starter for this lesson.
- You might like to include some work on mentally doubling numbers as this will help the children to find the perimeter of rectangles.

Focus
- Revise measuring accurately in centimetres and millimetres by doing some practical activities using lines and shapes.
- Arrange the class into groups. Hand out a number of small shapes to each group.
- Let the children draw around each shape to show its perimeter on paper. Then have them measure the perimeter in centimetres and/or millimetres. If you let the children measure in centimetres, you can also use this activity to revise converting from a larger to a smaller unit of measure.
- <u>Think and share</u>: Turn to **Pupil book page 134**. Read the information to revise the concept of perimeter. Answer the questions about the shapes with the class and let them work out the perimeter of each

shape, explaining how they did this (especially how they worked out any unknown dimensions using the known properties of shapes). (The perimeters are, from left to right: 16 m (all sides are the same length), 60 cm, 5.4 m (opposite sides are equal in length) and 17 cm (the shape has a line of symmetry so the unknown sides are 4 m and 2 cm).) The first question asks why you can't measure the perimeters of these shapes. (The shapes are not drawn accurately, so measuring the perimeter of the shape would not give an accurate answer.) The second question asks how you can work out the perimeter of a shape when you cannot measure the length of its sides. (You can work out any unknown dimensions and then add the lengths of all the sides.)
- Let the children work together to answer the questions about the 'formula'. Ask some children for their ideas. (The letter P stands for perimeter, l stands for length and w stands for width. You can only use the formula for the perimeter of a rectangle. The formula for the perimeter of a square can be written $P = 4 \times l$, where l is the length of each side.)
- For question 1, ask the children to share their estimates and record these. Then let a few children measure the perimeter of the board. Record their measurements.
- For question 2, the children can work independently to calculate the perimeter of each shape.

Follow-up
- Use the questions on **Workbook page 76** to check that the children are able to work out unknown side lengths and use them to calculate perimeter as well as to work back from a given perimeter to find the lengths of sides.
- Check the answers with the class and resolve any misconceptions or confusion.

Interesting mistakes
- When children have to work back from a given perimeter to find the length of two opposite sides of a rectangle (for example, in **Workbook page 76** question 2), they may forget that the perimeter is $2 \times (l + w)$ and not halve the length when they work it out.
- It is useful to get the children to draw a rough sketch and to label the sides with their lengths so they can double check that they get the correct perimeter. This mistake is likely to crop up again when you work with triangles, so it's important to deal with it here.

Answers for Pupil book page 134

❶ Individual answers.

❷ a 24 cm   b 12 cm   c 52 m
   d 41 cm   e 44 cm   f 16 cm
   g 37.4 m   h 11.8 m

More perimeter

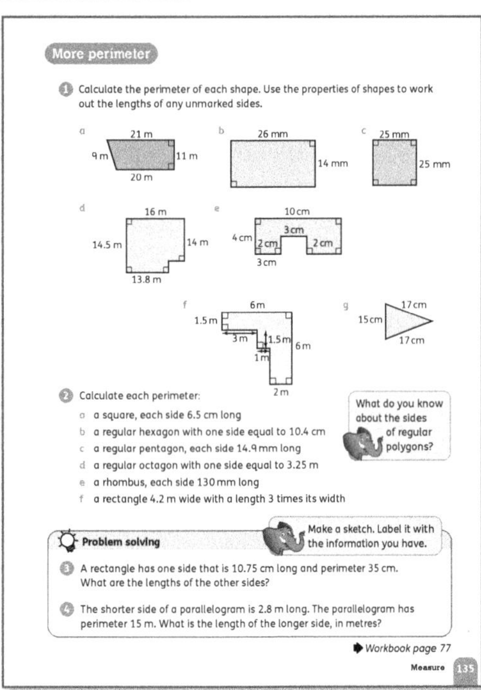

Materials

Rulers; rods of different lengths or drinking straws; composite shapes drawn on squared paper.

Warm-up

Use a suitable 'Mental problem-solving' activity (pages 23–24) as a starter for this lesson. You might like to include some further work on mentally doubling numbers as the children will be using this skill.

Focus

- Revise the concept of finding perimeter of a 'composite shape'.
  - Display a shape like this one for the class:

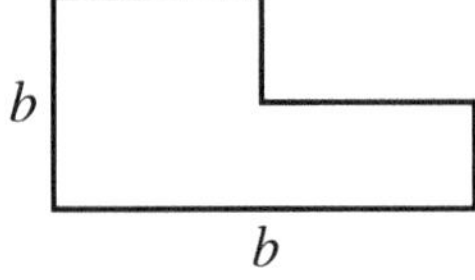

  - Discuss how you can work out the perimeter of the shape if you only know the lengths of a and b.
  - To do this, allocate lengths to a and b. For example, say: *I know that a is 10 cm and b is 15 cm. What is the perimeter of this shape?* Let the children discuss this in groups then take responses.
- Use rods or drinking straws to show why the perimeter is $2 \times (a + b)$ even though there is a 'chunk' out of the shape, for example:
  - Ask the children to model a similar shape using rods or straws. Let them physically move the straws so they can see that the two cut out pieces equal the other length and width.

- Model this for the class after they have experimented with their own shapes. For example:

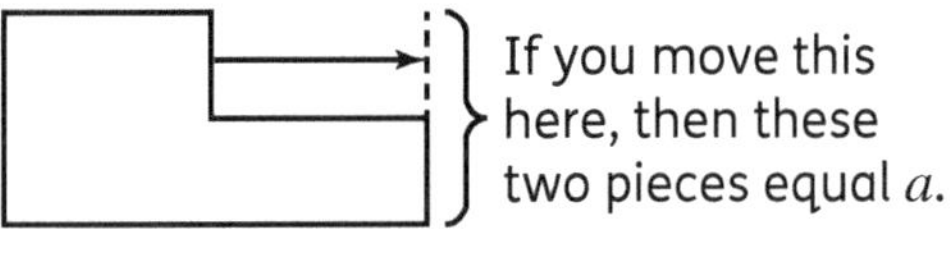

If you move this here, then these two pieces equal b.

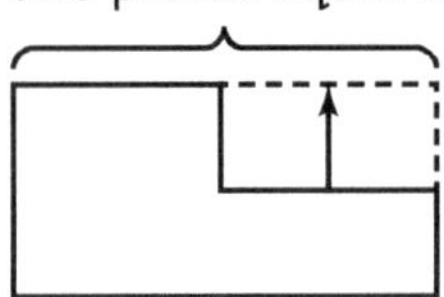

- You can demonstrate this using a computer program as well, but physically moving the pieces helps the children make connections and allows those who are less visual learners to experience the concept in a different way.
- Next ask the children to build a shape where there are 'extra lengths'. For example:

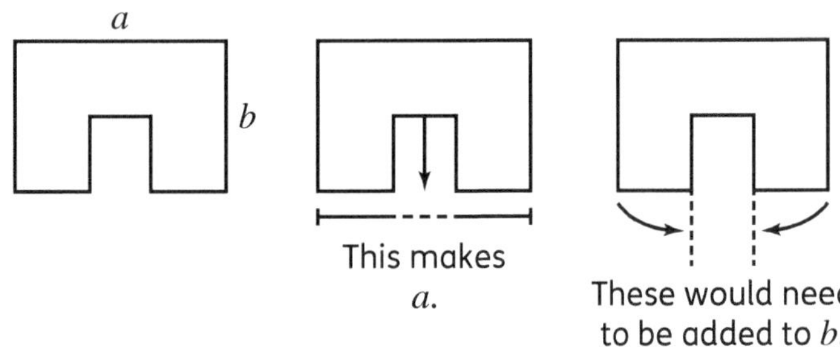

- Explain that in this case, we have to know the lengths of the shorter vertical sides to find the perimeter.
- Let the children work independently to complete questions 1 and 2 on **Pupil book page 135**.
- Problem solving: In questions 3 and 4, the children can sketch diagrams and label the given lengths, to help them find the unknown lengths.

Follow-up

Use the investigation on **Workbook page 77** to explore the idea that shapes with the same perimeter can have very different areas. The children will see that the rectangle with the greatest area is actually a square (a rectangle with 4 equal sides). This concept will be explored further in Unit 14.

Support

Give the children a number of fairly simple composite shapes drawn on squared paper. Ask them to draw a rectangle with the same perimeter. You could extend this by asking them to draw a different composite shape with the same perimeter, or with a perimeter twice as long as the one they've been given.

Challenge

The children could investigate different rectangles with an area of 24 squares and discuss how the perimeter varies with different side lengths.

Interesting mistakes

In working with perimeter of compound shapes, children may need additional support to work out the unknown lengths. Encourage them to separate the shape into two sets of lines. Look first at horizontal lines. Draw a complete length and use it to work out the lengths of the unknown bits. Repeat for vertical lines.

Answers for Pupil book page 135

1. **a** 61 m   **b** 80 mm   **c** 100 mm
   d 61 m   **e** 28 cm   **f** 24 m
   g 49 m
2. **a** 26 cm   **b** 62.4 cm   **c** 74.5 mm
   d 26 m   **e** 520 mm   **f** 33.6 m
3. 6.75 cm
4. 4.7 m

Answers for Workbook page 77

1. 13 cm by 1 cm; 12 cm by 2 cm; 11 cm by 3 cm; 10 cm by 4 cm; 9 cm by 5 cm; 8 cm by 6 cm; 7 cm by 7 cm
2. **a** Rectangle size followed by area in cm²: 13 cm by 1 cm: 13; 12 cm by 2 cm: 24; 11 cm by 3 cm: 33; 10 cm by 4 cm: 40; 9 cm by 5 cm: 45; 8 cm by 6 cm: 48; 7 cm by 7 cm: 49
   b All of them have different areas.
   c Two shapes could have the same perimeter but may not have the same area.

Revisit area

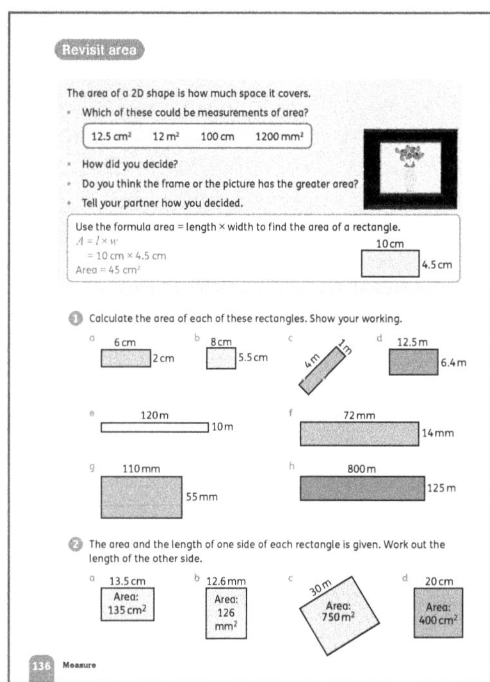

Materials

Masking tape or painter's tape; cuboid-shaped boxes or containers; stickers.

Warm-up

- Revise finding the area of rectangles and squares with the class.
- Draw rectangles in different orientations on the board. Label some of the dimensions, such as some or all of the sides.
- Ask: *With the information you have, for which rectangles can you find the area? For which rectangles do you need more information to find the area?*

Focus

- Ask the children to find the area of the classroom and objects within the room (for example, the top of a desk, the cover of a book). If you have a tiled floor in the classroom or elsewhere in the school, mark out a number of composite shapes with masking tape or painter's tape (both peel off easily). Aim for shapes that cover complete tiles or half tiles. For example:

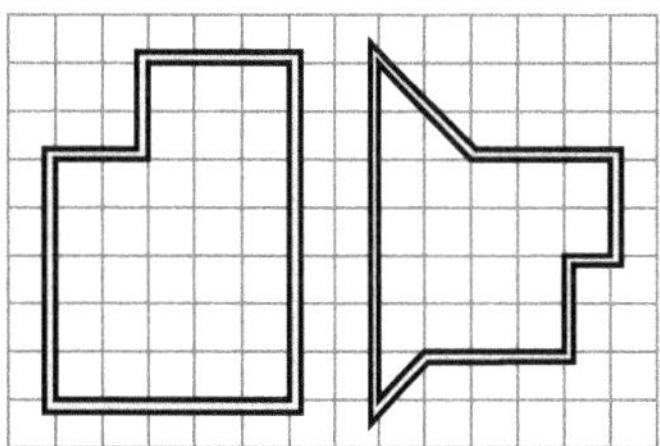

- Let the children measure the perimeter of each shape and work out its area in tiles.
- If you have time, give each group of children an area (for example, 8 tiles) and ask them to use tape to mark out three different shapes with the same area but different perimeters.
- Round off the activities with a number talk (pages 17–18) about perimeter and area. Ask the children what is similar and what is different about the concepts; how they work out each measurement and where it would be useful to know either perimeter, or area or both in everyday contexts.
- Turn to **Pupil book page 136**. Let the children work in pairs to read through the explanation and answer the questions. Ask for feedback from different children. The example box at the top of the page asks: Which of these could be measurements of area? (12.5 cm², 12 m², 1200 mm²) How did you decide? (Area is measured in square units, so I chose the ones with square units.) Do you think the frame or the picture has the greater area? (Individual answers.) Tell your partner how you decided. (For example, I visualised the frame cut into 4 separate rectangles, and how they would fit on the picture.)
- The children can do question 1 independently. Check that they use the formula and show their working.
- For question 2, you may need to do the first example with the class to show them how to use inverse operations (division) to find the side length. Use a fact family to show this, for example: $2 \times 3 = 6$, so $6 \div 2 = 3$ and $6 \div 3 = 2$. Explain that if $l \times w = A$ then it follows that $A \div w = l$ and $A \div l = w$.

Challenge

- Write an area, for example 36 cm², on the board. Ask the children if they can construct shapes with this area (on a cm² grid if possible).
- Encourage the children to move away from rectangles and try to construct triangles or other shapes, including composite shapes. Compare the shapes the children draw.

Support

- Give each group a set of cuboid boxes or containers. Ask the children to choose the face that they think has the largest (or smallest) area.
- Place a sticker on this face, or mark it somehow.
- Let the children measure the length and width of all the faces and calculate their area to see whether the face they selected originally is actually the one with the largest (or smallest) area.

Interesting mistakes

The children may forget to include the units when they calculate area. Continue to remind them to do this and refer back to a squared grid to reinforce the idea of a square unit.

Answers for Pupil book page 136

1   a   12 cm²   b   44 cm²
    c   4 m²   d   80 m²
    e   1200 m²   f   1008 mm²
    g   6050 mm²   h   100 000 m²
2   a   10 cm   b   10 mm
    c   25 m   d   20 cm

Area of composite shapes

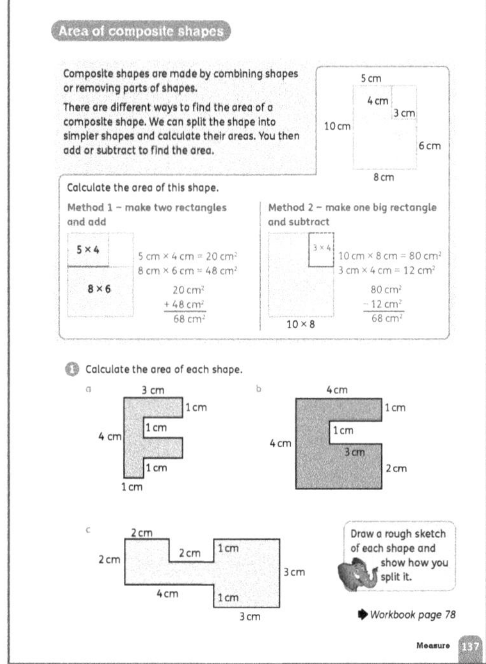

Materials

Rulers; squared paper.

Warm-up

As a mental starter for this lesson revise multiplication facts and square numbers because the children will use these to calculate area of shapes.

Focus

- This lesson offers a good opportunity for the children to talk through things on their own, to make decisions and to ask questions about how they can use what they already know to solve area problems involving more complicated shapes.
- Give the children time to talk through, in pairs, the examples on **Pupil book page 137** and the different methods of finding the area of a composite shape.
- Then hold a class discussion, focusing on what the children found challenging and what they learnt.
- Check that they have grasped the idea by drawing a compound shape on the board and asking them how they would work to find the area of the shape. If necessary, explain that the only skill they need is knowing how to find the area of a rectangle.
- Build on the children's ideas to decide how to break the shape down into rectangles, find the area of each and then find the total area.
- Use **Workbook page 78** to consolidate finding the area of simple composite shapes before moving back to **Pupil book page 137**.
- For question 1, explain that they can draw and label a rough sketch using a ruler, but not using exact measurements.

Challenge

Ask the children to design a poster all about area. It could show the methods they would use to find the area of compound shapes and how they can be used in real-life situations.

Support

Let the children work in pairs or groups to prepare a chart to teach others how to decompose shapes to work out the area in parts. They can provide diagrams drawn on squared paper to show examples. This work will help them to consolidate what they have done and to formalise and express their understanding.

Interesting mistakes

As with perimeter problems, the children may need additional support to work out the unknown lengths. Encourage them to separate the shape into two sets of lines. Look first at horizontal lines. Draw a complete length and use it to work out the lengths of the unknown bits. Repeat for vertical lines.

Answers for Pupil book page 137

1   a   8 cm²   b   13 cm²   c   15 cm²

Answers for Workbook page 78

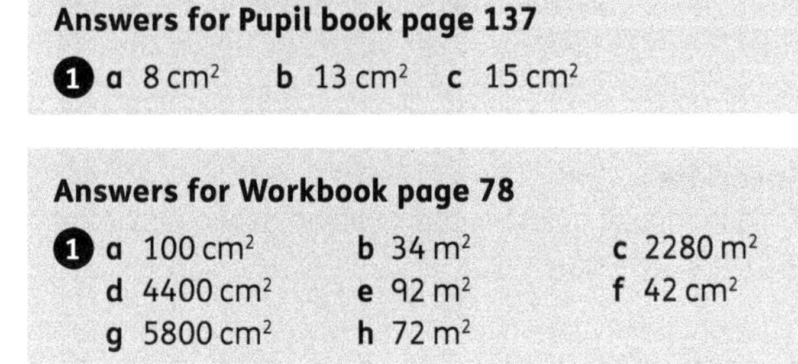

1   a   100 cm²   b   34 m²   c   2280 m²
    d   4400 cm²   e   92 m²   f   42 cm²
    g   5800 cm²   h   72 m²

Area of parallelograms

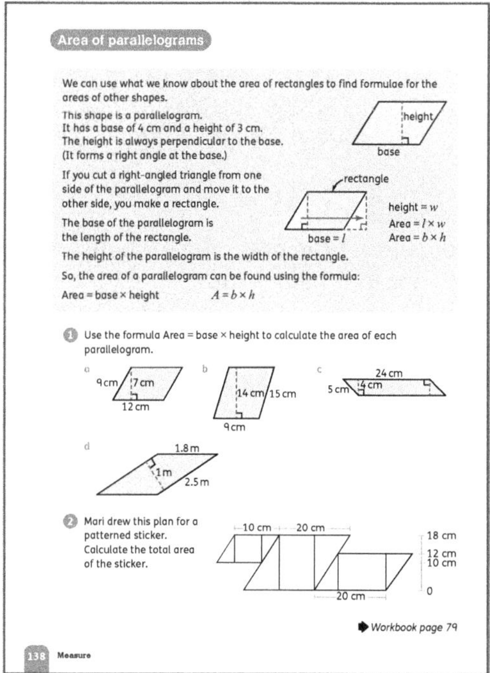

Materials

Centimetre squared paper; scissors; rulers; dynamic geometry software (optional).

Warm-up

Use any mental multiplication activity as a starter for this lesson.

Focus

- Use the example on **Pupil book page 138** as a class practical activity. Let the children draw the given parallelogram on squared paper, then let them work in pairs to cut the right-angled triangle from one side and move it to the other. Keep one parallelogram intact to show that they have the same area.
- Make sure the children realise that 'base' is another name for the length of the rectangle and that the height is a perpendicular the same length as the width of the rectangle.
- You can also use dynamic geometry software to show this.
- If necessary, ask the children to draw any other parallelogram on the squared paper and repeat the process.
- The challenge for the children in question 1 is to recognise which length is the base and which is the height. You may need to talk through this with the class before they do the calculation work.
- For question 2, let the children work in pairs to discuss and solve the problem.

Follow up

Use **Workbook page 79** to consolidate the children's understanding of area of a parallelogram. Discuss how they worked out the answers to question 2.

Interesting mistakes

- When calculating the area of a parallelogram the children may use the slant height rather than the correct perpendicular height, particularly when the lengths of all sides are given and the shape is not in an obvious orientation (as in shape d in question 1).
- Encourage the children to turn the shapes if they need to and let them use a set square to find the perpendicular pair of measurements. Remind them that they can use the properties of shapes and that they can write in measurements (such as opposite side lengths) to make it easier to see what to do.

Answers for Pupil book page 138

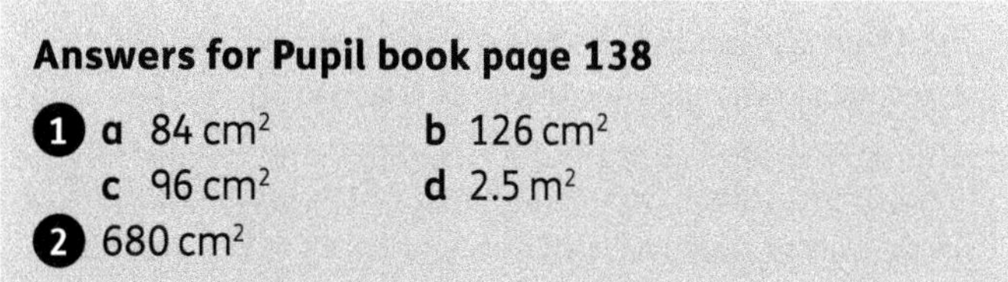

1. a 84 cm^2 b 126 cm^2
   c 96 cm^2 d 2.5 m^2
2. 680 cm^2

Answers for Workbook page 79

1. a 72 cm^2 b 144.4 cm^2 c 7.5 cm
   d 6 cm   e 16 cm   f 16.5 cm
2.

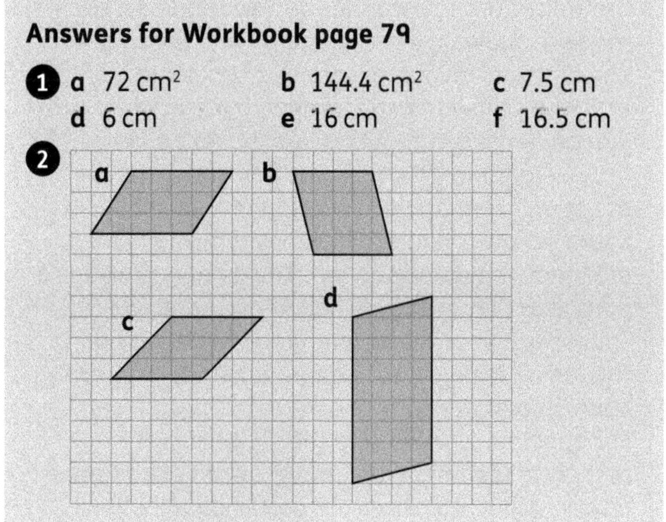

Area of triangles

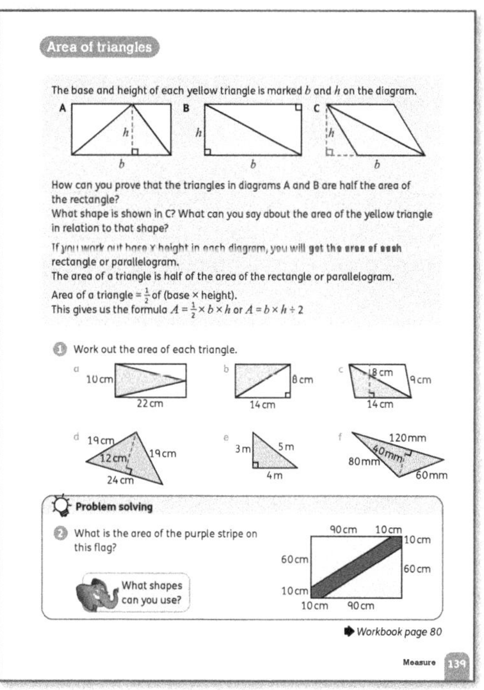

Materials

Squared paper; scissors.

Warm-up

Use any suitable mental calculation using halving and/ or multiplying as a starter for this lesson.

Focus

- Put the children into groups and let them work together to read through the explanation and examples on **Pupil book page 139**. The aim is to allow them to see how the formula for the area of a triangle is derived. If the children are not convinced, let them draw the shapes on squared paper and then cut them out to see that the triangles are half the area of the larger shapes in each example. The explanation box at the top of the page asks: How can you prove that the triangles in diagrams A and B are half the area of the rectangle? (In A, the yellow triangle is split into two triangles, each made by drawing in the diagonal of a rectangle. The diagonal divides a rectangle exactly in half. In B, the yellow triangle is made by drawing in the diagonal of a rectangle.) What shape is shown in C? (parallelogram) What can you say about the area of the yellow triangle in relation to that shape? (In C, the yellow triangle is made by drawing in the diagonal of a parallelogram. The diagonal divides a parallelogram exactly in half (by symmetry).)
- Talk about what the children found interesting in their discussions. Also, explore anything they found confusing or difficult.
- For question 1, make sure the children can recognise the height and the base of each triangle, emphasising that these are perpendicular. If necessary, go through this orally, identifying the height and base in each before asking the children to calculate the areas.
- <u>Problem solving:</u> For question 2, let the children work in pairs to develop and use a strategy to solve the problem. They will probably find it useful to copy and label the diagram.

Follow-up

Use **Workbook page 80** as an additional problem-solving activity. Make sure the children realise they can use the squares on the grids to calculate heights and base lengths.

Challenge

- Let the children investigate overlapping areas and how to solve problems involving overlaps. They can use overlapping stickers or draw overlapping shapes on squared paper and use the diagrams to investigate different methods. You may need to show them an example and ask: *What is the total area covered by these two overlapping stickers? How can you work this out? What do you need to know to work it out?* (For example, if you know the area of each sticker

you can add these and then subtract the area that overlaps.)

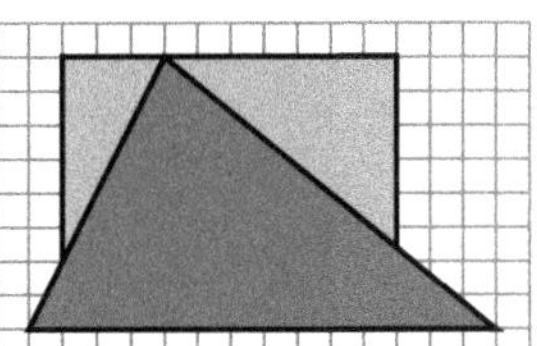

Interesting mistakes

- When calculating the area of a triangle the children may use the slant height rather than the correct perpendicular height. For example, in the example below, they might work out 17×8 and divide it by 2, or 17×15 and divide it by 2.

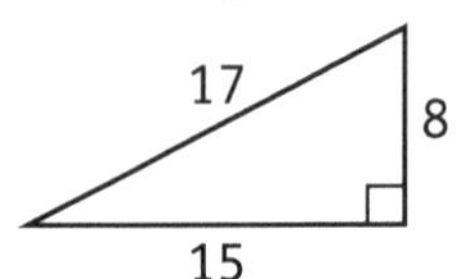

- Encourage the children to draw in the other sides of the rectangle so they are clear about which numbers to use.

Answers for Pupil book page 139

1 a 110 cm²   b 56 cm²   c 56 cm²
    d 144 mm²   e 6 cm²   f 2400 mm²

2 1600 cm²

Answers for Workbook page 80

1 a 6 cm²; 13.5 cm²; 3.375 cm²; 5.625 cm²
    b 10 cm²; 13.5 cm²; 20.625 cm²; 18.375 cm²

2 $a = 4$ cm; $b = 9$ cm; $c = 5$ cm; $d = 21$ cm

Estimate area using a grid

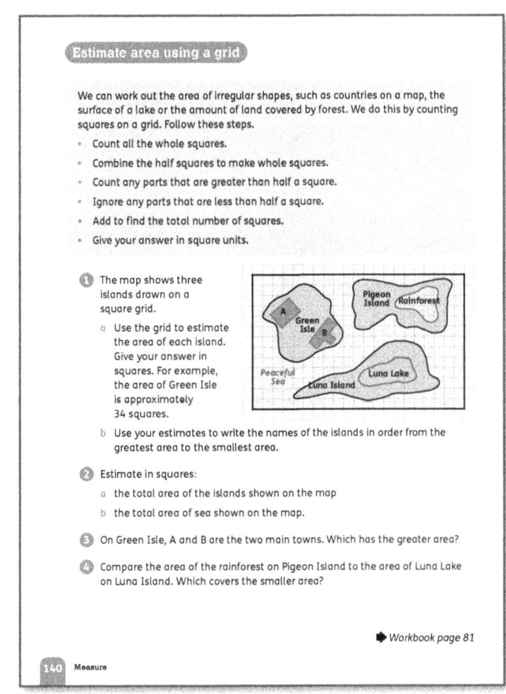

Materials

An atlas or large world map; a grid of square centimetres printed on an acetate sheet; sheets of shapes.

Warm-up

Use any of the 'Mental problem-solving' activities (pages 23–24) as a starter for this lesson.

Focus

- Revise finding the area of shapes using a grid of square centimetres. If possible, prepare an acetate sheet with a grid for each group. Let the children superimpose this on a sheet of shapes to find the dimensions of the sides and the area in square centimetres.
- Show the children an irregular shape drawn on a grid. Try to include some half squares, more than half squares and less than half squares in the area of the shape. Ask the children how we can find its area. They should remember earlier work on areas and say that you can count the squares. Discuss which squares to include.
- Point out that all complete squares should be counted. Count these, marking them as you go. Record the number of complete squares.
- Next, find the half squares. Count these, marking them as you go. Establish that the area is half the number of 'half squares' (in other words, two half squares make one complete square). Record the area covered by half squares.
- Next, count the areas that cover more than half a square. Mark these as you go and record the number. Tell the children you will ignore any areas covering less than half a square. Cross out these on the diagram.
- Total the areas you have recorded to find the estimated area of the shape. Record this and write the correct units alongside it.
- Use **Workbook page 81** to consolidate these ideas.
- Let the children work on their own to complete the questions before comparing their answers in groups. Encourage them to discuss and explain any differences.

Follow-up

- Let the children work in pairs to read through the information and complete questions 1–4 on **Pupil book page 140**.
- Hold a class discussion about the answers, bearing in mind that they are likely to be slightly different. Encourage the children to explain how they reached their answers, particularly where these are different from the others.

Challenge

Let the children do a mini-project to find out how the area of large irregular spaces, such as countries, is calculated. They can use reference books or online resources to find their answers. Let them prepare a short talk or slideshow to share their findings.

Interesting mistakes

The children may need additional support with the idea that they can simply ignore areas of less than half a square when they estimate the area of irregular shapes. Stress that they are estimating, and if they joined the areas they have ignored to the areas they have counted (including those greater than half a square), the areas usually balance out and give a fairly accurate estimate.

Answers for Pupil book page 140

1 a Pigeon Island is 37 squares; Luna Island is 41 squares

   b Luna Island; Pigeon Island; Green Isle

2 a 112 squares   **b** 161 squares

3 A

4 rainforest

Answers for Workbook page 81

1 A = 10 cm^2; B = 8 cm^2; C = 12 cm^2

2 Here are some example answers.

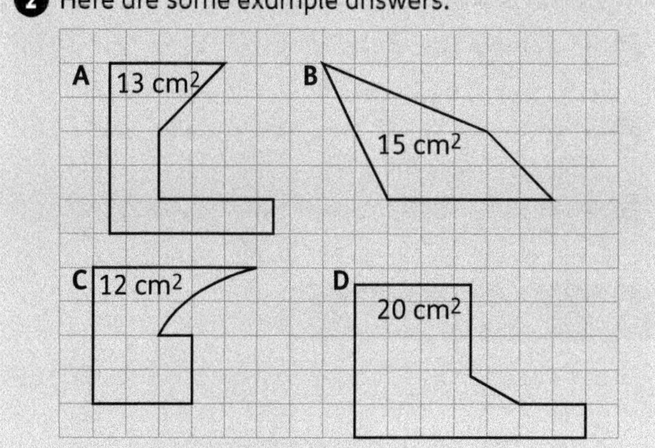

Mixed area problems

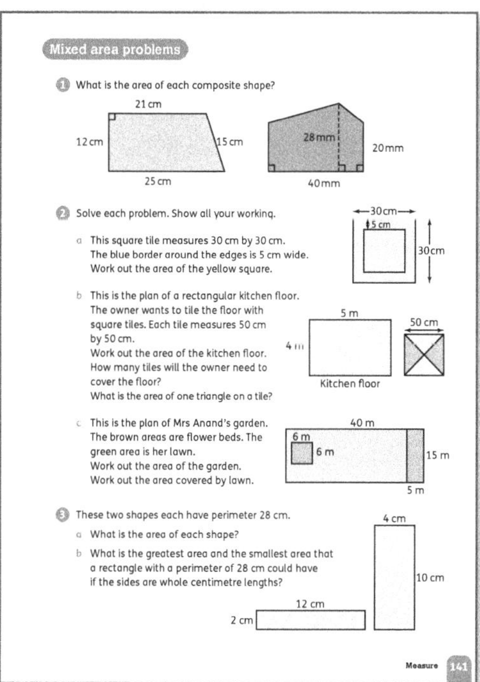

Materials

Rulers; squared paper.

Warm-up

Use any mental multiplication activity as a starter for this lesson.

Focus

Turn to **Pupil book page 141**. The children can work through questions 1–3 in pairs or groups.

Follow-up

You could split up the problems, giving different problems to different groups. If you do this, the children can report back on how they solved each problem so that they are exposed to different methods of working.

Support

Let the children draw shapes on squared paper if they need to and allow them to model the problem situations using scaled drawings.

Answers for Pupil book page 141

1   a   green shape: 276 cm²

    b   brown shape: 960 cm²

2   a   400 cm²   b   20 m²; 80 tiles; 625 cm²

    c   600 m²; 489 m²

3   a   24 cm²; 40 cm²   b   49 cm²; 13 cm²

Volume and capacity

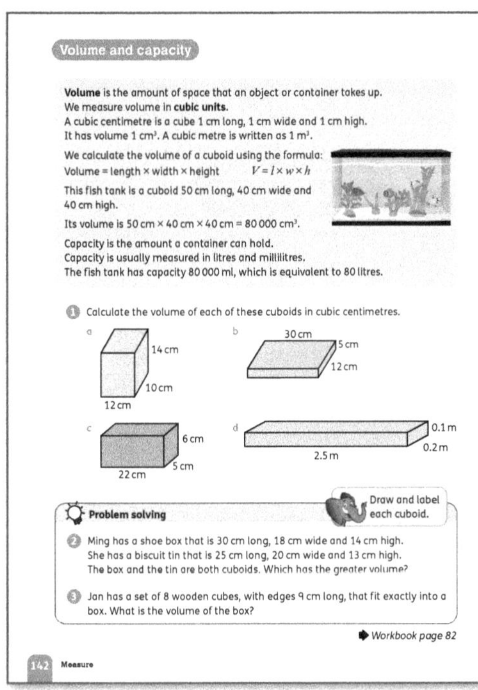

Materials

1 cm cube to show its actual size; measuring jug marked in millilitres; short video on the concept of volume (optional); worksheet on volume (see Support section).

Warm-up

- Make up a set of multiplication facts with 3 numbers (for example: 2 × 4 × 5, 3 × 6 × 3). Ask the children to work out the answers as a starter for this lesson.
- Remind them to think about the numbers and the most useful strategy for working.

Focus

- You may wish to show the class a short video to revise the concept of volume. There are a number of free videos available online. As with all media, check that it is appropriate and at the correct level before showing it to the class.
- Turn to **Pupil book page 142** and read through the information with the class. Remind them that they worked with both concepts in previous levels.
- Show them a 1 cm cube and remind them that this is the unit of volume; (use the term 'cubic unit'). Explain that if you filled a hollow cube this size with water, it would hold 1 ml. Show them the measuring jug and remind them that we use litres and millilitres to talk about liquid capacity. Let the children work independently to complete question 1 on **Pupil book page 142**. For question 1d, the children should convert the length, width and height measurements to centimetres before calculating the volume.
- <u>Problem solving:</u> Encourage the children to share how they solved questions 2 and 3.

Follow up

The children can work in pairs on the questions on **Workbook page 82**, finding unknown lengths given the volume. Check their answers as a class.

Support

- Give the children a worksheet similar to the one below. Let them label the shaded blocks on each cuboid as length, width or height and complete the rest of the information.
- They can keep this on their desk for reference as you work through the lessons on volume.

1   Fill in the correct name for the shaded part of this cuboid.

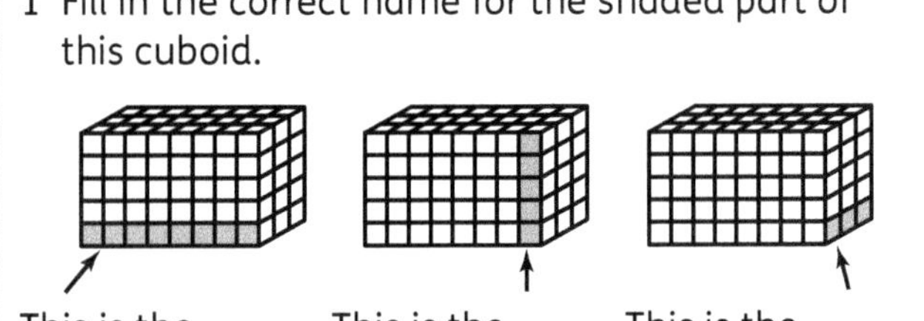

This is the _____.   This is the _____.   This is the _____.

2   Complete this table using the information in the pictures.

Length	Height	Width	Total number of blocks

3   Complete this sentence.

You can work out the number of blocks in a cuboid if you know the length, width and height. This is how you do it: _________________

Interesting mistakes

- It is useful to ask children to write down what they think volume means to check their understanding.
  - If they say 'It's how loud something is', you will need to address the mathematical vocabulary.
  - If they say 'It's how much you can fit into something', they have mistaken it for 'capacity'.
- Explain that volume is the amount of space something takes up. Use a closed but empty carboard box to show that it takes up space even though it is empty. If you fold the box down to a flat piece of card, it takes up less space, so its volume is reduced, but it still has a height, a length and a width (and it cannot hold anything).

Answers for Pupil book page 142

1   a   1680 cm³   b   1800 cm³
    c   660 cm³   d   0.05 m³

2   the shoe box (7560 cm³; the tin has volume 6500 cm³)

3   5832 cm³

Answers for Workbook page 82

1   a   6 cm   b   4.5 cm   c   4 cm
    d   16 cm   e   12 cm   f   2 cm

2   length = 18 m, width = 10 m, height = 2 m;
    length = 8 m, width = 9 m, height = 5 m;
    length = 6 m, width = 6 m, height = 10 m

3   length = 40 mm, width = 10 mm, height = 20 mm

Calculate volume

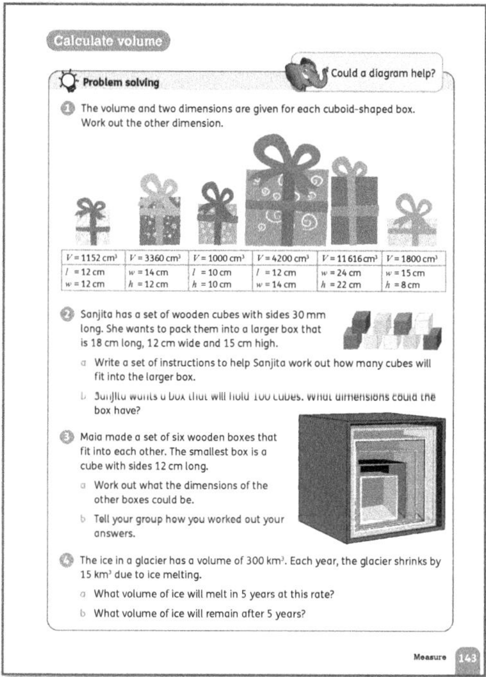

Materials

Calculators.

Warm-up

As a mental starter for this lesson, prepare some target numbers (for example: 1000, 48, 320, 160, 212), and ask the children to work in groups to make up four different 3-number multiplication facts with that product.

Focus

- <u>Problem solving</u>: Use questions 1–4 on **Pupil book page 143** to assess how well the children have understood their work on volume and to see that they can apply the ideas in different contexts.
- Ask the children to explain how they worked out their answers when you check the work.
- For question 1, let the children use a calculator to do the divisions and to multiply to check that the three numbers they end up with will give the correct volume.
- Note that question 4 uses cubic kilometres. These are not common units, but they do apply to masses of ice in glaciers and the volume of lava that is expelled from volcanoes.
- If the children ask, you could explain that there are 1000 metres in a kilometre, so a cubic kilometre is a cube 1000 m × 1000 m × 1000 m; in other words, a volume of 1000 000 000 m³ (a thousand million cubic metres).
- Ask the children what units they could use to measure the volume of cubes or 3D shapes smaller than a 1 cm cube. Label the sides of the 1 cm cube as 10 mm each and ask them to calculate its volume in mm³. (1000 mm³)

Challenge

Give the children a set of cubes or blocks. Tell them each cube represents a volume of 1 cubic metre. Give the groups a volume (base this on how many blocks you have), such as 36 cubic metres. Challenge the children to build the highest stable tower they can with that volume.

Interesting mistakes

Children often leave off units when they work with volume. Keep asking questions such as: *A volume of 24 what?* as a reminder. Some children might count only the blocks they can see and give that total as the volume of a shape, particularly when they realise that an empty box has volume so they think the sides are somehow the volume. Continue to model volume using cubes and real objects to help them understand the concept.

Answers for Pupil book page 143

1 8 cm; 20 cm; 10 cm; 25 cm; 22 cm; 15 cm

2 **a** First, calculate the volume of a wooden cube:
$3 \times 3 \times 3 = 27$ cm^3
Then calculate the volume of the box:
$18 \times 12 \times 15 = 3240$ cm^3.
Finally, divide the volume of the box by
the volume of one wooden cube to get the
number of cubes that will fit in the box:
$3240 \div 27 = 120$.
b Individual answers, for example, a box with
30 cm of length, 30 cm of width and 3 cm of
height (volume must equal 2700 cm^3).

3 **a** For example, cubes with side lengths 13 cm,
14 cm, 15 cm, 16 cm and 17 cm
b Individual answers.

4 **a** 75 km^3 **b** 225 km^3

End-of-unit check

Use some or all of these questions to assess how well
children have understood the concepts in this unit.
- Provide different shapes. *What is the perimeter of
this shape?*
- Provide a shape with its perimeter given. *How do you
know how long this side of the shape is?*
- Show a composite shape with some sides unlabelled.
*Which lengths do you need to work out the perimeter?
How can you find them?*
- *What information do you need to calculate the
perimeter/area of a rectangle?* (length and width)
- *What information do you need to calculate the
perimeter/area of a square?* (length of one side)
- *What is the area of a 3 cm × 5 cm rectangle?* (15 cm^2)
What is the area of a 30 m × 50 m rectangle?
(1500 m^2)
- *What pattern do you notice about these two answers?
Explain the pattern.* (The unit is the square of the unit
of the lengths. The value for the area of the larger
rectangle is 100 times the value for the smaller
rectangle, because the values for the length and
width are both 10 times and 10 × 10 = 100.)
- *What is the formula for calculating the area of a
rectangle?* (area = length × width)
- Show a composite shape. *How could you divide this
shape up to find its area?* (Split it into rectangles.)
- Give the children a piece of squared paper. *Draw a
triangle with an area of 16 square units. Explain how
you did it.* (The area of the triangle is half the area of
a rectangle with the same length and width, so the
height and length of the triangle need to multiply
together to make 32.)

- *A regular hexagon has sides of 5.5 cm. What is its
perimeter?* (33 cm)
- Prepare some tables like the one below and ask the
children fill in the missing values (shown in brackets).
Make sure they need to carry out a range of
calculations to find the missing perimeter, area and/
or one side length. *Complete this table. All the shapes
are rectangles.*

Length	Width	Perimeter	Area
12.5 cm	9 cm	(43 cm)	(112.5 cm^2)
(7 cm)	10 cm	34 cm	(70 cm^2)
8 cm	(9 cm)	(34 cm)	72 cm^2

- Prepare diagrams like the ones below. *What are
the unknown values?* (12 cm, 6 cm, 5.5 cm,
are = 104.5 cm^2)

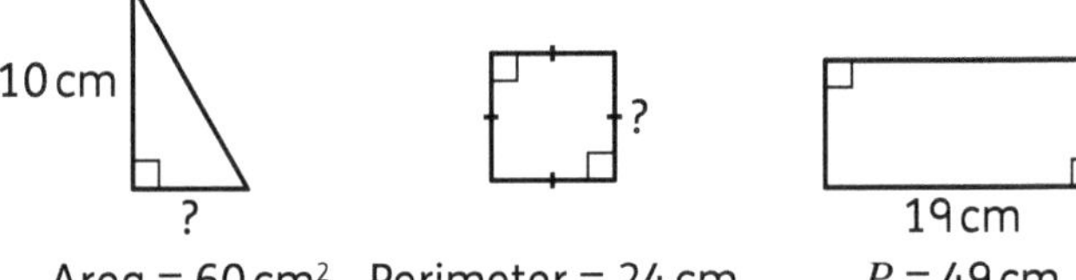

- Show children some 3D shapes. *Which of these takes
up most space? How do you know?*
- *A cuboid is made up of 7 layers. Each layer is made
of 19 blocks, each with a volume of 1 cm^3. What is
the volume of the cuboid? How did you work it out?*
(7 × 19 × 1 cm^3 = 133 cm^3)
- *If you turn a cuboid so its length becomes its height,
how does this affect the volume? Why?* (It stays the
same. The volume is the product of the three side
lengths and it does not matter which order is used to
multiply them.)
- *Two different cuboids each have a volume of 16 cm^3.
What could the dimensions of the cuboids be?* (1 cm
by 1 cm by 16 cm, 1 cm by 2 cm by 8 cm, 1 cm
by 4 cm by 4 cm, 2 cm by 2 cm by 4 cm, or other
dimensions with non-integer lengths)
- *A wooden cube has sides that are 19 cm long. Eight
of these cubes fit together to make a cuboid. What is
the volume of the cuboid?* (54 872 cm^3) *How could the
cubes be arranged to make the cuboid?* (1 by 1 by 8
cubes, 1 by 2 by 4 cubes, 2 by 2 by 2 cubes)
- *What is the difference between area and volume?*
(Area is the space taken up by a 2D shape. Volume is
the space taken up by a 3D shape.)
- *What is the volume of a cement brick that is 12.5 cm
wide, 10.5 cm high and 20 cm long?* (2625 cm^3)
- Show a box. *What measurements do you need to find
the volume of this box?* (length, width and height)
- *Estimate the volume of our classroom in metres cubed.*

UNIT 14 Algebra

Learning objectives

- Use simple formulae.
- Generate and describe linear number sequences.
- Find and use a position-to-term rule.
- Recognise the use of letters to represent quantities that vary in addition and subtraction calculations.
- Express unknown number problems algebraically.
- Find pairs of numbers that satisfy an equation with two unknowns.
- Enumerate possible combinations of two variables.

Key words

number sequence   term-to-term rule
position-to-term rule   variable   formula
formulae   substitute

Unit introduction

Materials

Movement on squared grid worksheet (see right).

Teaching guidance

- Give the children this problem involving 'number sequences' and generalisations:
  - The diagram below shows how a person walks from point A to point B. The person always walks south or west (in other words, they do not ever backtrack).
  - On day 1, they walked single blocks, starting with one block west, then one block south. They covered a distance of 20 blocks.
  - The next day they tried a new route, walking double blocks, starting with two blocks south and then two blocks west.

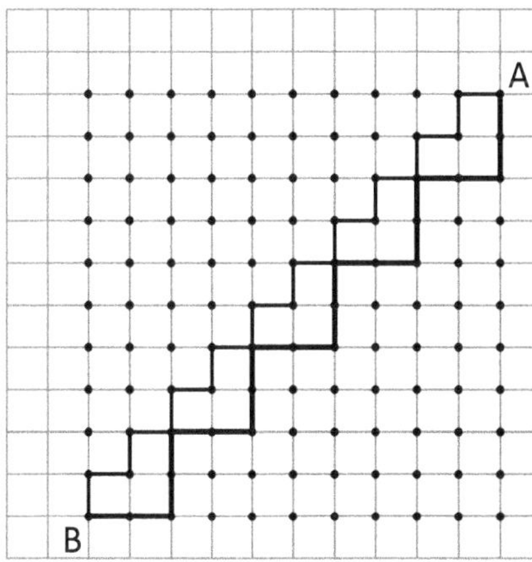

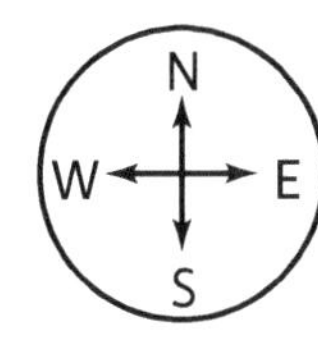

- Ask: *Will they always cover a distance of 20 blocks between A and B travelling only south and west? Give reasons for your answers.*
- (The person will have to travel 10 blocks south and 10 blocks west, or a total of 20 blocks, regardless of the order in which they choose to do the south and west parts of the walk.)
- Investigate how this works for different numbers of blocks, for example:
  - If there were 9 blocks on the vertical axis and 15 on the horizontal, the person would always have to walk 24 blocks to get from A to B travelling south and west or north and east.
- Let the children try to express this as a general rule, such as distance = $n + w$, where n is the number of blocks from north to south and w is the number of blocks from west to east.

Number sequences and rules

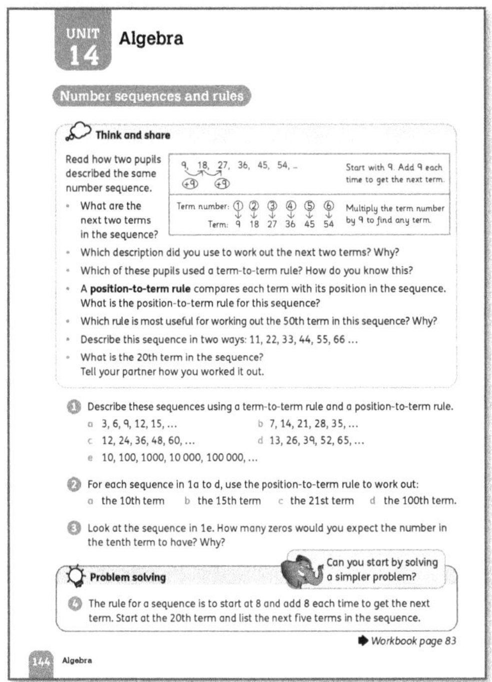

Materials

Calculators.

Warm-up

- Use a number of different skip-counting activities as a mental starter for this lesson. Vary the counting interval and the starting numbers. Include counting forwards and backwards.

Focus

- The children have already worked with number sequences and 'term-to-term rules' in Level 5.
- Think and share: Use the Think and share activity on **Pupil book page 144** to hold a number talk (pages 17–18) with the class. Make sure the children understand the definition of 'position-to-term' rule.

The children are asked the following questions. What are the next two terms in the sequence? (63 and 72). Which description did you use to work out the next two terms? Why? (Individual answers.) Which of the pupils used a term-to-term rule? How do you know this? (The first pupil did, because they added 9 to each term to get the next term.) What is the position-to-term rule for sequence? (Multiply the term number by 9.) Which rule is more useful for working out the 50th term in this sequence? Why? (The position-to-term rule is more useful because it allows you find the 50th term with one calculation instead of finding all the terms before it.) Describe the sequence 11, 22, 33, 44, 55, 66 … in two ways. (Term-to-term rule: start with 11 and add 11 each time to get the next term. Position-to-term rule: multiply the term number by 11 to find any term.) What is the 20th term in the sequence? (220; either continue adding 11 until, you have reached the 20th term or multiply 20 by 11).

- Complete questions 1–3 on **Pupil book page 144** orally with the class to ensure they are clear about the two different types of rules and can apply them.
- Problem solving: Use question 4 to revise the problem-solving strategy of solving a simpler problem. Show the class that by using a table and working out the first few terms, they can then find a rule to work out the 20th term (the term number × 8) and list the next terms in the sequence.

Follow-up
- When you are confident that the children can use the two types of rules correctly, turn to **Workbook page 83.**
- Let the children complete the questions independently. Allow them to use calculators to check their work.

Challenge
You can find a range of games involving patterns and missing numbers online. As with all online resources, check that they work and are appropriate before using them in class.

Support
As the focus is on patterns and the difference between terms, rather than calculation skills, encourage the children to use calculators to check pattern work and to generate patterns of their own.

Interesting mistakes
- Working with number patterns gives a good basis for introducing the concept of 'variables' and more formal algebra. However, children often do not see the connection between the physical models used to make patterns in earlier levels (for example: matchsticks, tiles, cubes, diagrams) and the mathematical language used to describe patterns in general terms. One of the reasons for this is that children will 'see' and therefore 'count' physical representations of patterns in different ways. More often than not, these ways are different from the more formal ways used in textbooks.

- Have ongoing discussions and number talks (see pages 17–18) about patterns and how they work, including linking them to concrete examples where you can help the children to recognise that patterns can be viewed and described in different ways.

Answers for Pupil book page 144
1 a term-to-term rule: +3; position-to-term: ×3
   b term-to-term rule: +7; position-to-term: ×7
   c term-to-term rule: +12; position-to-term: ×12
   d term-to-term rule: +13; position-to-term: ×13
   e term-to-term rule: ×10; position-to-term: 10^{term}

2 a 30; 70; 120; 130   b 45; 105; 180; 195
   c 63; 147; 252; 273   d 300; 700; 1200; 1300

3 10 zeros. The number of zeros is equal to the term number.

4 160, 168, 176, 184, 192

Answers for Workbook page 83
1 a 2, 4, 6, 8, 10, 12
      term-to-term rule: + 2
      position to term rule: × 2
   b 2.5, 5, 7.5, 10, 12.5, 15
      term-to-term rule: + 2.5
      position to term rule: 2.5
   c 19, 38, 57, 76, 95, 114
      term-to-term rule: 19
      position to term rule: × 19
   d 8, 11, 14, 17, 20, 23
      term-to-term rule: + 3
      position to term rule: × 3 + 5

2 −4, −1, 2, 5, 8, 11, 14, 17, 20, 23

3 4, 10, 16, 58, 88, 148

More number sequences

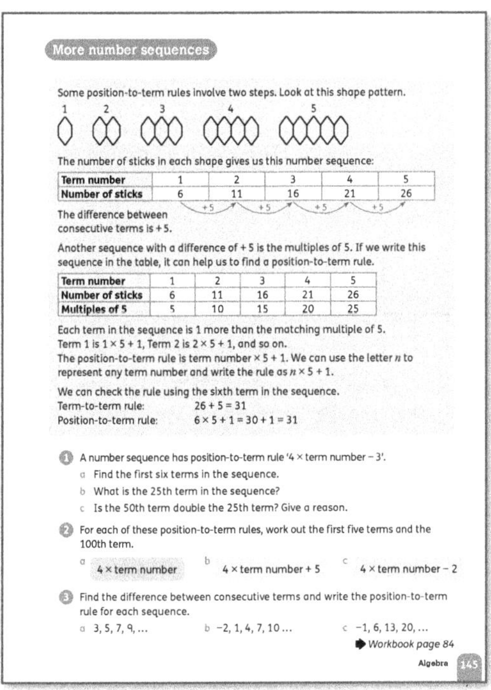

More number sequences

Some position-to-term rules involve two steps. Look at this shape pattern.

The number of sticks in each shape gives us this number sequence:

Term number	1	2	3	4	5
Number of sticks	6	11	16	21	26

The difference between consecutive terms is + 5.

Another sequence with a difference of + 5 is the multiples of 5. If we write this sequence in the table, it can help us to find a position-to-term rule.

Term number	1	2	3	4	5
Number of sticks	6	11	16	21	26
Multiples of 5	5	10	15	20	25

Each term in the sequence is 1 more than the matching multiple of 5.
Term 1 is $1 \times 5 + 1$, Term 2 is $2 \times 5 + 1$, and so on.
The position-to-term rule is term number $\times 5 + 1$. We can use the letter n to represent any term number and write the rule as $n \times 5 + 1$.

We can check the rule using the sixth term in the sequence.
Term-to-term rule: $26 + 5 = 31$
Position-to-term rule: $6 \times 5 + 1 = 30 + 1 = 31$

1 A number sequence has position-to-term rule '4 × term number − 3'.
   a Find the first six terms in the sequence.
   b What is the 25th term in the sequence?
   c Is the 50th term double the 25th term? Give a reason.

2 For each of these position-to-term rules, work out the first five terms and the 100th term.
   a 4 × term number   b 4 × term number + 5   c 4 × term number − 2

3 Find the difference between consecutive terms and write the position-to-term rule for each sequence.
   a 3, 5, 7, 9, …   b −2, 1, 4, 7, 10 …   c −1, 6, 13, 20, …

➡ *Workbook page 84*

Algebra 145

Materials
Calculators; examples of complex shape patterns like these:

Warm-up
- Continue to do different skip-counting activities as a mental starter for this lesson. Vary the counting interval and the starting numbers. Include counting forwards and backwards.
- You can also display any number chart and play games in which the children have to find numbers based on counting backwards and/or forwards in given amounts.

Focus
- Display your shape patterns and allocate one to each group. Explain that these are complex patterns but that even the most complex pattern has simpler patterns within it.
- Ask the children to find at least one simpler pattern in the complex one they are looking at.
- Let different groups give feedback to the class and make sure everyone can see the simpler units.
- Turn to **Pupil book page 145**. Examine the stick pattern with the class and use the table to show how the term-to-term rule works for this pattern. Remind the class that the term-to-term rule is useful for listing terms, but that it is not very helpful when you have to find the number of sticks in the hundredth term (for example).
- Use the second table to show how we can use the difference of 5 (in other words, multiples of 5) to work out a general position-to-term rule.
- The children should see that the number of sticks is one more than the related multiple of 5. Children may look at the first two terms and say that the rule could be multiply by 6 and subtract 1. This will give them 11, but it will not give them 16 or any other term number. Look back to the sticks and show that if you start with 1 stick, you can add 5 each time to make the pattern into this:

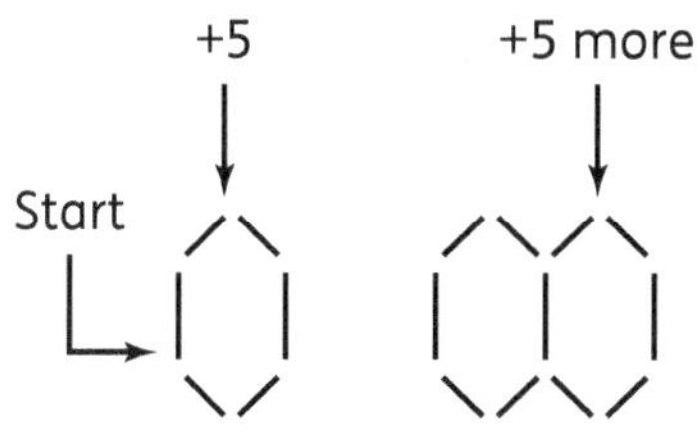

Make sure
- Make sure the children understand that n is a letter used to represent any term number. To work out the value of the term, you replace n with the term number. This is an important distinction because the children will later learn about the nth term, which is the value of any term (the general rule).
- Let the children work on the questions on **Pupil book page 145**.
- For question 1, the children can work in pairs.
- For question 2, the children can work independently to generate the terms, and then check their work with a partner.
- For question 3, remind the class that they should use tables to help them work out the difference and the rule.

Follow-up
- Use **Workbook page 84** to consolidate work on sequences and rules.
- Let the children work on their own to complete the questions.

Challenge
Let the children explore different shape patterns and use them to generate rules for the related number sequences. You will find many examples online, but you need to make sure that they are arithmetic sequences (in other words, they have a fixed difference between the numbers) and not geometric (pattern increases or decreases in a fixed ratio). Here are two examples you could start with:

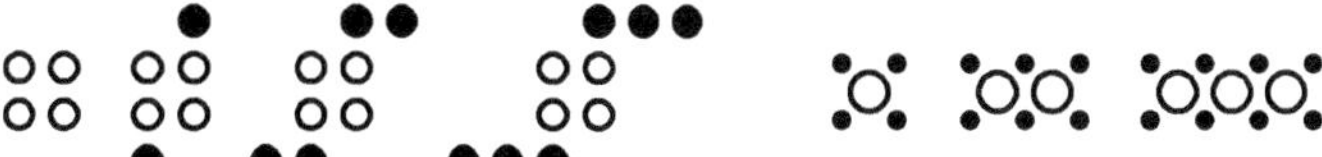

Support
- If children need additional support with the rules, it may help them to look at some incorrect examples and to use these to verbalise what went wrong. This may seem counter-intuitive, but many children will recognise their own thinking in the mistakes and learn from them. Here is an example:

The rule for a sequence is 'multiply the term number by 3 and then subtract 2'.

Three children were asked to write the first five terms.

Here are their answers.

Pupil A	Pupil B	Pupil C
–2, 1, 4, 7, 10	1, 4, 7, 10, 13	–1, 1, 3, 5, 7

Which child got the right answer?
Explain why the other two answers are wrong.
(Pupil B is correct. Pupil A used 0 as the term number for the first term instead of 1. Pupil C has multiplied each term number by 2 and then subtracted 3.)

Interesting mistakes

- Children sometimes think that sequences with numbers are different from patterns with objects. Use the stick pattern in this lesson to show that the number pattern is directly related to the stick pattern.
- Children often need additional support to find the rule. Explain that it is important to identify the repeating unit and to work out how that relates to the numbers. It may seem tedious to use tables to do this, but it is the most effective way of seeing the solution.

Answers for Pupil book page 145

1 a 1, 5, 9, 13, 17, 21, 25   b 97
  c No, the fiftieth term is 197 and it is not the double of 97.

2 4, 8, 12, 16, 20 and 400; 9, 13, 17, 21, 25 and 405; 2, 6, 10, 14, 18 and 398

3 a $2n + 1$ b $3n - 5$ c $7n - 8$

Answers for Workbook page 84

1 a 5, 6, 7, 14, 19, 29   b 3, 7, 11, 39, 59, 99
  c 1, 4, 9, 100, 225, 625   d 6, 9, 12, 33, 48, 78

2 a It starts with 1 and 3 dots are added each time.
  b 1, 4, 7, 10, 13, 16, 19   c $3 \times$ term number $- 2$
  d 58

Use letters for unknown numbers

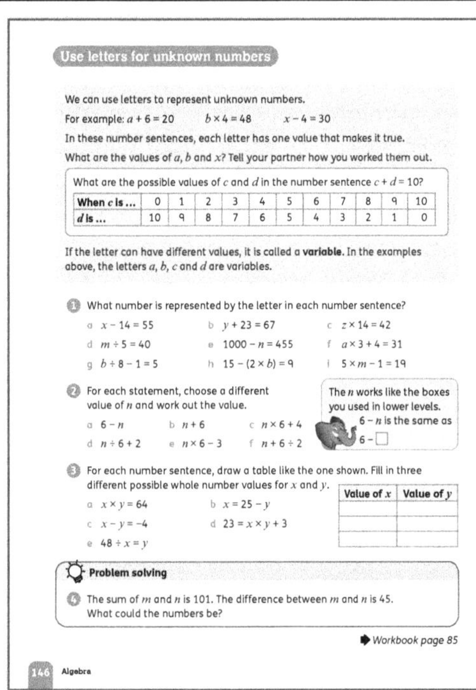

Materials

Calculators.

Warm-up

Use 'Letter-number code' (page 24) as a mental starter for this lesson.

Focus

- The children have previously worked with unknown values, sometimes in the form of a shape (such as $\square + 9 = 10$) and they know how to use inverse operations to work out the value. This lesson introduces the idea that a letter can stand for more than one value: in other words, the value of the letter can change or vary.
- The mathematical name for these letters is a variable. For example, if we say that $a + b = 4$, then a could be 0, 1, 2, 3, or 4 (negative values and fractional values are also possible in this case). The value we assign to a will affect the value of b.
- Work through the examples and explanation on **Pupil book page 146** with the class. The example box at the top of the page asks: What are the values of a, b, and x? Tell your partner how you worked them out. ($a = 14$, I worked out what to add to 6 to make 20; $b = 12$, I worked out 48 divided by 4, or what to multiply 4 by to make 48; $x = 34$, I worked out what number subtract 4 gives 30.)
- When you deal with the table, ask the children whether these are all the possible values for a and b when $a + b = 10$. (No, values could include decimals, fractions, and negative numbers.) Use a number line to show that $-7 + 17$ would also give a result of 10.
- The children can do question 1 orally, using mental strategies as far as possible but, if necessary, let them use jottings to find the solutions.
- For question 2, the children can work independently, or they could assign values to n and ask their partner to find the solution.
- For question 3, remind the class that the value they use for one variable affects the value of the other.
- Problem solving: The children can work in pairs to discuss and solve question 4.

Challenge

Explain to the children that we can represent two consecutive numbers using x and $x + 1$.

Ask:

- *Can you explain why this works for any two consecutive numbers?* (The second number is always one more than the first.)
- *How could you represent two consecutive odd numbers?* ($2x + 1$, $2x - 1$)
- *If $x + (x + 1) + (x + 2) = 48$, what is the value of $(x + 1)$?* (16)

Follow-up

- Use **Workbook page 85** to consolidate work on finding the value when there are two variables in an equation. Let the children consult if they need to.
- Check their work as a class. Discuss any mistakes and clarify any misconceptions.

Interesting mistakes

- Children sometimes get the idea that two variables have to have two different values. For example, if $a + b = 10$, they assume that $a = 5$ and $b = 5$ is not a possible solution.
- Using tables of values and organised lists will help the children to recognise that the variables can be equal.

Answers for Pupil book page 146

1 a 69   b 44   c 3
   d 200   e 545   f 9
   g 48   h 3   i 4

2 a–f   Individual answers.

3 a–e   Individual answers.

4 28 and 73

Answers for Workbook page 85

1 a For example:

Value of a	Value of b
2	20
5	8
4	10

b For example:

Value of a	Value of b
2	11
1	22
5.5	4

c For example:

Value of a	Value of b
2	4
1	8
1.6	5

d For example:

Value of a	Value of b
1	34
10	25
20	15

2

Value of a	1	2	3	4	6	9	12	18	36
Value of b	36	18	12	6	6	4	3	2	1

3 a

Value of x	Value of y
2	25
9	88
10	97
12	115
7	10x

b

Value of x	Value of y
8	4
53	19
23	9
71	25
y	2

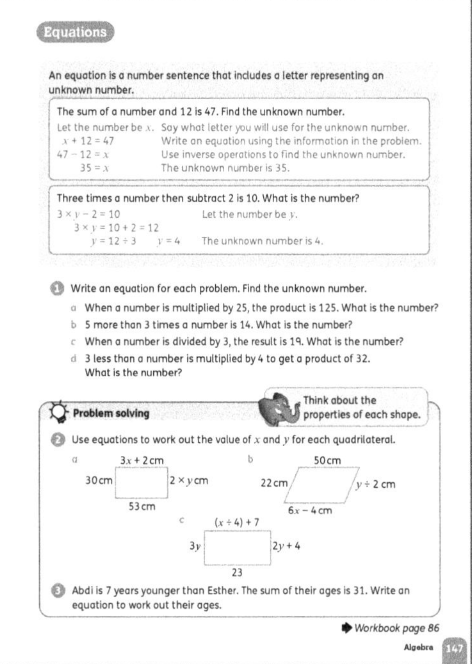

Equations

An equation is a number sentence that includes a letter representing an unknown number.

The sum of a number and 12 is 47. Find the unknown number.

Let the number be x.	Say what letter you will use for the unknown number.
$x + 12 = 47$	Write an equation using the information in the problem.
$47 - 12 = x$	Use inverse operations to find the unknown number.
$35 = x$	The unknown number is 35.

Three times a number then subtract 2 is 10. What is the number?

$3 \times y - 2 = 10$ Let the number be y.
$3 \times y = 10 + 2 = 12$
$y = 12 \div 3$ $y = 4$ The unknown number is 4.

1 Write an equation for each problem. Find the unknown number.

   a When a number is multiplied by 25, the product is 125. What is the number?
   b 5 more than 3 times a number is 14. What is the number?
   c When a number is divided by 3, the result is 19. What is the number?
   d 3 less than a number is multiplied by 4 to get a product of 32. What is the number?

Problem solving

Think about the properties of each shape.

2 Use equations to work out the value of x and y for each quadrilateral.

a $3x + 2$ cm, 30 cm, $2 \times y$ cm, 53 cm

b   50 cm, 22 cm, $y \div 2$ cm, $6x - 4$ cm

c $(x \div 4) + 7$, $3y$, $2y + 4$, 23

3 Abdi is 7 years younger than Esther. The sum of their ages is 31. Write an equation to work out their ages.

Workbook page 86

Algebra   147

Materials

Calculators; matching cards for expressions (as follows).

Warm-up

Use any of the 'Mental problem-solving' activities in the Activity bank (pages 23–24) as a mental starter for this lesson.

Focus

- Prepare a set of matching cards to revise and teach the operations words needed to make expressions and equations. For example:

The sum of two numbers	The difference between two numbers	The sum of two consecutive numbers	The product of two numbers
$x + y$	$x - y$	$x + x + 1$	$x \times y$
The product of a number and 7	6 more than a number	A number divided by 10	Twice a number
$x \times 7$	$x + 6$	$\dfrac{x}{10}$	$2 \times x$

- Display a word card showing a number sentence that includes a letter representing an unknown number. Ask the children to find the matching expression.
- Repeat, showing the expression and asking the children to say what each could mean.
- Turn to **Workbook page 86**. Discuss what an equation is with the class. Remind them that they have already worked with equations using numbers and letters, for example, in unknown angle problems. Talk through question 1a, as an example.
- Let them work in pairs to complete the questions.
- Turn to **Pupil book page 147**. Explain that we can write

equations and solve them to find unknown values. This is a very useful problem-solving strategy, so it's important that the children understand the method.

- Work systematically through the examples with the class. Talk about each step and what you are doing. If necessary, draw balance scales to show that what you do to one side of the equation, you must do to the other to keep both sides equal. In example 2, you add 2 to the left-hand side to 'get rid' of the –2. That means you have to add 2 to the right-hand side as well.
- Complete question 1 orally with the class. Give the children time to think about each statement and then ask for contributions.
- Record these for everyone to see. Encourage the children to share any different ways of writing the equations.
- <u>Problem solving</u>: For question 2, let the children work in pairs to make equations to find the unknown values in the shape problems.
- <u>Problem solving</u>: Let the children try question 3 on their own.

Challenge

- Provide some challenging open-ended problems for the children to solve, such as the following.
- The perimeter of a parallelogram is given as $2 \times (a + b) = 60$ cm. If a is a 2-digit whole number and b is a 1-digit whole number, what are the possible values of a and b?

Support

If children need additional support to solve equations, show them how to use bar models (page 21) to represent different types of problems, for example:

Two numbers have a sum of 54 and a difference of 12. What are the numbers?	The sum of two numbers is 42. One number is $\frac{1}{5}$ of the size of the other. What are the numbers?

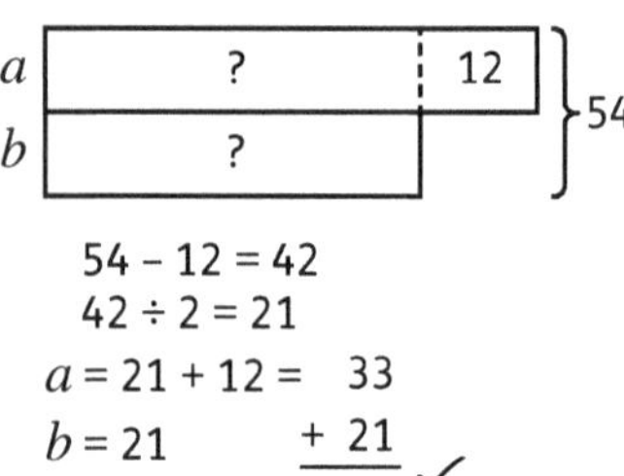

$$54 - 12 = 42$$
$$42 \div 2 = 21$$
$$a = 21 + 12 = \quad 33$$
$$b = 21 \qquad + 21$$
$$\overline{\quad 54 \quad} \checkmark$$

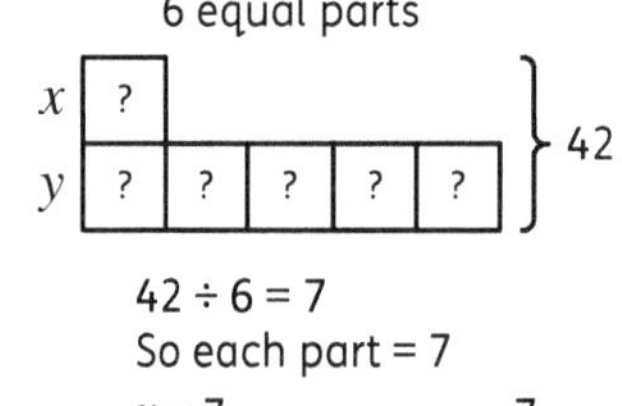

6 equal parts

$$42 \div 6 = 7$$
So each part = 7
$$x = 7 \qquad\qquad 7$$
$$y = 5 \times 7 = 35 \quad + 35$$
$$\overline{\quad 42 \quad} \checkmark$$

Formulae

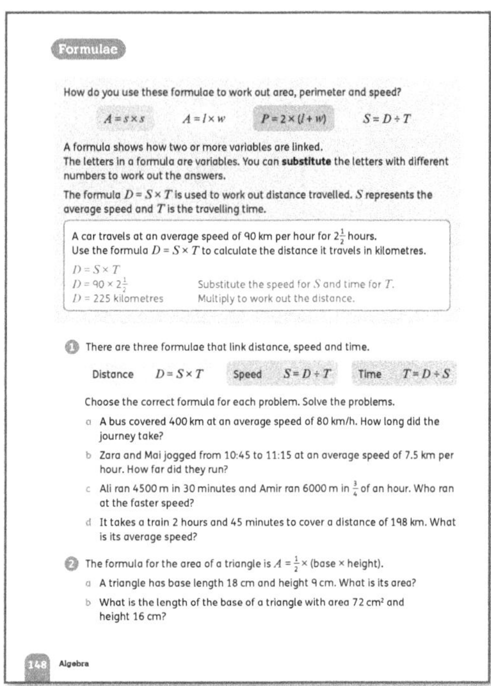

Formulae

How do you use these formulae to work out area, perimeter and speed?

$$A = s \times s \qquad A = l \times w \qquad P = 2 \times (l + w) \qquad S = D \div T$$

A formula shows how two or more variables are linked. The letters in a formula are variables. You can **substitute** the letters with different numbers to work out the answers.

The formula $D = S \times T$ is used to work out distance travelled. S represents the average speed and T is the travelling time.

A car travels at an average speed of 90 km per hour for $2\frac{1}{2}$ hours. Use the formula $D = S \times T$ to calculate the distance it travels in kilometres.

$$D = S \times T$$
$$D = 90 \times 2\tfrac{1}{2} \qquad \text{Substitute the speed for } S \text{ and time for } T.$$
$$D = 225 \text{ kilometres} \qquad \text{Multiply to work out the distance.}$$

1 There are three formulae that link distance, speed and time.

Distance $\quad D = S \times T \qquad$ Speed $\quad S = D \div T \qquad$ Time $\quad T = D \div S$

Choose the correct formula for each problem. Solve the problems.

a A bus covered 400 km at an average speed of 80 km/h. How long did the journey take?

b Zara and Mai jogged from 10:45 to 11:15 at an average speed of 7.5 km per hour. How far did they run?

c Ali ran 4500 m in 30 minutes and Amir ran 6000 m in $\frac{3}{4}$ of an hour. Who ran at the faster speed?

d It takes a train 2 hours and 45 minutes to cover a distance of 198 km. What is its average speed?

2 The formula for the area of a triangle is $A = \frac{1}{2} \times (\text{base} \times \text{height})$.

a A triangle has base length 18 cm and height 9 cm. What is its area?

b What is the length of the base of a triangle with area 72 cm² and height 16 cm?

148 Algebra

Warm-up

Use any of the 'Mental problem-solving' activities (pages 23–24) as a starter for this lesson.

Focus

- Turn to **Pupil book page 148**. Give the children time to read the formulae and explain how they use them. (You substitute the values of the letters you know and work out the answers.)
- Explain that a formula is a rule for working something out. To use a formula you replace all but one of the letters with numbers ('substitute') and calculate the unknown value. Use area as an example to show that you need to know l and w to work out A. Similarly, if you know A and l, you can work out w.
- Work through the distance formula example with the class. If necessary, do some additional step-by-step examples to work out distance, time and speed. Remind the class that speed is a rate, so it has two units (kilometres per hour).

- Let the children work on their own to complete questions 1 and 2 on **Pupil book page 148**.

Answers for Pupil book page 148

1   a $T = 400 \div 80 = 5$ hours
    b $D = 0.5 \times 7.5 = 3.75$ km
    c   Ali: $S = 4.5 \div 0.5 = 9$ km/h
        Amir: $S = 6 \div 0.75 = 8$ km/h
        So, Ali ran faster than Amir.
    d $S = 198 \div 2.75 = 72$ km/h
2   a $A = 81$ cm^2 b   9 cm

Solve problems with variables

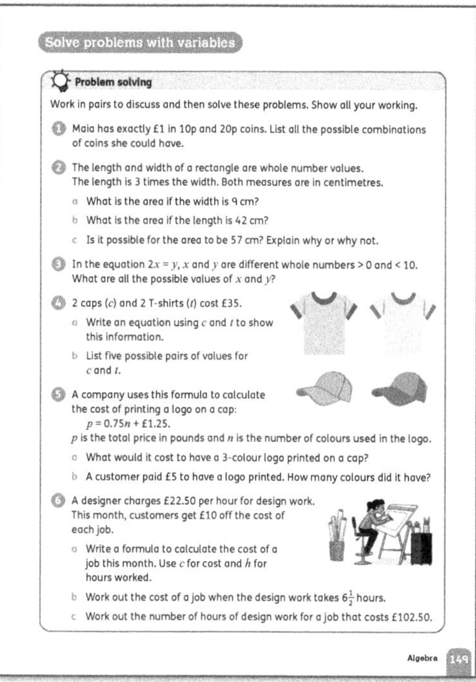

Materials

Calculators; squared paper for drawing bar models (page 21).

Warm-up

Choose any 'Calculation skills' activity with decimals (pages 24–26) as a starter for this lesson.

Focus

- Problem solving: No new concepts are taught in this lesson. Questions 1–6 require the children to apply what they have learnt to solve a range of different problem types.
- Turn to **Pupil book page 149**. Let the children spend some time discussing the problems with their partners and deciding what strategies or methods they will use to solve them.
- You could take feedback from the class before asking the children to work through and solve the problems.

Challenge

- Use the formula $P = 2 \times (l + w)$ to investigate whether the largest rectangle you can make for a given perimeter will always be a square. (The answer is 'yes', but the children can work this out by choosing different perimeters and then calculating the area for different combinations of l and w. Remind them that when we talk about the largest rectangle we mean the rectangle with the greatest area.

Answers for Pupil book page 149

1   Two 10p coins and four 20p coins, four 10p coins and three 20p coins, six 10p coins and two 20p coins, and eight 10p coins and one 20p coin.
2   a   243 cm^2 b   588 cm^2
    c   No; areas must be in cm^2.
3 $x = 1$ and $y = 2$, $x = 2$ and $y = 4$, $x = 3$ and $y = 6$, $x = 4$ and $y = 8$
4   a $2c + 2t = 35$
    b   Any two amounts of money that add to £17.50; for example, $c = £10$ and $t = £7.50$.
5   a   £3.50   b   5 colours
6   a $p = 22.5h - 10$ b   £136.25
    c   5 hours

End-of-unit check

Use some or all of these questions and problems to assess how well the children have understood the concepts in this unit.

- Provide a sequence and ask:
  - *What rule was used to make this sequence?*
  - *Is there any other way of carrying on this sequence? How?*
  - *What comes next in this sequence? Why?*
- *The sum of two positive whole numbers is 23. What could the numbers be?* (For example: 1 and 22, 5 and 19, 11 and 12)
- *If $2x + 3 = 11$, what is x?* ($x = 4$)
- *A cashier gives a customer £2.70 in 50p and 20p coins. How many of each coin could there be?* (1 50p coin and 11 20p coins, 3 50p coins and 6 20p coins, 5 50p coins and 1 20p coin)
- *I have to organise 72 crayons into boxes. Some boxes hold 8 crayons and others hold 12. How many different ways are there to organise the crayons?* (4 ways: $9 \times 8 + 0 \times 12$, $3 \times 8 + 4 \times 12$, $6 \times 8 + 2 \times 12$, $0 \times 8 + 6 \times 12$) *What is the smallest number of boxes I could use to organise all the crayons?* (6 boxes)
- *The formula for working out the volume of a cuboid is volume = length × width × height. What do you need to know to work out the length of a cuboid? Why?* (You need to know volume, width and height. The values can be substituted into the formula to find the length.) *What is the volume of a cuboid that is 12 cm long, 3 cm high and 5 cm wide?* (180 cm^3)

Learning objectives

- Use the language of probability and proportion to describe and compare possible outcomes.
- Identify when two events can happen at the same time and when they cannot, and know that the latter are mutually exclusive.
- Model probabilities through experiments using a large number of trials.

Key words

probability   probability scale   mutually exclusive
experiment   outcome   possible outcomes

Unit introduction

Teaching guidance

- Put the children into groups of three or four and give each group the same set of statements. Ask the children to decide whether they agree or disagree with each statement. They must be prepared to justify their choices to the rest of the class.
- Here are some statements you can use or adapt:
  - *If the weather forecast says there is a 100% chance of rain next Monday, it will definitely rain on that day.*
  - *You should choose heads in a coin flip because heads are easier to get than tails when you flip a coin.*
  - *When you spin a 1–6 spinner, it is harder to get a 6 than any other number.*
  - *If you play a game properly and follow the rules, it is a fair game.*
  - *When the weather forecast says there is a 40% chance of rain, it won't rain.*
  - *You always see someone you know on your way to or from school.*
  - *When you take a test, you have an equal chance of passing or failing.*
- After the children have talked about the statements and decided whether they agree or not, hold a class discussion about their reasons. If someone disagrees, ask them to justify their own position.

Describe probability

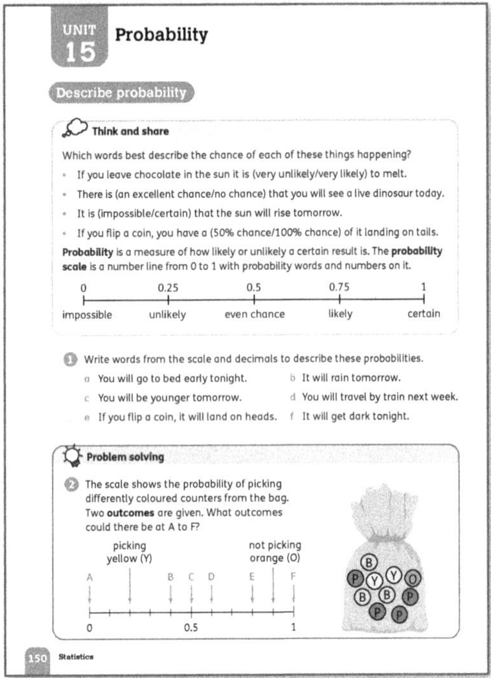

Materials

Large display of a 'probability scale' (see **Pupil book page 150**).

Warm-up

Do any activity that involves writing one number as a fraction of another as a mental starter for this lesson.

Focus

- <u>Think and share:</u> Turn to **Pupil book page 150**. Display a large probability scale. Revise how the probability scale works and use it to explain that the words we use to talk about how likely things are to happen are describing the chance or probability of it happening. Discuss the answers to the questions. (If you leave chocolate in the sun it is very likely to melt. There is no chance that you will see a live dinosaur today. It is certain that the sun will rise tomorrow. If you flip a coin you have a 50% chance that it will land on tails.) Point out the words on the scale and use the slider to show that events can fit anywhere on the scale.
- Ask the class what it means if there is 'zero chance' that something will happen. Point out that things with zero chance of happening are impossible.
- Then show them that the scale of probability extends from 0 to 1. Events that are certain have a probability of 1. All other events are given a value between 0 and 1 to show the chance that they will happen.
- Remind the class that we can use numbers from the probability scale to describe chance more accurately. For example, when you flip a coin you have an equal chance of getting heads or tails. That is a probability of $\frac{1}{2}$. If you spin a 1–6 spinner with six equal sections, the probability of each number is the same: you have

a $\frac{1}{6}$ chance of getting 1 and a $\frac{1}{6}$ chance of getting 6. The probability of spinning an even number is $\frac{3}{6}$ or $\frac{1}{2}$.

- It may be interesting to have a class discussion to clarify that a probability of $\frac{1}{2}$ for heads does not mean that you will get heads every one out of two times you flip the coin. It does mean that if you flip the coin a large number of times, you will end up with a more or less equal number of heads and tails.
- Make sure the children can confidently use the vocabulary of probability and relate probabilities to numbers.
- Let the children complete question 1 orally before they record their answers.
- <u>Problem solving:</u> The children can work in pairs to try to solve question 2. If necessary, check their understanding of the term 'outcome'.

Interesting mistakes

- The main focus at this level is on using everyday terms to discuss the chance of something happening. Children whose first language is not English may be unfamiliar with the terms 'certain', 'likely', 'even chance', 'unlikely' and 'impossible', and you may need to teach the words and their meanings before you start the activities.
- As you work through the unit, check to make sure the children remember and use the terms.

Answers for Pupil book page 150

1 a Individual answers.   b Individual answers.
   c impossible; 0   d Individual answers.
   e Equally likely; 0.5   f Certain; 1
2 A: picking a green; B: picking a pink;
   C: picking a pink or orange; D: picking a pink or a yellow; E: not picking a yellow; F: not picking a black

What is the probability?

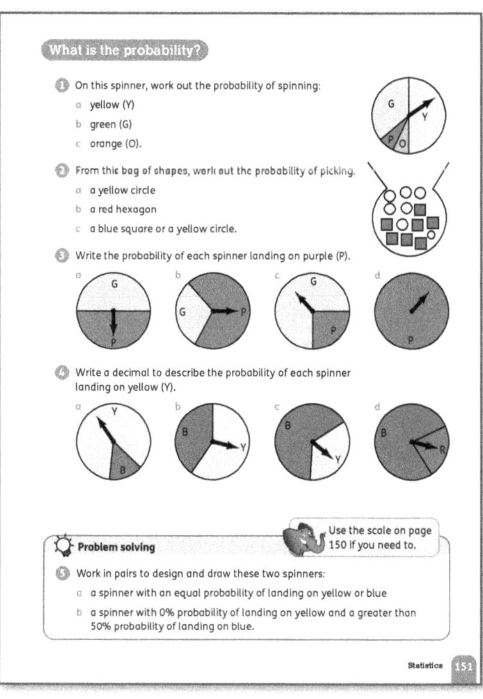

Materials

Protractors; ten coloured cards (five red, five white) in a bag or box; selection of coloured cubes (or balls or beads) in a box or bag; colouring pencils.

Warm-up

Do any activity that involves identifying a shaded fraction of a shape as a mental starter for this lesson.

Focus

- Put the ten coloured cards in a bag or box. Tell the class you are going to pull out and replace a card ten times. Ask them how many times out of ten they expect you will get a red card.
- As you pull out the cards, let the children tally how many times you draw red or white.
- Discuss the results and whether or not this is what the children expected. Reinforce the idea that you are equally likely to draw out red or white.
- Put a mixed selection of cubes (or balls or beads) into a bag. Choose one white, three black and six green, for example. Repeat the activity above. Discuss why white is less likely than black or green to be drawn from the bag.
- Turn to **Pupil book page 151**.
- Use question 1 with the class to revise how to find the fraction of the spinner that is shaded in each colour.
- For questions 2, remind the class to give the answers as a fraction.
- For questions 3 and 4, the children should be able to work out the fraction for each sector of the circles, but if not, they can measure in degrees and write the fractions with denominator 360.
- <u>Problem solving:</u> For question 5, the children can work in pairs. They should be able to design and draw their spinners without using a protractor. For 5b, compare their answers and check the probability of each colour.

Support

You may need to help the children work out what fraction of a circle each sector represents before they work through the activities.

Challenge

- Give the children some more challenging probability problems, for example:
  The numbers 1–30 are written on cards and placed in a box.
  If a number is drawn at random from the box, what is the probability that it will have the digit 1 in it? $\left(\frac{12}{30}\right)$
  Does the probability of getting a number with a 1 in it double if you put the numbers from 1–60 in the box? (No, for 1–60 the probability is $\frac{15}{30}$*)*
  Explain your answers. (Numbers with the digit 1 in 1–30 are: 1, 10, 11, 12, 13, 14, 15, 16, 17, 18, 19, 21. Numbers with the digit 1 in 31-60 are: 31, 41, 51)

Interesting mistakes

- When working with spinners, children may determine the probability using the number of sectors rather than the size of the angles at the centre. Use circles with divisions round the outside and ask the children to colour different fractions to make sectors to show that the size of the sector represents a different fraction of the circle. For example:
  - Give the children a circle marked in twelfths.
  - Ask them to colour $\frac{1}{4}$, then $\frac{1}{3}$ and then $\frac{1}{6}$ and then three separate pieces of $\frac{1}{12}$.
- Discuss the results, making sure they understand that the fraction they have coloured is the probability of the spinner landing on that colour.

Answers for Pupil book page 151

1 a 0.5   b 0.375   c 0.125

2 a 0.5   b 0   c 1

3 a $\frac{1}{2}$ b $\frac{2}{3}$ c $\frac{1}{4}$ d 1

4 a $\frac{5}{6}$ b $\frac{2}{3}$ c $\frac{1}{3}$ d 0

5 a, b Individual answers.

Can it happen at the same time?

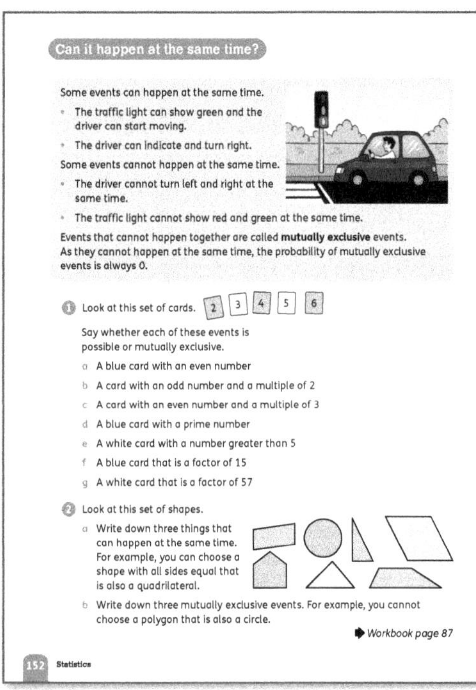

Warm-up

Use any suitable 'Mental problem-solving' activity (pages 23–24) as a starter for this lesson.

Focus

- Read through the explanation text on **Pupil book page 152** with the class to teach the term 'mutually exclusive'. Explain that this means that one thing happening excludes the other from happening.

- Ask the children to make up some examples of two mutually exclusive events, for example:
  - You cannot stand and lie down at the same time.
  - You cannot flip a coin and get heads and tails.
  - You cannot score a goal and score no goals at the same time.
- Focus on the word 'and'. In probability we say, *The probability of A **and** B happening is 0 when they are mutually exclusive. The probability of A **or** B is not mutually exclusive as they can happen independently.*
- Do question 1 orally with the children. Give them time to think about each statement and then, using a show of hands, say whether it is possible or mutually exclusive. If the children disagree, let them explain their thinking so they learn from their mistakes.
- Let the children work independently on question 2.
- Then ask them to read their statements aloud so others can check that they are correct.

Follow-up

- Use **Workbook page 87** to assess the children's understanding of mutually exclusive events.
- Check the answers with the class.

Support

- If children need additional support with this concept you could ask them to draw rough Venn diagrams to work out whether events are mutually exclusive.
- If there is no event that goes into the circle overlap, then the events are mutually exclusive. For example, ask: *Can you spin a number that is both odd and even? Can you spin a number that is both odd and prime?*

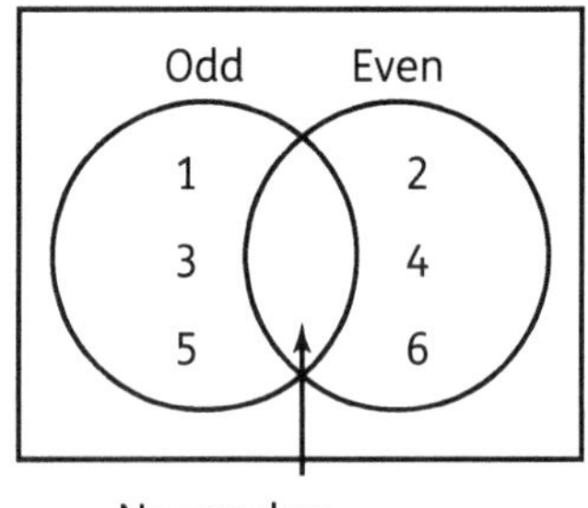

No overlap
mutually exclusive

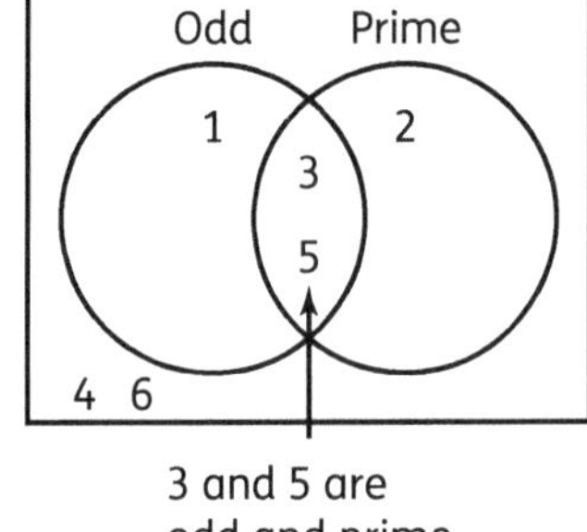

3 and 5 are
odd and prime

Answers for Pupil book page 152

1 a possible
b mutually exclusive
c possible
d possible
e mutually exclusive
f mutually exclusive
g possible

2 a For example, triangle and green; parallelogram and yellow; circle and blue
b For example, circle and green; trapezium and blue; triangle and blue

Answers for Workbook page 87

1 picking a sock from a drawer and picking a blue sock: not mutually exclusive

scoring two goals and winning a match: not mutually exclusive

having an odd number of number of pupils and an even number of pupils in a class: mutually exclusive

wearing a red sock and wearing a grey sock: not mutually exclusive

eating toast and drinking orange juice: not mutually exclusive

getting an A in a test and passing a test: not mutually exclusive

rolling a number cube with the numbers 1 to 6 and getting a square number and an odd number: not mutually exclusive

getting heads and tails on a single coin toss: mutually exclusive

turning left and right when you walk out of a doorway: mutually exclusive

picking a shape with five equal sides that is also a hexagon: mutually exclusive

2 a heads on the other/tails on the other
b an odd number

3 a odd and even **b** even
c odd **d** a prime square number

Experiment with probability

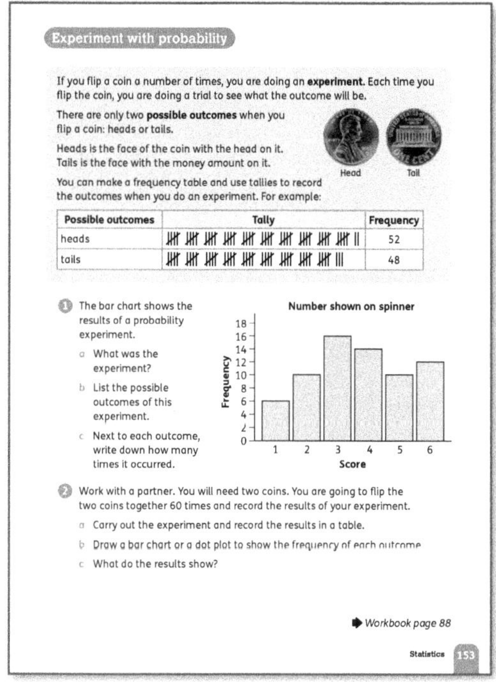

Materials

Coins; (see Follow up): card; scissors; bag or box.

Warm-up

Use any suitable 'Mental problem-solving' activity (pages 23–24) as a starter for this lesson.

Focus

- Ask the class what are the 'possible outcomes' when you flip a coin. (Only two are possible – heads or tails.)
- Turn to **Pupil book page 153**. Read through the information with the class making sure they understand the language used and remember how to draw up a frequency table and use it to tally and record results.
- For question 1, let the children work in pairs to discuss the bar chart and answer the questions. Take feedback when they have finished.
- For question 2, make sure each pair of children has a coin and then let them complete the 'experiment' on their own. Hold a class discussion around the results. Make sure they do not argue that the results show it's harder to throw heads or harder to throw tails. This is not statistically true.

Follow-up

- Tell the children to turn to **Workbook page 88**. Explain that they are going to do an experiment with 120 trials.
- Let everyone make the card and cut out the pieces before they start. Give the children time to do the experiment and record their results. Help anyone who is unsure of what to do.
- After the children have completed their graphs and commented on their own results, let them work in groups to compare their results.

Challenge

Work with the class to combine the results from all the experiments to see whether you then get outcomes that are more equal (the same number of 1, 2, 3, 4, 5 and 6). Do this by getting groups to combine their own results into one frequency table and then combine these as a class.

Interesting mistakes

- Children may need additional support to recognise that the probability of an event may not match the actual outcome. For example, the probability of a coin landing on heads or tails is equally likely, but you may throw heads five times in a row.
- Let the children carry out experiments and trials to show that results vary and that the predicted probability may not occur all the time. In statistics, the more trials you do in an experiment, the closer you get to the theoretical probability. So, the more times you flip a coin, the closer you will get to a probability of $\frac{1}{2}$.

Answers for Pupil book page 153

1 **a** It recorded the results of spinning a 6-sided spinner.
  b 1, 2, 3, 4, 5, 6
  c 1 = 6 times; 2 = 10 times; 3 = 16 times; 4 = 14 times; 5 = 10 times; 6 = 12 times
2 **a, b, c** Individual answers.

Answers for Workbook page 88

1 **a, b, c** Individual answers.
2 Individual answers.

End-of-unit check

Use some or all of these questions and problems to assess how well the children have understood the concepts in this unit.

- *Explain in your own words what 'probability' means.* (For example: How likely something is to happen.)
- *Tell me something that is certain/impossible.* (For example: Certain: New Year's Day will be on 1 January. Impossible: I will go on holiday to Mars.)
- Give some different events. *What is the probability that these events will happen?*
- *What does a probability of $\frac{3}{4}$ mean?* (The event is likely to happen.)
- *How does the probability scale work?* (It is a number line from 0 to 1, going from impossible (probability of 0) to certain to happen (probability of 1). Other events have a probability between 0 and 1, where closer to 0 means that it is unlikely to happen and closer to 1 means that it is likely to happen. Halfway along the scale, events are equally likely to happen as not happen.)
- Show a spinner. *What is the probability of the spinner landing on (give a colour on the spinner)?*
- *What is a probability experiment?* (repeating something a number of times to see how often different things happen)
- *I have 3 red beads and 5 blue beads in a bag and I pick one without looking. Which colour am I most likely to pick? Why?* (blue, because there are more blue beads than red beads)
- *What are the possible outcomes when you spin a spinner with a square, a red dot and a yellow sun on it?* (square, red dot, yellow sun) *What outcome has a probability of 0 on this spinner?* (for example: green triangle, purple line)
- *What does it mean if events are mutually exclusive?* (They cannot happen at the same time.) *Give an example of two mutually exclusive events.* (for example, rolling a 3 and a 4 on a dice in one roll)

Answers to Mixed practice 3

Answers for Pupil book pages 154–156

1 **a** 0–9.99   **b** 9
  c 40%   **d** 10–19.99
  e 40–49.99, because it is a group with price range £9.99.
2 **a** 08:25 to 08:43   **b** 7800 m
  c 18 minutes   **d** 9 minutes
  e 09:10
  f No. The graph stops at 09:49.
3

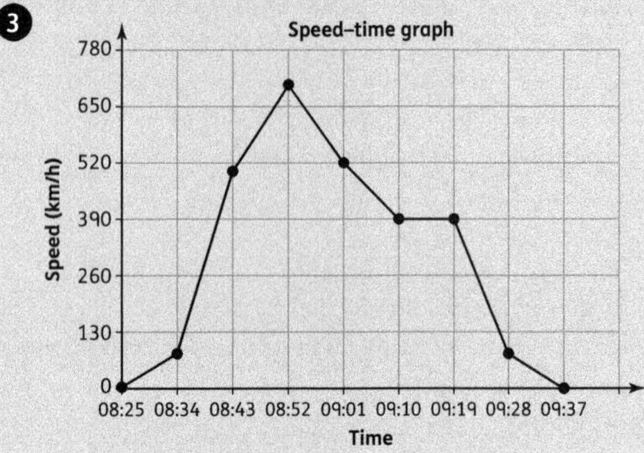

  b 463 kilometres
4 **a** A = reading; B = looking out the window; C = complaining about being bored; D = eating; E = games/movies
  b 25%   **c** 28 minutes   **d** $\frac{1}{3}$
  e Divide 360° by 120 to get 3°. Each minute is represented by 3°.
5 **a** 2 : 2 : 1   **b** 15 : 12 : 8   **c** $\frac{1}{2}$
6 **a** mode = 15; median = 15; range = 6
  b mode = 10; median = 12; range = 8
  c 15
7 **a** 2.5   **b** 25 cm by 10 cm
8 **a** 2   **b** 5.8 cm   **c** 6.16 cm²
9 **a** It doesn't affect the perimeter.
  b It makes it smaller.
10 617.5 cm²
11 **a** 5, 8, 11, 14, 17, 20   **b** add 3
  c $3n + 2$ **d** 62
12 **a** $n + 7 = 43$; $n = 36$ **b** $n + 12 = 103$; $n = 91$
  c $2n + 3 = 55$; $n = 26$ **d** $0.5n - 17 = 12$; $n = 58$ or $17 - 0.5n = 12$; $n = 10$
13 $x = 64$; $y = -44$; $z = 8$
14 **a** 1 and 3; 2 and 6; 3 and 9
  b 16.5
  c 0.25
15 **a** 3 (1, 2 and 3)   **b** $1 : \frac{1}{5}$; $2 : \frac{3}{5}$; $3 : \frac{1}{5}$
16 lemon = 8; lime = 12; orange = 18; mango = 2